HUMAN MUSCLE FATIGUE

When human muscle fatigues, athletic performance becomes impaired. For those individuals suffering muscle or metabolic diseases, the effects of muscle fatigue can make everyday tasks difficult. Understanding the scientific processes responsible for skeletal muscle fatigue is therefore central to the study of the physiology of sport, exercise and disease.

Written by a team of leading international exercise scientists, this book explores the mechanisms of muscle fatigue and presents a comprehensive survey of current research on this important topic. Examining the wide variety of protocols, assessment methods and exercise models used to study muscle fatigue, the book explores the differential effects of fatigue as influenced by:

- age
- gender
- fitness and training
- the use of ergogenic aids
- medical conditions including neuromuscular diseases, muscular dystrophies and myopathies.

Human Muscle Fatigue covers both clinical and applied approaches in sport and exercise physiology and devotes an entire section to the conceptual framework underpinning research in this area, helping readers from a wide range of backgrounds to engage with the topic. Accessible and detailed, this book is a key text for students and practitioners working in exercise and sports science, medicine, physical therapy and health.

Craig A. Williams is Associate Professor and Co-Director of the Children's Health and Exercise Research Centre, School of Sport and Health Scien___   ___ ___ University of Exeter, UK. Craig's research interests are ___ ___ ___ ___ hildren's fatigue and high-intensity exercise. Sél___ ___ ___ ___ member of the Laboratory of Exercise Biolog___ ___ ___ ___ ___ ___ ___ ice. His current research interests include the u___ ___ ___ ___ ___ ___ ___ ___ opy to investigate the muscle metabolic respo___ ___ ___ ___ ___ ___ ___ ___ muscle fatigue and recovery from high-intensity

HUMAN MUSCLE FATIGUE

Edited by Craig A. Williams and Sébastien Ratel

Routledge
Taylor & Francis Group

LONDON AND NEW YORK

First published 2009
by Routledge
2 Park Square, Milton Park, Abingdon, Oxon OX14 4RN

Simultaneously published in the USA and Canada
by Routledge
270 Madison Ave, New York, NY 10016

Routledge is an imprint of the Taylor & Francis Group, an informa business

Typeset in Sabon and Futura by
Wearset Ltd, Boldon, Tyne and Wear
Printed and bound in Great Britain by
CPI Antony Rowe, Chippenham, Wiltshire

British Library Cataloguing in Publication Data
A catalogue record for this book is available from the British Library

Library of Congress Cataloging in Publication Data
Human muscle fatigue/edited by Craig A. Williams and Sébastien Ratel.
p. cm.
1. Muscles–Physiology. 2. Fatigue. 3. Sports–Physiological aspects. 4. Exercise–Physiological
aspects. I. Williams, Craig A., 1965– II. Ratel, Sébastien.
QP321.H856 2009
612.7'4–dc22   2008055417

ISBN10: 0-415-45327-5 (hbk)
ISBN10: 0-415-45328-3 (pbk)
ISBN10: 0-203-88548-1 (ebk)

ISBN13: 978-0-415-45327-1 (hbk)
ISBN13: 978-0-415-45328-8 (pbk)
ISBN13: 978-0-203-88548-2 (ebk)

This book is dedicated to the late memory of two extraordinary family members. Firstly, my grandmother Chrissie Williams who taught me from an early age the value of education and who was an exceptional teacher. And secondly to my father-in-law Ted Wragg, an outstanding academic and role model.

Craig A. Williams

To my wife, Anne, for all her love and support through the years. To my wonderful daughter, Camille, the light of my life, you are the best – keep smiling.

Sébastien Ratel

CONTENTS

FIGURES

TABLES

CONTRIBUTORS

David G. Allen (MBBS, PhD), Professor, School of Medical Sciences, University of Sydney F13, NSW 2006, Australia.

David Bendahan (PhD), Directeur de Recherche, Centre de Résonance Magnétique Biologique et Médicale (CRMBM, UMR CNRS 6612), Faculté de la Timone, Université de la Méditerranée, Marseille, France.

David Bishop (PhD), Facolta di Scienze Motorie, Universita di Verona, Verona, Italia, School of Human Movement and Exercise Science, University of Western Australia, Crawley, Western Australia, Australia.

Gijs Bleijenberg (PhD), Professor, Centre Chronic Fatigue, Radboud University Nijmegen Medical Centre, the Netherlands.

Gregory C. Bogdanis (PhD), Department of Sports Medicine and Biology of Physical Activity, University of Athens, Athens, Greece.

Damien M. Callahan (MSc), Muscle Physiology Laboratory, Department of Kinesiology, University of Massachusetts, Amherst, USA.

Linda H. Chung (MSc), Muscle Physiology Laboratory, Department of Kinesiology, University of Massachusetts, Amherst, USA.

Marine Dididze (MD, PhD), The Miami Project to Cure Paralysis and Department of Neurological Surgery, University of Miami Miller School of Medicine, Miami, USA.

Pascale Duché (PhD), Professor, Laboratoire de Biologie des Activités Physiques et Sportives (BAPS, EA 3533), UFR STAPS, Université Blaise Pascal, Clermont-Ferrand, France.

Stephen A. Foulis (MSc), Muscle Physiology Laboratory, Department of Kinesiology, University of Massachusetts, Amherst, USA.

Alexander C.H. Geurts (PhD), Professor, Department of Rehabilitation, Radboud University Nijmegen Medical Centre, the Netherlands.

Ronald Haller (MD), Professor, Departments of Neurology and Internal Medicine, the University of Texas, Southwestern Medical School, Texas, USA.

Jane A. Kent-Braun (PhD), Professor, Muscle Physiology Laboratory, Department of Kinesiology, University of Massachusetts, Amherst MA 01003, USA.

Brittany C. Lee (MS), Departments of Radiology and Molecular Physiology and Biophysics, Vanderbilt University Medical School, Nashville, USA.

Nicola A. Maffiuletti (PhD), Senior Research Fellow, Neuromuscular Research Laboratory, Schulthess Clinic, Zürich, Switzerland.

Nancy J. Olsen (MD), Professor, Department of Medicine, University of Texas Southwestern Medical School, Dallas, USA.

George W. Padberg (PhD), Professor, Department of Neurology, Radboud University Nijmegen Medical Centre, the Netherlands.

Jane H. Park (PhD), Professor, Departments of Radiology and Molecular Physiology and Biophysics, Vanderbilt University Medical School, Nashville, USA.

Sébastien Ratel (PhD), Laboratoire de Biologie des Activités Physiques et Sportives (BAPS, EA 3533), UFR STAPS, Université Blaise Pascal, Clermont-Ferrand, France.

David W. Russ (PT, PhD), Assistant Professor, School of Physical Therapy, Ohio University, Athens, USA.

Christine K. Thomas (PhD), The Miami Project to Cure Paralysis and Department of Neurological Surgery, University of Miami Miller School of Medicine, Miami, USA.

Baziel G.M. van Engelen (PhD), Professor, Department of Neurology, Radboud University Nijmegen Medical Centre, the Netherlands.

John Vissing (D.M.Sci), Professor of Neurology, Neuromuscular Research Unit, the Copenhagen Muscle Research Center and the Department of Neurology, National University Hospital, Rigshospitalet, Copenhagen, Denmark.

Nicole B.M. Voet (MD, PhD Student), Resident Rehabilitation Medicine, Department of Rehabilitation, Radboud University Nijmegen Medical Centre, the Netherlands.

Craig A. Williams (PhD), Associate Professor, Children's Health and Exercise Research Centre, School of Sport and Health Sciences, University of Exeter, Exeter, United Kingdom.

Håkan Westerblad (MD, PhD), Professor, Department of Physiology and Pharmacology, Karolinska Instituet, Stockholm, Sweden.

Inge Zijdewind (PhD), Assistant Professor, Department of Medical Physiology, University Medical Center Groningen, Groningen, the Netherlands.

Machiel J. Zwarts (PhD), Professor, Department of Neurophysiology, Radboud University Nijmegen Medical Centre, the Netherlands.

PREFACE

This book has been written for scientists, physicians, lecturers, coaches and postgraduate students interested in the study of human muscle fatigue in sport, exercise and disease. The enormity of such a task can be judged by the fact that whole conferences have been devoted to the topic and voluminous texts have been written to try to bring together all the different disciplines affected by this phenomenon. Historically, some of the greatest scientists and researchers have devoted their expertise to the direct and indirect study of this topic. Eighteenth- and nineteenth-century scientists such as Galvani, DuBois-Reymond, Faraday and Mosso all aided the development of electrophysiology, the modern-day equivalent of neuroscience. Indeed it was Glisson, a century earlier, who established the concept of irritability, the foundation for electrophysiology. At around the same time as Glisson, scientists such as Swammerdam, Stensen and Goodard furthered the work of the Greek physician Galen by investigating the shortening and lengthening hypothesis of muscles. The greater understanding of how muscles worked lead to many studies of ergonomics, calculating how intensive physical labour was and when fatigue occurred.

Other scientists wrote important text books, such as Bainbridge's *Physiology of Muscular Exercise*, published in 1931, which differentiated central and peripheral fatigue – or as he describes it, fatigue partly within the nervous system and the active muscles. Important articles began to appear about the sense of effort, published in the journal *Brain*, by Waller in 1891. If we fast forward to the twentieth century, a laboratory in the United States took fatigue as part of its name, the "Harvard Fatigue Laboratory" directed by David Bruce Dill, which played a leading role not just in the development of studying fatigue, but in the pursuit of physiology in general. Since that time, other notable scientists such as Hill, Merton, Bigland-Ritchie and Edwards have contributed significantly to furthering our understanding of fatigue.

Although fatigue has been shown to have been addressed by researchers for more than several centuries, human muscle fatigue is currently still a subject of considerable interest. Over the last decade, the number of scientific publications on the subject has increased, especially with respect to age, gender, training and muscle diseases. This text book is devoted to providing the most recent and advanced information on human muscle fatigue among various populations and muscle pathologies. Contributing authors are international leaders in their fields and have provided a range of theoretical and practical information on fatigue. In each chapter, authors have suggested journal papers which shaped the topic area, and although there is always some degree of bias to these selections, we hope that readers will be encouraged to follow up and read authors' choices.

This book is divided into three parts. Part I provides a framework for subsequent sections of the book by reviewing the different definitions of muscle fatigue, the various tools and techniques of investigation and the cellular mechanisms of muscle fatigue. The main objectives are to clarify the meaning of muscle fatigue, indicate how fatigue can be identified, measured and classified, and understand how the various metabolic changes affect muscle performance in normal and diseased subjects.

In Part II, five chapters are devoted to identifying and explaining the differences in fatigability between individuals according to age, gender, training background, daily levels of physical activity, and explaining how the use of ergogenic aids may help to understand the mechanisms of fatigue during physical activity. This section highlights the task- and muscle-specific nature of fatigue, and how these characteristics may contribute to discrepancies with regard to the role of age, gender and training in muscle fatigue.

In Part III, the last four chapters describe muscle fatigue in a number of muscle and metabolic diseases such as myasthenia gravis, muscular dystrophies, dermatomyositis, polymyositis, body myositis, metabolic myopathies, etc. The objectives of these chapters are to provide an overview on the prevalence and assessment of muscle dysfunction and fatigue in each muscle disorder, and to suggest methods for alleviation of fatigue and weakness in order to improve muscle strength, endurance and overall quality of life.

We hope that this book will offer readers a better insight into the complexity of the subject by exploring the mechanisms of fatigue in relation to age, gender, training background and muscle disorders.

Craig A. Williams and Sébastien Ratel
November 2008

PART I

THE CONCEPTUAL FRAMEWORK OF MUSCLE FATIGUE

CHAPTER 1

DEFINITIONS OF MUSCLE FATIGUE

Craig A. Williams and Sébastien Ratel

INTRODUCTION

The aim of this chapter is to clarify the meaning of muscle fatigue, indicate how fatigue can be identified, measured and classified, and thus applied in the context of exercise, disease and sport. In the literature there exist many definitions for fatigue, and although measurement techniques have significantly improved over the last century to observe this phenomenon, its aetiology is still a matter of considerable debate. Whether in the context of athletic competition, manual labour or patients with various neuromuscular diseases or chronic fatigue-type syndromes, fatigue is a commonly experienced phenomenon with important consequences. The study of fatigue has a long and illustrious research history, but advancements in its understanding have been made, particularly when fatigue has been investigated in different populations, under differing sport, ergonomic and disease conditions. For example, findings in neuromyopathies, i.e. myasthenia gravis, have presented scientists and clinicians with the opportunity to understand factors controlling force generation and to improve habitual performance.

In this chapter, to understand more fully the context and aetiology of fatigue, three important parameters are considered. These include:

1 the status[1] of the individual;
2 the task[2] required of the individual; and
3 the location of the fatigue.

Although the consequences of fatigue will differ depending on the three above parameters, the eventual and observable decline on performance allows clinicians and exercise scientists to explain the causal factors. This challenge cannot be underestimated because, despite several centuries of research into fatigue, the complexity of the problem is one of the few agreements amongst researchers investigating fatigue.

The search for the causality of fatigue can be divided into investigations which examine predominant factors at the central nervous system, the neuromuscular junction or the muscle unit. The former can be termed as central fatigue, while the latter two can be termed as peripheral fatigue. The complexity of establishing fatigue mechanisms is highlighted when causal factors can occur as a result of alterations at the peripheral *as well as* central areas, often synergistically, thus preventing the explicit identification of a single causal mechanism. The majority of research has tended to focus on peripheral factors and will predominantly be the focus of this book; this does not, however, discount the importance of central mechanisms.

DEFINITIONS OF FATIGUE

The following box represents a variety of definitions for fatigue, and the list is by no means comprehensive. Critical to the definition are three unifying points:

1   There is a decline in one or more of the biological systems.
2   The decline is reversible.
3   The decline may or may not occur before an observable performance or task failure occurs.

The first point usually refers to the decline in force, velocity or power of the biological system, usually with reference to muscle performance. The second point, reversibility, distinguishes fatigue from injury or disease when muscle performance might be impaired for a period of time or significantly debilitated. The latter point is an important consideration as it establishes the decline in performance often observed during a maximal "all-out" effort compared to the fatigue which might be progressively experienced during a prolonged exercise task.

Of course, the definition of any variable is important because without agreement on the definition, the observation and measurement of fatigue becomes extremely difficult. Part of this difficulty can be explained because fatigue occupies multiple roles and mechanisms, which can operate at many levels, from the CNS to the intrinsic muscle fibre itself (Westerblad *et al.*, 1991; Fitts, 1994; Enoka, 1995).

Definitions of fatigue

A reversible state of force depression, including a lower rate of rise of force and a slower relaxation.

(Fitts and Holloszy, 1978)

The failure to maintain a required or expected force.

(Edwards, 1981)

Muscle fatigue is a decline in the maximal contractile force of the muscle.

(Vøllestad, 1997)

The inability to maintain of a physiological process to continue functioning at a particular level and/or the inability of the total organism to maintain a predetermined exercise intensity.

(Fifth International Symposium on Biochemistry of Exercise, 1982)

Reduction in the maximal force generating capability of the muscle during exercise.

(Miller *et al.*, 1995)

Any reduction in the force-generating capacity (measured by the maximum voluntary contraction), regardless of the task performed.

(Bigland-Ritchie and Woods, 1984)

A loss of maximal force generating capacity.

(Bigland-Ritchie *et al.*, 1986)

A condition in which there is a loss in the capacity for developing force and/or velocity of a muscle, resulting from muscle activity under load which is reversible by rest.

(NHLBI, 1990)

Any reduction in a person's ability to exert force or power in response to voluntary effort, regardless of whether or not the task itself can still be performed successfully.

(Enoka and Stuart, 1992)

Any exercise-induced reduction in the maximal capacity to generate force or power output.

(Vøllestad, 1997)

Intensive activity of muscles causes a decline in performance, known as fatigue.

(Allen and Westerblad, 2001)

> Performing a motor task for long periods of time induces motor fatigue, which is generally defined as a decline in a person's ability to exert force.
>
> (Lorist *et al.*, 2002)
>
> The development of less than expected amount of force as a consequence of muscle activation.
>
> (McCully *et al.*, 2002)
>
> Fatigue is known to be reflected in the EMG signal as an increase of its amplitude and a decrease of its characteristic spectral frequencies.
>
> (Kallenberg *et al.*, 2007)

Other terms, often used interchangeably with fatigue, are exhaustion and weakness. However, these two terms are describing different functions and should not be used synonymously. Exhaustion refers to "the moment in time when the expected force level cannot be maintained" (Vøllestad, 1995: p. 186). Fatigue, as defined as a decline in the maximal force of the muscle, differentiates itself from exhaustion in that it is possible for fatigue to be observed during submaximal levels without a noticeable effect on performance. This interchanging of terms is partially due to the negative ramifications often associated with fatigue. However, the consequences of fatigue should also be viewed from a positive or protectionist perspective, as much as a negative one, as the body prohibits a metabolic crisis and conserves the integrity of the muscle fibre. As such, fatigue can be viewed as a consequence of one or several "fail-safe" mechanisms in the organism that call for temperance before damage occurs (Edwards, 1983). Another term associated with fatigue and often thought to represent a modified form is "plasticity". This term is particularly important to rehabilitative therapists in that fatigue is a manifestation of plasticity that modifies contractile properties in order to improve the efficiency of contractions (Sargeant, 1994).

Interestingly, in exercise and sports science there has been some interest in exhaustive tests, often in the guise of timed rides or runs to exhaustion. This endurance time is then associated with the fatigability. In such studies, interventions are often applied and the time-to-exhaustion exercise repeated, and consequently findings with a demonstrated longer time to exhaustion have been used to imply improvements in fatigability. However, there are two weaknesses in this approach. First, tests which use time to exhaustion are notoriously unreliable and possess a high coefficient of variation (Jeukendrup *et al.*, 1996), as well as being ecologically invalid. It is rare in a sport or clinical situation that an individual would need to perform a task "for as long as possible". Second, this approach assumes there is a relationship between the decline in maximal performance and the time to exhaustion. Several studies have shown a large variability between time to exhaustion and declines in maximal voluntary contractions (MVC) (Vøllestad *et al.*, 1988).

Weakness, which often results in observable decrements in force or power generation, is more symptomatic of maintenance of low force or power over a sustained period of time, but is independent of exercise. Weakness may occur due to the sustained atrophy of muscles experienced during ageing. Under these circumstances the inability to delay muscle fatigue during a task requiring absolute power or force is reduced. Interesting however, is the observation that relative to the maximal force or power, atrophied muscle shows a greater resistance to fatigue (Larsson and Karlsson, 1978; Overend et al., 1992). Although the mechanism for this observation is not entirely clear, increases in capillary density and the concentration of mitochondrial and metabolic enzymes and shifts in fibre types are thought to be responsible.

Therefore, we are generally left with two commonly accepted definitions. The first is that fatigue is an exercise-induced reduction in the ability of muscle to produce power or force, irrespective of task completion (Bigland-Ritchie and Woods, 1984). And second, fatigue is considered as the inability to maintain a required or expected force or power output (Edwards, 1981) under maximal or submaximal sustained contraction conditions, i.e. the time to task failure or muscle endurance. As Enoka and Duchateau (2008) concisely describe, the distinction between these two definitions is that the former one does not consider muscle fatigue to be the point at which task failure occurs or the time period of muscle exhaustion.

In their review, Enoka and Duchateau (2008), rather than focusing on a global mechanism for muscle fatigue which they argue does not exist, elaborated on causal mechanisms specific to individual tasks which are responsible for fatigue. By critiquing the experimental approaches that focus on identifying the mechanisms which limit task failure, Enoka and Duchateau presented several lines of evidence which highlighted the contributions of previously underrated factors. These factors included synergist, antagonist and postural muscle activity as limiting factors to performance across a range of physical activities. The results from these studies were collected by manipulating protocols common to fatigue studies, such as interrupting a fatiguing exercise bout with brief maximal voluntary or electrically simulated contractions, or by measuring the decline in maximal power immediately after fatiguing exercise both pre- and post-intervention study. Clearly, not only which definition of fatigue is accepted is important, but the associated methodologies, including the protocol, are also critical to elucidate the rate-limiting factors of fatiguing muscle performance.

Most important to those researchers accepting the first definition of fatigue is that the maximal force or power produced by the muscles displays a transient decline soon after the commencement of the sustained physical exercise. However, what cannot be ignored is the importance of the different variables constituting the task itself, e.g. the intensity of the exercise, the amount of muscle mass involved, the type of contraction, the recovery periods between successive bouts of exercise and the environmental conditions in which the exercise takes place. Although we have stated that a general lack of agreement for the definition of fatigue is problematic, consensus can be reached. We suggest that in order to address a specific question related to fatigue, especially given its multi-dimensional nature, provided the definition complements the quantification of the developing fatigue, both definitions allow fatigue to be investigated from the broadest perspective.

IDENTIFICATION AND MEASUREMENT OF FATIGUE

Although fatigue is a common occurrence in a sporting or clinical setting, it is clear that the observed symptoms of fatigue reveal differing associated mechanisms and processes. The lack of a commonly agreed definition and the numerous methodological techniques to investigate fatigue have no doubt contributed to the differing opinions of the differing mechanisms and processes. It is also clear that results from maximal voluntary contraction studies must be carefully interpreted when compared to the electrical stimulation of single-muscle fibres. However, a common unifying measurement is the assessment of force, velocity or power under different exercise-induced or artificially simulated exercise conditions.

All definitions of fatigue necessitate a decline in force, velocity or power, and both animal and human studies have identified these changes and their associations with fibre composition. In animal studies, preparations using isolated single living and skinned muscle fibres, whole muscle *in-vitro* and anaesthetised *in-situ* have been utilised (Fitts, 1994; Allen and Westerblad, 2001; Gandevia, 2001). Studies with human participants have used *in-vivo* tests or isolated organelles from muscle biopsies (Karlsson and Saltin, 1970; Hermansen and Osnes, 1972; Gollnick *et al.*, 1991; Brickley *et al.*, 2007). These have been mostly related to examining the effects on metabolic by-products, e.g. H^+, La^- and Ca^{2+}.

In both animal and human studies, force, velocity and power results have confirmed similar loss of contractile function with fatigue, regardless of muscle type. As described earlier, a commonly used protocol to examine fatigue involves completing a series of MVC during the fatiguing bout of exercise. By observing the decline in the maximal force generation of the muscle, it is possible to quantify the extent of the fatigue. However, as the capacity to maintain an MVC under isometric conditions can be limited by motivation (Vøllestad, 1997), it is also possible to stimulate the muscle electrically with a tetanic frequency. The use of isometric contraction reduces the number of extraneous variables compared to dynamic contractions, including stabilising the joint complexes and synergistic activity. Another extension of this protocol involves measuring the MVC immediately after the fatiguing bout of exercise.

Measurement of fatigue during dynamic exercise is a more complicated situation because of the inherently greater number of extraneous variables to contend with. Models of both dynamic-, constant- and intermittent-type exercise have been utilised, mostly related to understanding the processes of the metabolic pathways and the mechanisms of fatigue (Spriet *et al.*, 1989; Balsom *et al.*, 1992; Gaitanos *et al.*, 1993; Nevill *et al.*, 1996). For further information on this topic, the reader is directed to a review by Cairns *et al.* (2005). Although the manifestation of fatigue can be difficult to quantify during isometric and dynamic human movement, the eventual effects of fatigue and exhaustion will result in significant deterioration of muscle performance, which is akin to the deterioration of technique in sports performance or basic daily functions, e.g. walking, lifting and carrying.

MECHANISMS OF FATIGUE

Although, because of the confounding variables in which fatigue manifests itself, Edwards (1983) states that it might not be appropriate to ask the question "What is the cause of fatigue?" (p. 20), it has not stopped various mechanistic models being proposed. One of the most well-known ones is a model by Edwards, known as the "chain of command" (Figure 1.1). In this model the various possible locations for the underlying fatigue are identified. They range from the brain and the "psyche" to the muscle sarcolemma to the eventual generators of force and power, the cross-bridge proteins. It is important to recognise that as with all models, there are a number of inherent assumptions. These include that much of the original research on which this schematic is based originated from isolated muscle preparations. Therefore, for these results to be extrapolated and applied to whole muscle groups or human performance, other parallel sites of fatigue must be eliminated as an explanatory variable before fatigue at the contractile proteins can be recognised. Also, there is an assumption that what is occurring within the whole muscle is reflective of the metabolic changes in individual muscle cells. Edwards himself states that this assumption cannot be taken for granted when investigating fatiguing or diseased muscle.

In the chain of command schematic, locating fatigue is often conceptualised as finding the weak link in the chain. Thus, if the weak link where the fatigue occurred can be identified, then it would be possible to strengthen that link, delay fatigue and enhance performance. Rather than the linear chain-type approach of investigating fatigue, a more integrationist model has been proposed (Edwards, 1983). This model, utilising catastrophe theory, accounts for the changing scenario of losses of energy and excitation/activation losses and their effects on force production. It provides a more complex theoretical mapping of what is likely occurring when

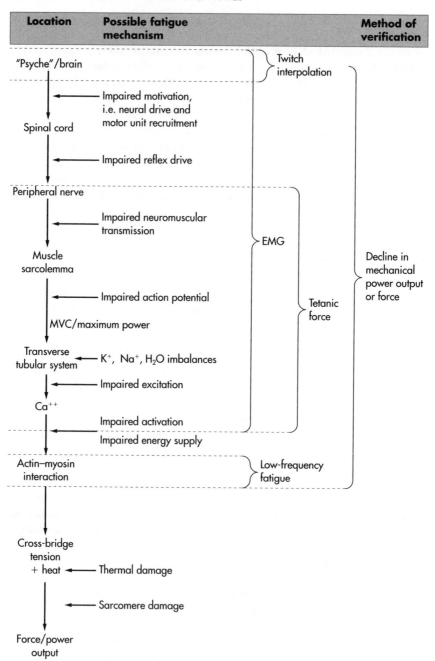

Figure 1.1 Chain of command mechanism of fatigue (adapted from Edwards, 1983 with permission)

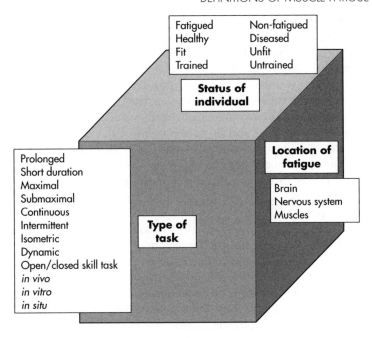

Figure 1.2 A multi-factorial perspective of fatigue

there is a decline in force or power output. As a three-dimensional sche-matic, the rapidly unfolding pathways in the model provide a continuum of the relationship between energy supply and the excitation/activation processes. By adding other extraneous variables which impact on fatigue, a multi-factorial picture of this complex phenomenon is built up (Figure 1.2). Hence, the interactions between the type of task, the status of the individual and the location of the fatigue will exemplify the relationships between the energy exchange, the central or peripheral fatigue and the reduction in the force-generating capabilities of the actin and myosin cross-bridges.

CONCLUSION

Although it might be almost impossible to identify the single most import-ant limiting factor of fatigue, this should not deter scientists and clinicians from attempting to resolve many of the issues which confound this concept. Fatigue is commonly defined as any reduction in the maximal capacity to generate force. Although most definitions of fatigue focus on force production, fatigue not only impedes a fibre's capacity for maximal force generation, but – importantly – the maximum speed of shortening or lengthening, and consequently power output, will also be affected. The measurement of MVC (force), maximal power output (force *and* the speed of shortening/lengthening) or the use of electrical stimulation is fundamental to the assessment of central or peripheral fatigue. All investigations of fatigue can be categorised into three broad areas: fatigue as related to central fatigue; failure at the neuromuscular junction; and deficits at the muscle. Central fatigue is defined as a failure in locations found in the brain, spinal cord and up to the point of the excitation site of the motoneuron. Because it is comparatively easier to measure compared to central fatigue, peripheral fatigue (failure in the transmission of the neural signal or a failure of the muscle to respond to neural excitation) provides a larger data base of evidence. Although the approaches to studying central and peripheral fatigue produce a different set of challenges both mechanistically and methodologically, whatever approach is taken, the manifestations of fatigue are the same. There is a decline in efficiency of force production, resulting in an increase of the amount of effort to continue or in the continuation being unachievable because the necessary force cannot be sustained. The aetiology of fatigue has been advanced using a variety of clinical populations, sports performance and normal healthy adults. For example, the significance of substantial neuromuscular transmission failure occurring during normal human fatigue is contentious, but not when applied to the pathological disease, myasthenia gravis.

Whether it will ever be possible to identify *the* limiting cause of fatigue during a task is debatable. What is clear is that fatigue comprises a spectrum of events for which there is no single causative factor, with many factors occupying potential roles in its aetiology. These facts make fatigue such a complex and controversial concept.

FIVE KEY PAPERS THAT SHAPED THE TOPIC AREA

Study 1. Kronecker, H. and Stirling, W. (1898). The genesis of tetanus. *Journal of Physiology*, 1, 45, 384–421.

In a series of experiments designed to continue the earlier work of Ranvier on "red and pale coloured" muscle, Kronecker and Stirling used both rabbit and frog muscle to examine the tetanus curves, using induction shocks from a du Bois-Reymond induction machine. By manipulating the strength of the stimulation and its frequency, Kronecker and Stirling were able to confirm differences in the red and pale muscle of these two animals. The authors continued to experiment with stimulating muscle, discussing issues related to fatigue and concluded: "In ordinary movements requiring the least special skill we employ the simple voluntary tetanus, which economic Nature produces by stimuli repeated at the longest intervals practicable, and thus fatigue remains at a minimum" (p. 420).

Study 2. Sahlin, K., Tonkonogi, M. and Soderlund, K. (1998). Energy supply and muscle fatigue in humans. *Acta Physiologica Scandinavica*, 162, 261–266.

This review article focused on the limitation of energetic processes and the relation to muscle fatigue. In this paper, the evidence for energy deficiency as being a determinant of fatigue during high-intensity exercise was strongly championed. For prolonged exercise, fatigue coinciding with glycogen depletion and signs of energy deficiency at the single-fibre level were discussed. It is also stated, however, that the complexity is more pronounced during prolonged, compared to high-intensity, exercise and clear lines of reasoning are proposed that energy deficiency is not the sole explanation. The authors conclude that metabolic factors are likely to play an important role in physical performance *in vivo*, but there is no doubt that conditions exist where fatigue could not be explained by metabolic changes.

Study 3. Edwards, R.H.T. (1983). Biochemical basis of fatigue in exercise performance: catastrophe theory of muscular fatigue. In H.G. Knuttgen, J.A. Vogel and J. Poortmans (eds), *Biochemistry of Exercise V*. Champaign: Human Kinetics, pp. 3–28.

In this chapter, muscle fatigue is defined as that discussed under the auspices of the Ciba Foundation symposium held in 1980, i.e. "an inability to maintain force or power output during sustained or repetitive contractions". Furthermore, R.H.T. Edwards proposes the catastrophe model of fatigue that is illustrated by a three-dimensional graphical representation by the inter-dependence of primarily energy-dependent mechanisms of fatigue with those involving excitation or excitation contraction (E–C) coupling. In certain types of exercise, the "catastrophic" theory can be illustrated by a failure in the excitation or E–C coupling, thereby leading to limit the extent of energy exchange within muscular cells.

Study 4. Blomstrand, E., Hassmén, P., Ekblom, B. and Newsholme, E.A. (1991). Administration of branched-chain amino acids during sustained exercise-effects on performance and plasma concentration of some amino acids. *European Journal of Applied Physiology*, 63, 83–88.

This paper continued the earlier work of Newsholme in the late 1980s related to mechanisms suggestive of causing central fatigue by measuring brain monoamine concentrations. Blomstrad *et al.* studied marathon runners and, by randomly dividing the group into an experimental group who received a drink with a mix of branched chain amino acids and a placebo group, the effects on running and mental performance were assessed. Although the results were varied and depended on how the group data were analysed, there were some intriguing results which added to the theory. Both Blomstrand and Newsholme produced many articles related to amino acid concentration and performance, and effectively provided much of the early evidence in support of the role of an increasing level of 5-hydroxytryptamine in the brain and fatigue.

Study 5. Bigland-Ritchie, B. (1981). EMG and fatigue of human voluntary and stimulated contractions. *Ciba Foundation Symposium*, 82, 130–156.

The use of MVC alongside electromyogram (EMG) recording allows for a combination of physiological variables to be investigated. In this paper Bigland-Ritchie, during a 60s MVC of the adductor pollicis muscle, found that the loss of force is accompanied by a parallel decline in both the integrated surface EMG and the single-muscle fibre spike counts recorded intramuscularly. This decline was not due to neuromuscular block, since the muscle mass action potential (M wave) evoked by single maximal shocks to the nerve were well maintained; nor did the size of the single-fibre spike change. It was concluded that this observation reflected a decline in the firing pattern of the motoneuron pool. The force of a sustained MVC continued to match that from maximal tetanic nerve stimulation; thus, all motor units remained active. Continuous nerve stimulation at the frequency required to match the voluntary force of unfatigued muscle led to a progressive failure of the M wave, and a more rapid force loss than in an MVC. Both were largely restored by reducing the stimulus frequency. The decline in neural firing rate was found to correlate well with the rate of muscle contractile slowing. It thus optimises force by maintaining a relatively constant degree of tetanic fusion, while avoiding peripheral failure of electrical propagation.

GLOSSARY OF TERMS

Ca^{2+}	calcium ion
EMG	electromyographic activity
K^+	potassium ion
MVC	maximal voluntary isometric contraction
Na^+	sodium ion

NOTES

1 "Status" refers to indices of health, levels of fitness, whether in a fatigued or non-fatigued state.
2 "Task" is defined not only as different types of isometric and dynamic exercise, but also similar exercise performed by various muscles which possess differing contractile properties.

REFERENCES

Allen, D.G. and Westerblad, H. (2001). Role of phosphate and calcium stores in muscle fatigue. *Journal of Physiology*, 536, 657–665.

Balsom, P.D., Seger, J.Y., Sjodin, B. and Ekblom, B. (1992). Physiological responses to maximal intensity intermittent exercise. *European Journal of Applied Physiology*, 65, 144–149.

Bigland-Ritchie, B. and Woods, J.J. (1984). Changes in muscle contractile properties and neural control during human muscular fatigue. *Muscle and Nerve*, 7, 691–699.

Bigland-Ritchie, B., Furbush, F. and Woods, J.J. (1986). Fatigue of intermittent submaximal voluntary contractions: central and peripheral factors. *Journal of Applied Physiology*, 61, 421–429.

Biochemistry of Exercise (1982). Proceedings of the 5th International Symposium on the Biochemistry of Exercise. Edited by H.G. Knuttgen, J.A. Vogel and J. Poortmans. Champaign: Human Kinetics.

Brickley, G., Green, S., Jenkins, D.J., McEinery, M., Wishart, C., Doust, J.H. and Williams, C.A. (2007). Muscle metabolism during constant and alternating intensity exercise around critical power. *International Journal of Sports Medicine*, 28, 303–315.

Cairns, S.P., Knicker, A.J., Thompson, M.W. and Sjøgaard, G. (2005). Evaluation of models used to study neuromuscular fatigue. *Exercise and Sport Sciences Reviews*, 33, 9–16.

Edwards, R.H.T. (1981). Human muscle function and fatigue. In R. Porter and J. Whelan (eds), *Human Muscle Fatigue: Physiological Mechanisms*. London: Pitman Medical, pp. 1–18.

Edwards, R.H.T. (1983). Biochemical bases of fatigue in exercise performance: catastrophe theory of muscular fatigue. In H.G. Knuttgen, J.A. Vogel and J. Poortmans (eds), *Biochemistry of Exercise*. Champaign: Human Kinetics, pp. 3–28.

Enoka, R.M. (1995). Mechanisms of muscle fatigue: central factors and task dependency. *Journal of Electromyography and Kinesiology*, 5, 141–149.

Enoka, R.M. and Duchateau, J. (2008). Muscle fatigue: what, why and how it influences muscle function. *Journal of Physiology*, 586, 1, 11–23.

Enoka, R.M. and Stuart, D.G. (1992). Neurobiology of muscle fatigue. *Journal of Applied Physiology*, 72, 1631–1648.

Fitts, R.H. (1994). Cellular mechanisms of muscle fatigue. *Physiological Reviews*, 74, 49–94.

Fitts, R.H. and Holloszy, J.O. (1978). Effects of fatigue and recovery on contractile properties of frog muscle. *Journal of Applied Physiology*, 45, 899–902.

Gaitanos, G.C., Williams, C., Boobis, L.H. and Brooks, S. (1993). Human muscle metabolism during intermittent maximal exercise. *Journal of Applied Physiology*, 75, 712–719.

Gandevia, S.C. (2001). Spinal and supraspinal factors in human muscle fatigue. *Physiological Reviews*, 81, 1725–1789.

Gollnick P.D., Körge, P., Karpakka, J. and Saltin, B. (1991). Elongation of skeletal muscle relaxation during exercise is linked to reduced calcium uptake by the sarcoplasmic reticulum in man. *Acta Physiologica Scandinavica*, 142, 135–136.

Hermansen, L. and Osnes, J. (1972). Blood and muscle pH after maximal exercise in man. *Journal of Applied Physiology*, 32, 302–308.

Jeukendrup, A., Saris, W.H., Brouns, F. and Kester, A.D. (1996). A new validated endurance performance test. *Medicine and Science in Sports and Exercise*, 28, 266–270.

Kallenberg, L.A., Schulte, E., Disselhorst-Klug, C. and Hermens, H.J. (2007). Myoelectric manifestations of fatigue at low contraction levels in subjects with and without chronic pain. *Journal of Electromyography and Kinesiology*, 17, 264–274.

Karlsson, J. and Saltin, B. (1970). Lactate, ATP and CP in working muscles during exhaustive exercise in man. *Journal of Applied Physiology*, 29, 598–602.

Larsson, L. and Karlsson, J. (1978). Isometric and dynamic endurance as a function of age and skeletal muscle characteristics. *Acta Physiologica Scandinavica*, 104, 129–136.

Lorist, M.M., Kernell, D., Meijman, T.F. and Zijdewind, I. (2002). Motor fatigue and cognitive task performance in humans. *Journal of Physiology*, 15, 313–319.

Miller, R.G., Kent-Braun, J.A., Sharma, K.R. and Weiner, M.W. (1995). Mechanisms of human muscle fatigue: quantitating the contribution of metabolic factors and activation impairment. In S.C. Gandevia, R.M. Enoka, A.J. McComas, D.G. Stuart and C.K. Thomas (eds) *Advances in Experimental Medicine and Biology: Fatigue Neural and Muscular Mechanisms*. New York: Springer.

Nevill, M.E., Bogdanis, G.C., Boobis, L.H., Lakomy, H.K.A. and Williams, C. (1996). Muscle metabolism and performance during sprinting. In R.J. Maughan and S.M. Shireffs (eds), *Biochemistry of Exercise IX*. Champaign: Human Kinetics, pp. 243–259.

NHLBI (1990) Workshop summary: respiratory muscle fatigue: report of the Respiratory Muscle Fatigue Workshop Group. *The American Review of Respiratory Disease*, 142, 474–480.

Overend, T.J., Cunningham, D.A., Paterson, D.H. and Smith, W.D. (1992). Physiological responses of young and elderly men to prolonged exercise at critical power. *European Journal of Applied Physiology*, 64, 187–193.

Sargeant, A.J. (1994). Human power output and muscle fatigue. *International Journal of Sports Medicine*, 15, 116–121.

Spriet, L.L., Lindinger, M.I., McKelvie, R.S., Heigenhauser, G.J.F. and Jones, N.L. (1989). Muscle glycogenolysis and H^+ concentration during maximal intermittent cycling. *Journal of Applied Physiology*, 66, 8–13.

Vøllestad N.K. (1995). Metabolic correlates of fatigue from different types of exercise in man. In S.C. Gandevia, R.M. Enoka, A.J. McComas, D.G. Stuart and C.K. Thomas (eds), *Fatigue: Neural and Muscular Mechanisms*. New York: Plenum Press, pp. 185–194.

Vøllestad, N.K. (1997). Measurement of human fatigue. *Journal of Neuroscience Methods*, 74, 219–227.

Vøllestad, N.K., Sejersted, O.M., Bahr, R., Woods, J.J. and Bigland-Ritchie, B. (1988). Motor drive and metabolic responses during repeated responses during repeated submaximal contractions in man. *Journal of Applied Physiology*, 64, 4, 1421–1427.

Westerblad, H., Lee, J.A., Lannergren, J. and Allen, D.G. (1991). Cellular mechanisms of fatigue in skeletal muscle. *American Journal of Physiology*, 261, C195–209.

CHAPTER 2

MEASUREMENT METHODS OF MUSCLE FATIGUE

Nicola A. Maffiuletti and David Bendahan

OBJECTIVES

The main objectives of this chapter are to describe:

- the different methods allowing quantification of muscle fatigue, which include the tools and the procedures for the assessment of force–time and power–time curves (part I);
- the techniques and the procedures for the assessment of neuromuscular mechanisms (part II); and
- the metabolic (part III) mechanisms underlying muscle fatigue. The advantages and drawbacks of the different tools and techniques will be briefly discussed.

INTRODUCTION

Muscle fatigue (hereafter referred to as fatigue) could be defined as an exercise-induced reduction in muscle performance. Two different measurement models are generally used to quantify such impairment. The first consists of quantifying the reduction of power output during real exercise such as cycling or running (generally at a maximal intensity), where dynamic performance decline could be attributed to reduced force and/or velocity. The second model is based on the assessment of maximal isometric muscle force-generating capacity before and immediately after real or simulated exercise, to describe the decline in static muscle performance. Therefore, measurement methods of fatigue mainly consist of maximal force or power recordings obtained before, during or after exercise.

Because fatigue is a decline in muscle force or power induced by exercise, all fatigue models (Cairns *et al.*, 2005) entail the two following components: fatigue induction and fatigue quantification (Table 2.1). Accordingly, the two important factors that should be determined when designing a fatigue study are the type of exercise inducing fatigue and the measurement methods of fatigue, which include type and timing of the measurements. Depending on the study population and research question, investigation of the neuromuscular and metabolic mechanisms underlying fatigue (i.e. fatigue mechanisms) could also be considered (Table 2.1).

Fatigue is an important functional parameter for physical work and daily activities, and its assessment is relevant to practitioners in sport, occupational, geriatric and orthopaedic medicine. In all these settings, it is essential to respect the specificity between the activity of interest and the type of exercise inducing fatigue. The best option is represented by fatigue quantification during real exercise in specific conditions (e.g. cycling power loss for a road cyclist). However, it is quite challenging to assess accurately muscle force/power during the majority of sport, work and daily conditions, because of the acyclic nature of these activities. There is indeed the

Table 2.1 Important factors for the definition of a fatigue model

---

Type of exercise inducing fatigue (fatigue induction)
Real
- On the field
- In the laboratory
Simulated*
- On the field
- In the laboratory

Measurement methods of fatigue (fatigue quantification)
Type of measurements (fatigue outcomes and, accordingly, tools) – cf. part I
- Muscle power
- Muscle force
- Other variables (work, velocity, position, time-to-exhaustion)
Timing of measurements (and definition of fatigue)
   During exercise
   - Two measurements (e.g. initial versus final force/power): fatigue index
   - Multiple measurements (e.g. slope of force/power): fatigue time course
   Before and after exercise
   - Two measurements (e.g. before versus immediately after exercise): fatigue index
   - Multiple measurements (e.g. before, immediately after and minutes, hours or days after exercise): fatigue recovery
Fatigue mechanisms (physiological changes underlying fatigue)
- Neuromuscular techniques – cf. part II
- Metabolic techniques – cf. part III

Note
*Type of muscle action – cf. Table 2.2

Table 2.2 Type of muscle action and factors to control for fatigue induction (simulated exercise) and quantification

Open chain, single joint-muscle	vs.	*Closed chain*, multiple joints-muscles
Unilateral	vs.	*Bilateral* • Simultaneous/alternate actions
Voluntary	vs.	*Stimulated* • Stimulus type (electric, magnetic) • Stimulus site (cortex, nerve trunk, muscle) • Stimulus level (maximal, submaximal) • Number of stimuli (single, multiple)
Isometric • Muscle length	vs.	*Dynamic* • Action mode (concentric and/or eccentric, stretch-shortening cycle) • Velocity • Range of motion
Maximal • Rate of force development	vs.	*Submaximal* • Intensity
Sustained • Exhaustion	vs.	*Intermittent* • Duty cycle • Exhaustion

Other *important factors* to control:
• Subject characteristics (age, sex, physical activity level, training status);
• Familiarisation with both fatiguing exercise and assessments;
• Motivation during exercise and assessments.

necessity to select/conceive a fatiguing simulated exercise, which would mimic the type of muscle actions realised during real exercise (Table 2.2). It is worth noting that such specificity should also be respected when determining the characteristics of muscle actions for fatigue quantification (cf. next section).

In this chapter, we will enumerate the different methods allowing quantification of muscle fatigue, which mainly include the tools for the assessment of force–time and power–time curves (part I) and the techniques for the evaluation of neuromuscular (part II) and metabolic (part III) mechanisms of fatigue.

MEASUREMENT METHODS OF MUSCLE FATIGUE

Part I: tools and procedures for force/power measurement

In this section, we will present the methodology of performance decline assessment consecutive to real or simulated exercise, which includes the tools, test procedures, outcomes and some considerations (advantages, drawbacks) for different physical tasks. The choice of an appropriate

fatigue outcome and of the timing of measurements should be made in relation to the research question. As previously outlined, two models are generally used to assess exercise-induced performance decline: mechanical power
loss during dynamic exercise versus isometric force loss as a before-to-after
exercise ratio. Therefore, these two modalities differ in the main outcome
measure (power versus force) and in the number and timing of repeated
measurements (several, during exercise versus two, before and after exercise).

Mechanical power decline can be quantified during whole-, lower- or
upper-body short maximal exercise where dynamic contractions are performed by several muscles, mainly in closed-chain conditions. Maximal
power is attained within a few seconds and a gradual fall is subsequently
observed. These muscle actions are very similar to those realised during
sport, work or daily activities. On the other hand, isometric open-chain
testing (e.g. seated knee extension) has been, for technical reasons (standard conditions, high reliability), the most commonly used method to assess
force decline (or torque, hereafter referred to as force) of single-muscle
groups. However, isometric force measures underestimate significantly
functional impairment, i.e. the decline in peak muscle power is higher than
maximal isometric force, so that the former is considered to be more
appropriate to assess performance in dynamic exercise. Power assessment
may also give more information compared to isometric force because significant changes in the metabolic processes involved in energy release and
utilisation are more easily detected by muscle shortening (Vøllestad, 1997).

Friction-braked (Monark), air-braked (Repco, Kingcycle), electromag-
netically braked (Lode) and special constructed constant-velocity (McCartney et al., 1983) ergometers, as well as mobile systems (SRM, Powertap)
mounted on the drive train of the cyclist's own bike can be used to measure
cycling power against a resistance. Systematic and random errors associated to power measurements provided by these different ergometers have
been reviewed by Paton and Hopkins (2001). Typically, temporal changes
in maximal or mean power output are examined through one or several
repeated short-term all-out exercise bouts, where bout number and duration can be modulated according to anaerobic pathways contribution. The
30-s all-out Wingate test or repeated-sprint exercise (RSE) tests (e.g. five
6 s maximal sprints every 30 s, Bishop et al., 2001) are some examples of
fatigue assessment during cycling exercise. Concerning the outcome measures, it has been demonstrated that (i) power output is more reliably
assessed with measures of mean power as opposed to peak power during
RSE (Glaister et al., 2004); and (ii) although fatigue measures have lower
reliability than absolute power output, the per cent decrement recorded
over each effort is the more reliable technique (Fitzsimons et al., 1993).
Laboratory fatigue tests on cycle ergometers are largely used because of the
possibility to combine cardiorespiratory, blood pressure or blood lactate
assessments. Despite these advantages, cycle ergometry is not an exercise
mode specific to sports involving repeated sprints.

Non-motorised treadmills (Woodway) instrumented with a force transducer (strain gauge load cell) have therefore been developed (Lakomy, 1984), which offer the possibility of testing RSE using running as the exercise model (ecological validity). A non-elastic belt is used to tether the wall-mounted height-adjustable strain gauge to the subject's body via a harness. Treadmill speed is generally quantified using spring-loaded generators or reflective optical switches. Therefore, horizontal power can be calculated as the product of force and speed for average and instantaneous data. In these conditions, it has been demonstrated that measures of speed are more reliable than measures of power and force, and the use of average data further increases the reliability compared to instantaneous values (Tong *et al.*, 2001). However, reliability of fatigue indices for RSE on non-motorised treadmills is still poor (Hughes *et al.*, 2006).

Alternatively, the decline in mechanical power output with fatigue can also be quantified during daily activities, such as walking and stair climbing. Funato *et al.* (2001) recently described a self-driven instrumented treadmill which allows recording of both horizontal and vertical ground reaction forces, as well as horizontal pushing force on a handlebar during walking and running. Contrary to this high-tech approach, Margaria *et al.* (1966) proposed a formula to estimate mechanical power during stair climbing, which is simply based on the measurement of the time necessary to cover a given vertical distance at maximal velocity, two steps at a time, where:

$$\text{power} = \frac{(\text{body mass} \times 9.81 \times \text{vertical distance})}{\text{time}}$$

Unfortunately, there is a lack of information on the decline in walking and stair-climbing power in elderly and patient populations during daily life and as a result of a treatment or surgical intervention, which could be extremely relevant to geriatric, neurological and orthopaedic medicine.

Coming back to fatigue assessments for athletic populations, sport-specific ergometers such as wind-braked (Concept2), friction-braked (Gjessing) or air-braked (Rowperfect) rowing ergometers can be used, which generally provide online feedback about stroke parameters such as speed, pace and power. The most commonly adopted ergometer is probably the Concept2. Despite high test–retest reliability of rowing performance on this ergometer, power values provided by the machine are significantly underestimated (~7%) compared to the actual power produced by the rower (Boyas *et al.*, 2006). However, rowing ergometers are rarely used to evaluate fatigue during very short all-out exercise, but rather, they are used to simulate rowing competitions (mainly 2,000 m), where pacing strategy invalidates fatigue assessment.

Vertical jump power can be directly measured using fixed or portable force plates (Kistler, Bertec, AMTI), while it can also be estimated from flight time recordings (Bosco *et al.*, 1983) provided by contact mats (Ergojump) or photoelectric cells (Optojump) on the field. Typically, mechanical power associated with single jumps is recorded during a series of consecutive counter-movement jumps (15–60 s, at a frequency of about 1 Hz), in an attempt to mimic sport-specific fatigue (e.g. alpine skiing). The power decline is subsequently expressed as a percentage of the higher of the first 3–5 jumps to the last 3–5 jumps. Another estimate of muscle fatigue is the number of jumps in which power declines to 25–50% of the initial power. The main limitations of these types of fatigue tests are the control of posture, particularly in the fatigued state, which is necessary for the validity of jumping power assessments, and the maximality of jumps throughout the entire test duration, since some subjects may pace themselves to some extent during the test. Portable accelerometry-based systems are now available (Myotest), which could provide jumping power under field conditions; however, their validity and reliability remain to be ascertained.

In addition to the valid measurement of muscle torque, work or power, commercially available isokinetic dynamometers (Biodex, Humac Norm, Con-Trex) offer the possibility to test muscle fatigue (or "endurance", which is basically the opposite of fatigue), mainly in open-chain but also in closed-chain conditions. Isokinetic fatigue is often quantified as the percentage decline in gravity-corrected peak torque or average work (which additionally takes into account the range of motion) over a certain number of repetitions (usually 20–50), completed at a relatively fast concentric velocity (usually $180°·s^{-1}$). Thorstensson and Karlsson (1976) were the first to define a fatigue index as the torque of the three last contractions as a percentage of the initial three contractions out of 50 contractions. They also demonstrated a significant correlation between the fatigue index and the percentage of fast fibres in the vastus lateralis muscle of ten subjects. Recently, we (Maffiuletti *et al.*, 2007) and others (Pincivero *et al.*, 2001) have provided experimental evidence that the linear slope of the decline in torque or work output over the series of fatiguing contractions is a better indicator than the fatigue index in terms of test–retest reliability. However, reliability of isokinetic fatigue indexes is inferior to that of the absolute parameters (Maffiuletti *et al.*, 2007), similar to cycling and running ergometry. Therefore we recommend interpretation of all these fatigue outcome measures with caution. It should also be remembered that the use of isokinetic actions for fatigue assessment is questionable because they do not exist in normal movements.

Finally, strength training machines equipped with linear encoders, accelerometers and force transducers allow force and power quantification during static or dynamic actions in both open-chain and closed-chain conditions. In isometric conditions, the force transducer (generally an S-shaped strain gauge load cell) should be fixed so as to record the direct line of force. In dynamic conditions, some commercially available machines (Technogym) are equipped with encoders measuring linear displacement, which in turn allows calculation of work and power (Bosco *et al.*, 1995).

Alternatively, linear encoder systems (MuscleLab) or accelerometers (Myotest) can be applied to standard machines and even to barbells or free loads. In isometric conditions, the decline in isometric MVC force is very often considered as the main measure of interest (see next section), while fatigue-induced alterations in rate of force development could also be examined. In dynamic conditions, the decline in concentric power during a typical strength training session, both within the same set and within different sets, has become a routine measure in the control of training for high-level and recreational sportsmen.

Part II: techniques and procedures to evaluate neuromuscular mechanisms of fatigue

In this section we will present how surface EMG and muscle force recordings obtained before and immediately after exercise can be used to investigate central and peripheral mechanisms of fatigue. The classical approach consists of realising a series of voluntary and stimulated contractions of the agonist and antagonist muscles, with concomitant EMG and force recordings. It is essential that this series of measurements is completed immediately after (within 1–2 min) the end of the fatiguing exercise and its total duration does not exceed 5–10 min. Measurements can also be performed during exercise using short interruptions to investigate the time course of alterations, and/or hours or days after the fatiguing bout, to examine the recovery process (particularly for muscle damage studies). Such neuromuscular assessments provide some evidence of the central and peripheral alterations induced by exercise, according to their localisation within the neuromuscular system. Central fatigue is defined as a progressive reduction of voluntary activation of muscle during exercise (activation failure) that could be due to supraspinal and/or spinal mechanisms (Gandevia, 2001). On the other hand, peripheral fatigue is produced by changes at or distal to the neuromuscular junction, and could include transmission and/or contractile failure (excitation–contraction coupling failure). It should, however, be remembered that contrary to animal studies, these *in vivo* assessments do not allow isolation of single processes of fatigue.

Force recordings

Isometric muscle force, as assessed using force transducers or isokinetic devices, is the most widely employed parameter for muscle fatigue studies. MVC force has been defined as the most direct assessment of fatigue and the first choice method, because the output is an integrated result of the total chain of neuromuscular events (Vøllestad, 1997). Very short (3–5 s) actions are generally used to assess this variable, where MVC force is attained 400–600 ms after contraction onset and a plateau is seen in the subsequent period. Moreover, electrical or magnetic stimuli adequately superimposed to the MVC plateau allow investigation of the maximal evocable force (Table 2.3), and in turn of muscle activation failure (see below). It should, however, be pointed out that, although isometric MVC force assessment is highly reliable, easily administered, requires little skill involvement and is relatively inexpensive (Wilson and Murphy, 1996), several experimental recommendations should be observed to obtain real MVC and muscle activation scores, more particularly in the fatigued state. These criteria, which have been described in detail elsewhere (Wilson and Murphy, 1996; Gandevia, 2001) include: appropriate familiarisation and practice; visual feedback of performance with variations in gain; standardised verbal encouragement with clear instructions (hard versus hard and fast); several repeat trials; negligible pre-tension prior to the test; and eventually provision of rewards.

Muscle activation failure can be measured using twitch interpolation during an MVC, where one (or more) supramaximal stimulus is delivered to the motor axons innervating the muscle and voluntary versus evoked force outputs are compared. Supramaximality of stimulation should be ensured by using an intensity consistently higher (20–50%) than that eliciting the maximal M wave (i.e. compound muscle action potential) and peak twitch. Any additional force produced by the stimulation indicates incomplete motor unit recruitment and/or suboptimal firing frequency of active units. Two formulas are classically used to estimate the extent of activation: voluntary activation, where the size of interpolated twitch is divided by the

Table 2.3 Conditions, techniques and outcomes associated with muscle force recordings

Voluntary contraction
  Static
    • Without stimulation (MVC force)
    • With stimulation* (maximal evocable force (activation))
  Dynamic
    • Without stimulation (MVC force)
    • With stimulation* (maximal evocable force (activation))

Stimulated contraction
  Electrically
    • Nerve or muscle
      – Single* (twitch contractile properties)
      – Short train* (contractile properties)
      – Tetanic* (Force-frequency relation and low-to-high frequency ratio)
        – Low frequency
        – High frequency
  Magnetically
    • Cortex
      – Single* (superimposed twitch torque)
    • Nerve
      – Single* (twitch contractile properties)
      – Short train* (contractile properties)

Note
*Single, short train and tetanic stimuli superimposed to MVC allow investigation of muscle activation (twitch interpolation).

size of a control twitch produced by identical stimulation in a relaxed potentiated state (Thomas *et al.*, 1989), i.e. voluntary activation (%) = [1 − (interpolated twitch/control twitch)] × 100, and central activation ratio, which consists of a simple MVC-to-maximal evocable force ratio (Kent-Braun and LeBlanc, 1996), i.e. central activation ratio = MVC/ (MVC + interpolated twitch). Importantly, voluntary activation has been shown to be a more sensitive (Bilodeau, 2006) and valid (Place *et al.*, 2007) index of central fatigue than the central activation ratio. However, limitations in the twitch interpolation technique have been identified (Gandevia, 2001), which also include the contribution of intramuscular processes to superimposed force with fatigue (Place *et al.*, 2008).

On the other hand, force evoked by single, paired or trains of stimuli under resting conditions can be used to examine contractile impairments (Table 2.3) and, together with the analysis of the M wave recorded by surface EMG (see below), allows excitation–contraction coupling failure to be approached *in vivo* (Desmedt and Hainaut, 1968). For example, M wave amplitude and twitch force both decrease with the occurrence of transmission failure. With contractile failure however, no changes are observed in M wave amplitude, while twitch force is significantly decreased, i.e. failure of excitation–contraction coupling. Additionally, fatigue-related changes in twitch-time course parameters such as increased time to peak twitch (contraction time) and half relaxation time (Duchateau and Hainaut, 1985) could be compared to M wave changes to gain insight into excitation–contraction coupling impairments.

Alternative approaches to the study of excitation–contraction coupling alterations with fatigue are represented by the analysis of low-to-high frequency force ratio (Edwards *et al.*, 1977) (see the section below, "Electrical or magnetic stimulation?") and muscle twitch post-activation potentiation (Miller *et al.*, 1987). This latter refers to any enhanced contractile response which results from prior contractile activity and is generally computed as the ratio between the size of the twitch recorded before and immediately after an MVC. Since fatigue can coexist and therefore be masked by potentiation, we (Place *et al.*, 2007) and others (Kufel *et al.*, 2002) recently suggested that potentiated twitch or doublets are superior to nonpotentiated responses for detecting contractile fatigue. We would also like to remind that unpotentiated twitch force is an inappropriate measure of peripheral fatigue (Edwards *et al.*, 1977; Millet and Lepers, 2004).

EMG recordings

Electromyography (EMG) is the recording of electrical signals that are sent from motoneurons to muscle fibres (action potentials) while they propagate along the sarcolemma, from the neuromuscular junction to the extremities of the muscle fibres (Enoka, 2002). The classical configuration for surface EMG of a whole muscle consists of using two small electrodes positioned between the innervation zone and the tendinous insertion (bipolar), while single electrode (monopolar) or arrays and even grids of electrodes (multipolar) can be used (see www.seniam.org for recommendations about electrode placement and location).

Surface EMG reflects both central and peripheral neuromuscular properties, since its main characteristics, such as amplitude and power spectrum, depend on muscle fibre membrane properties and on the timing of motor unit action potentials (Farina et al., 2004). Researchers should be aware of the physiological and nonphysiological factors influencing surface EMG (Farina et al., 2004), as well as of the real information extracted from this signal. For example, global surface EMG underestimates motor unit activity due to overlapping of positive and negative phases of motor unit potentials (i.e. amplitude cancellation) (Keenan et al., 2005). It should also be considered that the difficulties in interpretation of isometric EMG amplitude are further amplified in the case of dynamic contractions (Farina, 2006) and spectral estimations such as mean and median frequency.

Although less used than the twitch interpolation, EMG amplitude obtained during an MVC (Table 2.4) is another technique to detect central alterations due to fatigue, which, contrary to twitch interpolation, allows changes in descending drive between the synergists of a muscle group to be distinguished (e.g. elbow flexors, knee extensors). In order to enable peripheral changes to be excluded from the interpretation of the data and therefore to provide an index of central activation failure, average rectified or root mean square EMG could be expressed relative to the size or surface of the M wave obtained with supramaximal nerve stimulation. However, due to the limitations outlined above, caution should be taken in using this latter technique for central fatigue assessment. The gold standard for activation failure is probably the comparison of the decline in MVC force with

Table 2.4 Conditions, techniques and outcomes associated with surface EMG recordings

---

Voluntary contraction
- Static (root mean square or average rectified value)
- Dynamic (root mean square or average rectified value)

Evoked contraction
- Mechanically
  - Stretch (stretch reflex)*
  - Tendon percussion (T reflex)*
- Electrically or magnetically
  - Motor cortex (motor-evoked potential)*
  - Motor nerve
    - Maximal (M wave)*
    - Submaximal (H reflex)*

Note
*The amplitude of these potentials should be normalised to the amplitude of the maximal M wave obtained in the same conditions.

the decline in maximal tetanic force (Merton, 1954; Bigland-Ritchie *et al.*, 1978), even if it is not feasible in the large majority of anaesthetised human subjects.

On the other hand, compound action potentials can be evoked using mechanical, electrical or magnetic stimuli (Table 2.4) and their main characteristics (peak-to-peak amplitude and duration, area, latency) can be compared before and after fatigue. The maximal M wave is evoked by the recruitment of all motor axons and therefore provides an estimate of the response given by the whole motoneuron pool. Its amplitude is a measure of transmission across the neuromuscular junction and muscle membrane excitability. Maximal M wave amplitude is also used as a normalisation standard for all evoked potentials, provided the former is obtained in the same conditions (muscle action and muscle length) as the latter. Stimulated potentials include motor-evoked potentials and Hoffmann (H) reflexes. Fatigue-induced changes in motor-evoked potential amplitude and silent period duration, which are obtained using transcranial magnetic stimulation (see below), allow investigation of corticospinal excitability alterations. H reflex, evoked by selective (submaximal) stimulation of Ia afferents contained in the corresponding mixed nerve, can be potentiated (Trimble and Harp, 1998) or depressed (Racinais *et al.*, 2007) following exercise, indicating altered motoneuron excitability and/or presynaptic inhibition of Ia afferents. Finally, spinal reflexes can also be evoked as a result of transient stretches, which can be produced in a reliable way using electromagnetic hammers or special ergometers (Avela *et al.*, 1999) to obtain, respectively, the tendon (T) reflex and the stretch reflex. Compared to the already challenging H reflex, the interpretation of these reflex responses is further complicated by the influence of γ fusimotor drive, which controls the sensitivity to stretch of muscle spindle primary endings (Pierrot-Deseilligny and Mazevet, 2000).

Electric or magnetic stimulation?

As discussed above, non-invasive electrical and magnetic stimuli can be applied at different levels of the neuraxis (from motor cortex to peripheral nerve trunk) or over the muscle, both at rest and during voluntary contractions (see Table 2.1), to investigate central and/or peripheral mechanisms of fatigue through the analysis of evoked EMG and force signals. Electrical stimulation, in which current is transmitted to a body organ through surface electrodes of different sizes and shapes, has been used in clinical and research settings for several decades. This modality is very effective for stimulation of superficial peripheral nerves. However, deep nerves cannot be easily excited because of the high resistance of some tissues (e.g. bones) and because of the discomfort associated with strong stimulations. Magnetic stimulators were therefore developed in the 1980s (Barker *et al.*, 1985) for painless cortical stimulation, which is in contrast to the sensations felt using electrical stimulation with electrodes positioned on the scalp. Since its introduction, transcranial magnetic nerve stimulation has been in widespread use for the investigation of motor-evoked potential characteristics and silent period duration, and more recently, also for the assessment of voluntary activation of upper limb muscles (Todd *et al.*, 2003). Interestingly, magnetic stimulation of peripheral nerves (e.g. phrenic, femoral) has also been shown to provide valid and reliable assessment of muscle contractile function, including fatigue (Polkey *et al.*, 1996; Vergès *et al.*, 2009) and voluntary activation level (O'Brien *et al.*, 2008), compared to classical electrical stimulation. Although magnetic stimulation presents some advantages with respect to electrical stimulation (e.g. lower discomfort, possibility to excite deep nerves), it should be remembered that magnetic stimulators are quite expensive and cannot deliver trains of stimuli at high intensity and high frequency, such as those generated by commercially available electrical stimulators.

Supramaximal trains are indeed required for investigating exercise-induced changes in the force–frequency relation, more particularly in the low-to-high frequency–force ratio (generally 10–20 Hz to 50–100 Hz) (Edwards *et al.*, 1977). Any increase in this ratio following exercise is commonly associated with long-lasting, low-frequency fatigue, which may be caused by excitation–contraction coupling impairment. On the other hand, a decrease of the low-to-high frequency ratio is associated with high-frequency fatigue, which is thought to reflect impairment of muscle excitation at the level of the neuromuscular junction. However, tetanic electrical stimulation of intact muscles is painful and unsuitable in clinical practice. Therefore, we recently proposed to use force traces of paired stimuli at both 10 Hz and 100 Hz as a surrogate for stimulation trains for the assessment of low- and high-frequency fatigue (Vergès *et al.*, 2009).

Surface electrical stimulation can be differentiated in nerve versus muscle stimulation according to the position and size of the stimulating electrodes. In the first case, small electrodes are positioned over the peripheral nerve trunk, at a site where it runs close to the skin (e.g. poplitea fossa for the tibial nerve, femoral triangle for the femoral nerve). Over-the-muscle electrical stimulation is performed with large electrodes placed in proximity to the muscle motor point, which activates intramuscular nerve branches and not the muscle fibres directly (Hultman *et al.*, 1983). Even if this latter modality is largely adopted for central fatigue (twitch interpolation) and contractile fatigue assessment, three important limitations have to be acknowledged: muscle activation is incomplete and superficial; M waves cannot be easily recorded; central contributions from the recruitment of spinal motoneurons could overestimate considerably the recorded force (Collins, 2007). Therefore, nerve and muscle electrical stimulation should be considered as two distinct techniques of stimulation, which cannot be used interchangeably for fatigue studies. Nerve stimulation should always be used, whenever possible, to assess neuromuscular function.

Part III: techniques and procedures to evaluate metabolic mechanisms of fatigue

Thirty years ago, Hoult *et al.* reported for the first time that phosphorus metabolites could be observed *in vivo* using ^{31}P Magnetic Resonance Spectroscopy (MRS), opening promising opportunities of understanding muscle energetics *in vivo* under strictly non-invasive conditions (Hoult *et al.*, 1974). From that time, MRS technology has rapidly evolved with the development of RF-surface coils in 1980 (Ackerman *et al.*, 1980), the availability of high-field, wide-bore superconducting magnets and methodological developments (dedicated pulse sequences, spatial localisation of NMR signal), making MRS a tool of choice for investigating muscle energetics non-invasively in animals and humans. So far, a large number of publications have been devoted to the investigation of muscle energetics in a variety of conditions ranging from diseases (Chance *et al.*, 1980; Cozzone *et al.*, 1996; Argov *et al.*, 2000) to training regimens (McCully *et al.*, 1988; Kent-Braun *et al.*, 1990; Hug *et al.*, 2004), as compared to normal conditions. Such studies have provided interesting information regarding not only pathologies and metabolic changes associated with training, but also about what normally occurs throughout muscle contraction in terms of balance between energy production and consumption. In this review, we intend to provide key information to the non-specialist in order to understand what ^{31}P MRS can tell us about muscle energetics. After a brief introduction regarding the MR techniques and the corresponding information obtained in exercising muscle, we will present the different ergometers used in order to measure the mechanical performance throughout MR investigation of muscle fatigue both in humans and in animals. The MR results related to muscle fatigue are beyond the scope of the present chapter.

Informational content of a ^{31}P MR spectrum

Measurement of phosphorylated compound concentrations in living cells is not easy. Traditional methods (such as percutaneous needle biopsy and freeze clamping) exhibit limitations, especially related to alteration of anatomic integrity and partial degradation of phosphorylated metabolites during extraction and analysis. In addition, repeated measurements cannot be performed on the same muscle, making the achievement of high time-resolution kinetics impossible. Compared to analytical methods, ^{31}P MRS offers the opportunity of measuring non-invasively and continuously with high time-resolution, the concentration of phosphorylated compounds involved in muscle energetics. In addition, comparison of direct biochemical measurements with ^{31}P MRS findings suggests that the two methods give comparable results (Tarnopolsky and Parise, 1999).

MR spectra are actually generated by placing samples in a powerful magnetic field and then exciting them with a radio-frequency energy. A typical ^{31}P MRS spectrum exhibits 6–7 peaks corresponding to phosphocreatine (PCr), inorganic phosphate (P_i), the three phosphate groups of ATP (in position α, β and γ) and phosphomonoesters (PME) (Figures 2.1 and 2.2). The P_i signal of ATP displays occasionally an up-field shoulder corresponding to NAD$^+$/NADH. Between the PCr and P_i signals, the phosphodiester signal is sometimes observed. This signal is usually assigned to glycerophosphorylcholine and glycerophosphoryethanolamine, which can be detected as a small peak in normal muscle spectra (mainly from the lower limb) and as a larger peak in patients with muscle dystrophy (Lodi *et al.*, 1997), indicating membrane breakdown. Given the low sensitivity of the technique, the free metabolically active ADP concentration, which is only a tiny fraction of its total intracellular concentration, cannot be measured. It can, however, be calculated using the creatine kinase equilibrium, where the total creatine content is taken as either 42.5 mM or considering that phosphocreatine represents 85% of the total creatine content (Kemp *et al.*, 2001). Similarly, AMP concentration can be calculated using the adenylate kinase equilibrium (Kemp *et al.*, 2001). In the absence of biochemical data, ATP is often assumed to be normal and is used as the equivalent of an internal standard in order to calculate the concentrations of other metabolites.

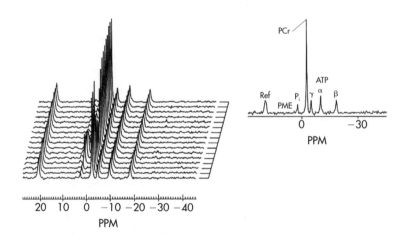

Figure 2.1 Typical series of ^{31}P MRS recorded in human forearm flexor muscles. MRS have been recorded at 4.7T (Biospec 47/30 Bruker) during a standardised rest–exercise–recovery protocol with a time resolution of 15 s. A single spectrum with the corresponding assignments is represented on the higher panel. Ref: reference compound (phenyl phosphonic acid), PME (phosphomonoesters), P_i (inorganic phosphate), PDE (phosphodiesters), PCr (phosphocreatine), phosphate groups of ATP in position γ, α and β

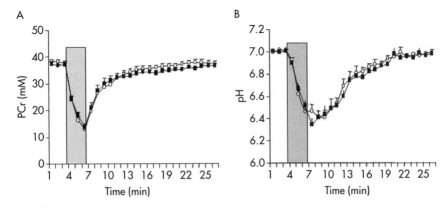

Figure 2.2 Typical PCr (A) and pH (B) time-dependent changes recorded throughout a standardised rest–exercise (shaded area)–recovery protocol in a group of 18 subjects. Measurements (black and white symbols) have been recorded seven days apart. Results are presented as means ±SD, and one can observe the very good reproducibility of measurements

Apart from the dynamic measurements of high-energy phosphate compounds, ^{31}P MRS offers the only non-invasive way of assessing intracellular pH. Indeed, under conditions of physiological pH, and considering one of the pKa, i.e. 6.75, two forms of P_i coexist ($H_2PO_4^-$ and HPO_4^{2-}). These two forms are exchanging so fast that only a single P_i signal is detected. However, the chemical shift of this single signal is a weighted average of both monobasic and dibasic forms. Due to this sensitivity of the P_i chemical shift to pH, it is possible, with appropriate calibration curves, to translate any P_i shift in terms of intracellular acidosis or alkalosis.

The quantitative measurements of high-energy phosphate compounds and pH allow computation of a number of derived metabolic parameters, such as the oxidative phosphorylation potential and the free energy of ATP hydrolysis *in vivo*. All those parameters reflect the regulation and control of energy metabolism, ion transport and muscular contraction.

Also, magnesium concentration has been calculated from changes in ATP chemical shift on the basis that the chemical shift of the resonance corresponding to the beta group is sensitive to the intracellular magnesium concentration (Gupta *et al.*, 1983; Ward *et al.*, 1996; Iotti *et al.*, 2000). Fat, fibrous tissue blood and extracellular fluid contribute no significant signal and mitochondrial metabolites are too tightly bound to interfere.

Given the low magnetic resonance sensitivity of ^{31}P (6% of proton) and the low tissue concentrations of some of the relevant metabolites, MR signals are time-averaged over a period ranging from a few seconds to several minutes, depending on the required signal-to-noise ratio and the desired time-resolution. In addition, the MR signal is detected with a surface coil over a relatively large muscle volume proportional to the surface coil radius, making this signal a weighted average of the muscle fibres existing within the sampling volume. This has to be taken as a comparative item with histological and biochemical analyses, which are often carried out on very small samples of tissues, which may not give a representative picture of the biochemical state of the muscle. However, care has to be taken not to turn this advantage into a drawback while sampling exercising and non-exercising muscles at the same time. MRS can be combined with MRI in order to achieve a proper localisation of the coil.

Technical considerations

The requirement for magnetic field homogeneity generally dictates that the muscle examined be positioned at magnet centre, and remains in a fixed position during data collection. In that respect, dedicated ergometers have been designed in each laboratory in order to investigate exercising muscles within superconducting magnets. So far, adductor pollicis (Miller et al., 1987), forearm and wrist flexor muscles (Chance et al., 1980; Arnold et al., 1984; Bendahan et al., 1990), calf (Ryschom et al., 1995; Lodi et al., 2002) and thigh muscles (Bernus et al., 1993; Rodenburg et al., 1994; Whipp et al., 1999; Francescato and Cettolo, 2001) have been investigated using ^{31}P MRS. These ergometers are intended to allow muscle exercise within a superconducting magnet and should provide accurate measurements of the corresponding mechanical performance so that metabolic changes can be properly analysed with respect to the mechanical measurements. Initially, basic ergometers have been designed for isometric contractions, but due to the absence of movement, mechanical work and power output could not be quantified (Miller et al., 1988; Bangsbo et al., 1993). Other ergometers, including bulbs (Quistorff et al., 1990; Saab et al., 2000), elastic bands (Francescato and Cettolo, 2001) or cable-and-pulley systems (Bendahan et al., 1990; Marsh et al., 1991; McCreary et al., 1996; Nygren and Kaijser, 2002; Raymer et al., 2004) have been designed for dynamic contractions of wrist and plantar flexor muscles. For the fewer ergometers developed for quadriceps, exercise was either limited to a single leg (Park et al., 1990; Weidman et al., 1991; Smith et al., 1998; Barker et al., 2006) or work-rate was not accurately determined (Gonzalez et al., 1993; Whipp et al., 1999). Indeed, no quantitative measurements of weight displacement, contraction frequency or displacement speed was performed, so the resulting power output was just roughly estimated (Whipp et al., 1999). Using a more elaborate ergometer, Rodenburg et al. reported both force and displacement speed values, but mentioned workload reduction due to temperature changes during a prolonged exercise (Rodenburgh et al., 1994). More recently, a new ergometer allowing dynamic and isometric knee extension exercise within a whole body 1.5 T Siemens Vision Plus magnet has been proposed. As reported by Layec et al. (2008), mechanical measurements obtained with this ergometer were highly reliable and accurate without any significant effect of magnetic field. In addition, it was possible to assess energy metabolism over a broad range of exercise intensities and for subjects with different heights. This ergometer allows dynamic (unilateral or bilateral) and isometric knee extension exercises using both legs. The isometric facility was particularly important, given that it allows a reliable standardisation of exercise on the basis of MVC measurements.

In addition to experiments conducted in humans, animal experiments can also be conducted as long as mechanical measurements can also be recorded and, more importantly, muscle can be exercised. Paradoxically, in animal research the benefit of MR being non-invasive is lost, because the experimental set-ups usually incorporate invasive systems to induce muscle contractions by direct nerve stimulation or to measure mechanical performance with a force transducer attached to the muscle tendon (Foley and Meyer, 1993; Hogan et al., 1998; Cieslar and Dobson, 2001). Surgical intervention is required to position the stimulation electrode on the nerve and to introduce the force transducer. In addition to the risk of disturbing muscle physiology, in particular the neurovascular supply, these interventions necessitate sacrificing the animal after the experiment and thus prohibit repetitive explorations in the same animal. Interestingly, Giannesini et al. (2005) have designed and constructed a new apparatus integrating two non-invasive systems for transcutaneous muscle stimulation and force measurement respectively (Figure 2.3). Briefly, this system integrates four distinct components allowing prolonged anaesthesia with control of the animal's body temperature, transcutaneous electrical stimulation of the gastrocnemius muscle, force output measurement and multimodal MR acquisition. Prolonged anaesthesia is maintained by gas inhalation via a custom-built facemask connected to an open-circuit gas anaesthetic machine. In order to maintain the rat at physiological temperature during anaesthesia, the cradle integrates an electric heating blanket in a feedback loop with a temperature control unit connected to a rectal probe. Transcutaneous stimulation of the gastrocnemius muscle is performed using two surface electrodes connected to an electrical stimulator. Electrodes are integrated in the cradle so that when the rat is placed in the cradle, one electrode is located above the knee, and the other at the heel. Force output is measured with a custom-built ergometer consisting of a foot pedal (20×42 mm) directly connected to a hydraulic piston. The foot pedal rotates freely on a nylon axis, which is situated at the level of the ankle joint. A hydraulic circuit filled with water connects the piston to a pressure transducer, which is placed outside the magnet. The pedal position can be adjusted such as to modify the angle between foot and lower hind limb. This adjustment enables the gastrocnemius muscle to be stretched passively in order to obtain maximum force production in response to electrical stimulation. MR acquisition is performed using a custom-built spectroscopy/imaging probe consisting of an elliptic ^{31}P MRS surface coil (10×16 mm) geometrically decoupled inside a 30 mm diameter Helmholtz imaging coil. Using this system, the failure of muscle performance has been compared among different stimulation protocols.

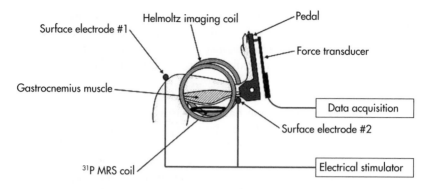

Figure 2.3 Schematic representation of the non-invasive setup allowing *in vivo* MR investigation of skeletal muscle function in rat gastrocnemius muscle

Reproducibility of measurements

For meaningful comparisons, data must be reproducible from study to study and it is important to assess repeatability in groups of subjects and patients. Foremost, it must be remembered that metabolic changes in a control population are highly heterogeneous (Miller *et al.*, 1987; Bendahan *et al.*, 1990; Miller *et al.*, 1995; Kemp *et al.*, 2001). The usual standardisation procedures, such as measurements of maximum voluntary contraction, do not reduce this between-subject variability (Bendahan *et al.*, 1990; Miller *et al.*, 1995; Kemp *et al.*, 2001). In that respect, it is clear that a set of factors which could influence metabolic changes in exercising muscle has to be taken into account and ad hoc standardisation procedures that would reduce the inter-subject variability have to be used for proper comparisons among various groups of subjects.

Between-subject variability

In a limited number of subjects, it has been shown that neither fasting nor carbohydrate loading significantly modified the extent of metabolic changes in exercising muscle (Miller *et al.*, 1995). Similarly, severe metabolic acidosis induced by ammonium chloride loading had no effect on changes recorded during exercise (Miller *et al.*, 1995). Ageing is another potential factor that should be taken into account. The studies based on measurements of enzymatic activities have suggested a decreased oxidative capacity with age (Kohrt *et al.*, 1991; Coggan *et al.*, 1992). Results from MRS investigations differed in whether physical activity has been taken into account or not (Russ and Kent-Braun, 2004). Ageing has been related to both a 50% reduction in the PCr recovery rate and a reduced mitochondrial content, with no alteration regarding the metabolic changes in exercise (Conley *et al.*, 2000). When physical activity has been taken into account, no deficit of aerobic capacity has been reported, whereas the PCr depletion was not modified and the pH was more alkalotic in older subjects (Kent-Braun and Ng, 2000), in agreement with previous studies conducted in sedentary and moderately active subjects (Chilibeck *et al.*, 1998). These results would imply a primary role of disuse in the decline of oxidative capacity rather than an inherent age-related defect. In addition, the recent observation of increased maximum aerobic capacity in older subjects as a result of training would be in keeping with the hypothesis that the decline in oxidative capacity would mainly result from a reduction in habitual physical activity rather than of ageing per se (Jubrias *et al.*, 2001). However, this conclusion must be moderated on the basis of two points. First, the oxidative capacity per mitochondrial volume measured in older subjects after a training period was still below the value reported in young adult muscle (Conley *et al.*, 2000); and adaptive mechanisms were distinct from what has been reported so far in young adult muscle, especially regarding resistance training (Jubrias *et al.*, 2001).

Another factor that should be taken into account regarding the between-subject variability is related to high body fat stores. Indeed, changes in fibre type proportions that could affect muscle energetics have been reported in obese subjects (Wade *et al.*, 1990) and high-energy phosphate metabolism has been investigated using ^{31}P MRS in order to test this hypothesis. Studies conducted in moderately obese women (BMI range: 27–30 kg·m^2) (Larson-Meyer *et al.*, 2000b) and in prepubertal girls with a familial predisposition to obesity (Treuth *et al.*, 2001) have clearly shown that muscle energetics was not altered. In addition, a weight-reduction programme had also no effect on mitochondrial function, thereby suggesting that a low substrate oxidative capacity of skeletal muscle is not involved in the pathogenesis of obesity on the contrary to what has been suggested earlier (Kriketos *et al.*, 1996).

Within-subject variability

Beside the between-subject variability, the within-subject variability is also important to analyse. Several studies have clearly demonstrated very low within-subject variability, as illustrated in Figure 2.2, and indicated a very high reproducibility (Miller *et al.*, 1995; Larson-Meyer *et al.*, 2000a; Taylor, 2000; Bendahan *et al.*, 2002). For instance, variation coefficients calculated from repeated experiments were 5–10% for metabolic indices at rest, such as ADP and PCr/P$_i$, and 3–15% for variables measured during exercise and in the recovery period (Larson-Meyer *et al.*, 2000a). The reliability of MRS parameters has also been investigated in children in response to three repeated exhaustive ramp exercise tests (Barker *et al.*, 2006). Indices recorded at rest, at the end of exercise and a threshold value (IT) related to the combined evolution of pH and P$_i$/PCr ratio were calculated. While the inter-observer variability was estimated as 5%, the coefficient of variations (CV) associated with the P_i/PCr ratios recorded at rest, at end of exercise and the IT values were 37%, 50% and 16%, respectively (Barker *et al.*, 2006). This indicates an acceptable reproducibility for IT values, but a poor reliability for P_i/PCr values. Values related to pH were also highly reproducible, with CV corresponding to 0.6% and 0.9% for resting and end-of-exercise values.

Standardisation procedures

This within-subject variability may also be used in order to understand muscle energetics and to initiate ad hoc standardisation procedures that could be used as reliable comparative methods between groups of subjects. It has been shown in normally active adults that an index of PCr consumption (($(PCr+P_i)$/PCr) (measured with ^{31}P MRS) was highly correlated with power output normalised to the volume of muscle (measured with MRI) in the plantar flexor compartment, indicating that combined MRS and MRI measurements could offer a way of reliably comparing subjects with different body size (Fowler *et al.*, 1997). Similarly, expressing metabolic changes such as PCr breakdown and intracellular acidosis in exercise muscle with respect to power output has been reported to offer an interesting standardisation procedure, independent of exercise protocols and anthropometric measurements (Mattei *et al.*, 2002). In agreement with such an approach, it has been shown, in exercising muscle, that the extent of PCr breakdown

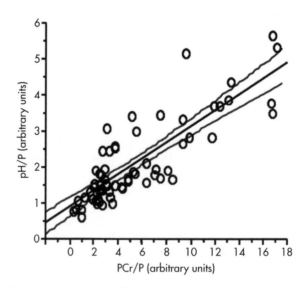

Figure 2.4 Linear relationship between PCr and pH values scaled to power output (P) and recorded in a group of subjects for different exercise intensities. Results are presented as arbitrary units. This linear relationship allows the standardisation of metabolic changes regardless of exercise intensity

was linearly linked to the intracellular acidosis, regardless of exercise protocols (Bendahan *et al.*, 1990; Mattei *et al.*, 2002), as illustrated in Figure 2.4. This suggests that various combinations of [PCr] and pH are multiple solutions for muscle fibres to reach a given level or range of ADP, which in turn will act as a regulator of energy production as previously described (Kemp *et al.*, 1993).

Future investigations

Employing non-MR-based techniques simultaneously with MRS can aid in data interpretation and will certainly broaden the scope of muscle investigation. Near-infrared spectroscopy (NIRS) provides data on tissue oxygenation, but reliability has still to be proven, mainly on the basis of comparative and/or combined analyses using MRS and NIRS (Hamaoka *et al.*, 1996). Electromyography is another technique of interest which can be used to study correlation between metabolic and electrical changes and provide interesting features regarding muscle fatigue (Bendahan *et al.*, 1996). Functional MRI can also be of interest for understanding muscle activation during exercise. Based on T2 changes due to uptake or redistribution of fluid within the exercising muscle, functional MRI is considered as a semi-quantitative method of assessing muscle recruitment during exercise (Meyer *et al.*, 2001).

CONCLUSION

It is concluded that maximal power (dynamic exercise) or MVC force (static exercise) can be used as a sort of gold standard to identify whether fatigue occurs or not. For fatigue studies based on power output assessment during dynamic exercise, mean power (cycling), torque slope (isokinetic) and speed (running) are better outcomes than peak power. The reliability of fatigue indexes obtained in these conditions should, however, be improved in the future. On the other hand, a series of isometric MVC and stimulated contractions (at rest and MVC-superimposed) completed before and immediately after exercise using concomitant force and EMG recordings can provide a valid and rapid assessment of central and peripheral fatigue induced by sport, work and daily activities.

MR spectroscopy and imaging can be used in order to investigate the metabolic and functional bases of muscle fatigue. The biological variability of results has to be acknowledged in order to determine invariant criteria which characterise "normality". A multimodal MR approach including MRS and MRI, together with other non-invasive techniques, should be chosen in order to thoroughly identify the exact mechanisms leading to muscle fatigue.

FIVE KEY PAPERS THAT SHAPED THE TOPIC AREA

Study 1. Vøllestad, N.K. (1997). Measurement of human muscle fatigue. *Journal of Neuroscience Methods*, 74, 219–227.

A short review focusing on the measurement methods of human muscle fatigue. After a quick introductory section, and basic definitions of fatigue (MVC force, maximal evocable force, maximal power output, muscle fatigue and central fatigue), the author presents the possible sites of fatigue from CNS processes to force/power output, and kindly relates these sites to the methods described in the last part of the review. These methods include direct assessments such as MVC force generation, power output, tetanic force and low-frequency fatigue, which should be privileged in fatigue studies, but also indirect assessments (twitch interpolation, endurance time, EMG), which are criticised for their limited value in measurement of human muscle fatigue.

Study 2. Gandevia, S.C. (2001). Spinal and supraspinal factors in human muscle fatigue. *Physiological Reviews*, 81, 1725–1789.

This excellent review covers all there is to know about central fatigue. After presenting the historical aspects and definitions related to central fatigue ("a progressive reduction in voluntary activation of muscle during exercise"), and providing evidence about the notion of "submaximal" voluntary activation during "maximal" efforts, the author describes in detail the spinal (e.g. altered input from muscle spindle, tendon organ and group III and IV muscle afferents) and supraspinal (e.g. changes in cortical excitability and inhibitability) mechanisms of muscle fatigue, as inferred from electrophysiological recordings (voluntary activation, evoked reflexes, motor-evoked potentials, etc.).

Study 3. Cairns, S.P., Knicker, A.J., Thompson, M.W. and Sjøgaard, G. (2005). Evaluation of models used to study neuromuscular fatigue. *Exercise Sport Sciences Reviews*, 33, 9–16.

This original review describes and appraises the different approaches (models) used to study neuromuscular fatigue in human subjects and muscle preparations. The authors suggest that the cause of fatigue has to be regarded as multi-factorial, and that the mechanisms of neuromuscular fatigue relate to the entire approach of how fatigue is studied (i.e. the fatigue model involved). They also propose that the diversity of fatigue mechanisms can be explained, in part, by the use of different fatigue measures (muscle force, displacement, work or power) and the timing of their measurement. Several methodological considerations are proposed, in line with the idea that fatigue assessment should depend on the specific question to be addressed. At the end of the paper, specific recommendations for models to study neuromuscular fatigue are provided, which are extremely helpful to researchers for designing rigorous and clever fatigue studies.

Study 4. Giannesini, B., Izquierdo, M., Le Fur, Y., Cozzone, P.J., Fingerle, J., Himber, J., Kunnecke, B., Von Kienlin, M. and Bendahan, D. (2005). New experimental setup for studying strictly noninvasively skeletal muscle function in rat using 1H-magnetic resonance (MR) imaging and ^{31}P-MR spectroscopy. *Magnetic Resonance in Medicine*, 54, 1058–1064.

This paper describes for the first time a new experimental setup allowing strictly non-invasive investigation of muscle function using MR techniques. The corresponding setup integrates two systems allowing muscle contraction by transcutaneous stimulation and force measurement with a dedicated ergometer. The experimental results provide a direct comparison between two different set-ups, and clearly demonstrate that longitudinal studies are now possible with this original setup.

Study 5. Mattei, J.P., Kozak-Ribbens, G., Roussel, M., Le Fur, Y., Cozzone, P.J. and Bendahan, D. (2002). New parameters reducing the interindividual variability of metabolic changes during muscle contraction in humans. A ^{31}P MRS study with physiological and clinical implications. *Biochimica et Biophysica Acta*, 1554, 129–136.

Considering that inter-individual variations in skeletal muscle metabolism can be a problem for reliable comparative analyses, the authors describe in the present study an original normalisation method based on relationships between metabolic and work-rate indices recorded during a ^{31}P MRS study. The study, conducted with a group of 65 subjects, demonstrates that the amplitude of mechanical performance accounts for 50% of the between-subjects variations. They report a linear relationship, independent of any anthropometric measurements, capturing more than 90% of the variability.

GLOSSARY OF TERMS

^{31}P MRS   31 phosphorus magnetic resonance spectroscopy
ADP   adenosine diphosphate
AMP   adenosine monophosphate
ATP   adenosine triphosphate
BMI   body mass index
CV   coefficient of variation
EMG   electromyographic activity
H reflex   Hoffmann reflex
$H_2PO_4^-$ ion dihydrogenorthophosphate
HPO_4^{2-} ion hydrogenorthophosphate

IT   intra threshold
MRI   magnetic resonance imaging
MRS   magnetic resonance spectroscopy
MVC   maximal voluntary isometric contraction
NAD   nicotinamide adenine dinucleotide
NIRS   near-infrared spectroscopy
NMR   nuclear magnetic resonance
PCr/P_i phosphocreatine/inorganic phosphate ratio
PCr   phosphocreatine
P_i inorganic phosphate
RF   radio frequency
T reflex   tendon reflex

ACKNOWLEDGEMENTS

The research activity related to muscle at CRMBM (UMR CNRS 6612, headed by Professor P.J. Cozzone) is conducted in collaboration with B. Giannesini, J. Gondin, S. Guis, Y. Le Fur, G. Layec, J.P. Mattéi, C. Vilmen, A. Tonson and K. Yashiro. Financial support has been obtained from CNRS (Centre National de la Recherche Scientifique), ANR (Agence Nationale pour le Recherche), AFM (Association Française contre les Myopathies) and PHRC (Programme Hospitalier de Recherche Clinique).

REFERENCES

Ackerman, J.J., Grove, T.H., Wong, G.G., Gadian, D.G. and Radda G.K. (1980). Mapping of metabolites in whole animals by ^{31}P NMR using surface coils. *Nature*, 283, 167–170.

Argov, Z., Lofberg, M. and Arnold, D.L. (2000). Insights into muscle diseases gained by phosphorus magnetic resonance spectroscopy. *Muscle and Nerve*, 23, 1316–1334.

Arnold, D.L., Matthews, P.M. and Radda G.K. (1984). Metabolic recovery after exercise and the assessment of mitochondrial function in vivo in human skeletal muscle by means of ^{31}P NMR. *Magnetic Resonance in Medicine*, 1, 307–315.

Avela, J., Kyrolainen, H. and Komi, P.V. (1999). Altered reflex sensitivity after repeated and prolonged passive muscle stretching. *Journal of Applied Physiology*, 86, 1283–1291.

Bangsbo, J., Johansen, L., Quistorff, B. and Saltin, B. (1993). NMR and analytic biochemical evaluation of CrP and nucleotides in the human calf during muscle contraction. *Journal of Applied Physiology*, 74, 2034–2039.

Barker, A., Welsman, J., Welford, D., Fulford, J., Williams, C. and Armstrong, N. (2006). Reliability of ^{31}P-magnetic resonance spectroscopy during an exhaustive incremental exercise test in children. *European Journal of Applied Physiology*, 98, 556–565.

Barker, A.T., Jalinous, R. and Freeston, I.L. (1985). Non-invasive magnetic stimulation of human motor cortex. *Lancet*, 1, 1106–1107.

Bendahan, D., Confort-Gouny, S., Kozak-Reiss, G. and Cozzone, P.J. (1990). Heterogeneity of metabolic response to muscular exercise in humans. New criteria of invariance defined by in vivo phosphorus-^{31}NMR spectroscopy. *FEBS Letters*, 272, 155–158.

Bendahan, D., Jammes, Y., Salvan, A.M., Badier, M., Confort-Gouny, S., Guillot, C. and Cozzone, P.J. (1996). Combined electromyography–^{31}P-magnetic resonance spectroscopy study of human muscle fatigue during static contraction. *Muscle and Nerve*, 19, 715–721.

Bendahan, D., Mattei, J.P., Ghattas, B., Confort-Gouny, S., Le Guern, M.E. and Cozzone, P.J. (2002). Citrulline/malate promotes aerobic energy production in human exercising muscle. *British Journal of Sports Medicine*, 36, 282–289.

Bernus, G., Gonzalez de Suso, J.M., Alonso, J., Martin, P.A., Prat, J.A. and Arus, C. (1993). ^{31}P-MRS of quadriceps reveals quantitative differences between sprinters and long-distance runners. *Medicine and Science in Sports and Exercise*, 25, 479–484.

Bigland-Ritchie, B., Jones, D.A., Hosking, G.P. and Edwards, R.H. (1978). Central and peripheral fatigue in sustained maximum voluntary contractions of human quadriceps muscle. *Clinical Science and Molecular Medicine*, 54, 609–614.

Bilodeau, M. (2006). Central fatigue in continuous and intermittent contractions of triceps brachii. *Muscle and Nerve*, 34, 205–213.

Bishop, D., Spencer, M., Duffield, R. and Lawrence, S. (2001). The validity of a repeated sprint ability test. *Journal of Science and Medicine in Sport*, 4, 19–29.

Bosco, C., Belli, A., Astrua, M., Tihanyi, J., Pozzo, R., Kellis, S., Tsarpela, O., Foti, C., Manno, R. and Tranquilli, C. (1995). A dynamometer for evaluation of dynamic muscle work. *European Journal of Applied Physiology and Occupational Physiology*, 70, 379–386.

Bosco, C., Luhtanen, P. and Komi, P.V. (1983). A simple method for measurement of mechanical power in jumping. *European Journal of Applied Physiology and Occupational Physiology*, 50, 273–282.

Boyas, S., Nordez, A., Cornu, C. and Guével, A. (2006). Power responses of a rowing ergometer: mechanical sensors vs. Concept2 measurement system. *International Journal of Sports Medicine*, 27, 830–833.

Cairns, S.P., Knicker, A.J., Thompson, M.W. and Sjøgaard, G. (2005). Evaluation of models used to study neuromuscular fatigue. *Exercise Sport Sciences Reviews*, 33, 9–16.

Chance, B., Eleff, S. and Leigh Jr, J.S. (1980). Noninvasive, nondestructive approaches to cell bioenergetics. *Proceedings of the National Academy of Sciences*, 77, 7430–7434.

Chilibeck, P.D., McCreary, C.R., Marsh, G.D., Paterson, D.H., Noble, E.G., Taylor, A.W. and Thompson, R.T. (1998). Evaluation of muscle oxidative potential by ^{31}P-MRS during incremental exercise in old and young humans. *European Journal of Applied Physiology and Occupational Physiology*, 78, 460–465.

Cieslar, J.H. and Dobson, G.P. (2001) Force reduction uncoupled from pH and $H_2PO_4^-$ in rat gastrocnemius in vivo with continuous 2-Hz stimulation. *Ameri-*

can Journal of Physiology Regulation, Integrative and Comparative Physiology, 281, R511–518.

Coggan, A.R., Spina, R.J., King, D.S., Rogers, M.A., Brown, M., Nemeth, P.M. and Holloszy, J.O. (1992). Histochemical and enzymatic comparison of the gastrocnemius muscle of young and elderly men and women. *Journal of Gerontology*, 47, B71–76.

Collins, D.F. (2007). Central contributions to contractions evoked by tetanic neuromuscular electrical stimulation. *Exercise Sport Sciences Reviews*, 35, 102–109.

Conley, K.E., Jubrias, S.A. and Esselman, P.C. (2000). Oxidative capacity and ageing in human muscle. *Journal of Physiology*, 526, 1, 203–210.

Cozzone, P.J., Vion-Dury, J., Bendahan, D. and Confort-Gouny, S. (1996). Future path of magnetic resonance spectroscopy in clinical medicine. *La Revue du Praticien*, 46, 853–858.

Desmedt, J.E. and Hainaut, K. (1968). Kinetics of myofilament activation in potentiated contraction: staircase phenomenon in human skeletal muscle. *Nature*, 217, 529–532.

Duchateau, J. and Hainaut, K. (1985). Electrical and mechanical failures during sustained and intermittent contractions in humans. *Journal of Applied Physiology*, 58, 942–947.

Edwards, R.H., Hill, D.K., Jones, D.A. and Merton, P.A. (1977). Fatigue of long duration in human skeletal muscle after exercise. *Journal of Physiology*, 272, 769–778.

Enoka, R.M. (2002). *Neuromechanics of Human Movement*. Champaign: Human Kinetics.

Farina, D. (2006). Interpretation of the surface electromyogram in dynamic contractions. *Exercise Sport Sciences Reviews*, 34, 121–127.

Farina, D., Merletti, R. and Enoka, R.M. (2004). The extraction of neural strategies from the surface EMG. *Journal of Applied Physiology*, 96, 1486–1495.

Fitzsimons, M., Dawson, B., Ward, D. and Wilkinson, A. (1993). Cycling and running tests of repeated sprint ability. *Australian Journal of Science and Medicine in Sport*, 25, 82–87.

Foley, J.M. and Meyer, R.A. (1993). Energy cost of twitch and tetanic contractions of rat muscle estimated in situ by gated ^{31}P NMR. *NMR in Biomedicine*, 6, 32–38.

Fowler, M.D., Ryschon, T.W., Wysong, R.E., Combs, C.A. and Balaban, R.S. (1997). Normalised metabolic stress for ^{31}P-MR spectroscopy studies of human skeletal muscle: MVC vs. muscle volume. *Journal of Applied Physiology*, 83, 875–883.

Francescato, M.P. and Cettolo, V. (2001). Two-pedal ergometer for in vivo MRS studies of human calf muscles. *Magnetic Resonance in Medicine*, 46, 1000–1005.

Funato, K., Yanagiya, T. and Fukunaga, T. (2001). Ergometry for estimation of mechanical power output in sprinting in humans using a newly developed self-driven treadmill. *European Journal of Applied Physiology*, 84, 169–173.

Gandevia, S.C. (2001). Spinal and supraspinal factors in human muscle fatigue. *Physiological Reviews*, 81, 1725–1789.

Giannesini, B., Izquierdo, M., Le Fur, Y., Cozzone, P.J., Fingerle, J., Himber, J., Kunnecke, B., Von Kienlin, M. and Bendahan, D. (2005). New experimental setup for studying strictly noninvasively skeletal muscle function in rat using

1H-magnetic resonance (MR) imaging and ^{31}P-MR spectroscopy. *Magnetic Resonance in Medicine*, 54, 1058–1064.

Glaister, M., Stone, M.H., Stewart, A.M., Hughes, M. and Moir, G.L. (2004). The reliability and validity of fatigue measures during short-duration maximal-intensity intermittent cycling. *Journal of Strength and Conditioning Research*, 18, 459–462.

Gonzalez de Suso, J.M., Bernus, G., Alonso, J., Alay, A., Capdevila, A., Gili, J., Prat, J.A. and Arus, C. (1993). Development and characterization of an ergometer to study the bioenergetics of the human quadriceps muscle by ^{31}P NMR spectroscopy inside a standard MR scanner. *Magnetic Resonance in Medicine*, 29, 575–581.

Gupta, R.K., Gupta, P., Yushok, W.D. and Rose, Z.B. (1983). On the noninvasive measurement of intracellular free magnesium by ^{31}P NMR spectroscopy. *Physiological Chemistry and Physics and Medical NMR*, 15, 265–280.

Hamaoka, T., Iwane, H., Shimomitsu, T., Katsumura, T., Murase, N., Nishio, S., Osada, T., Kurosawa, Y. and Chance, B. (1996). Noninvasive measures of oxidative metabolism on working human muscles by near-infrared spectroscopy. *Journal of Applied Physiology*, 81, 1410–1417.

Hogan, M.C., Ingham, E. and Kurdak, S.S. (1998). Contraction duration affects metabolic energy cost and fatigue in skeletal muscle. *American Journal of Physiology*, 274, E397–402.

Hoult, D.I., Busby, S.J., Gadian, D.G., Radda, G.K., Richards, R.E. and Seeley, P.J. (1974). Observation of tissue metabolites using ^{31}P nuclear magnetic resonance. *Nature*, 252, 285–287.

Hug, F., Bendahan, D., Le Fur, Y., Cozzone, P.J. and Grelot, L. (2004). Heterogeneity of muscle recruitment pattern during pedaling in professional road cyclists: a magnetic resonance imaging and electromyography study. *European Journal of Applied Physiology*, 92, 334–342.

Hughes, M.G., Doherty, M., Tong, R.J., Reilly, T. and Cable, N.T. (2006). Reliability of repeated sprint exercise in non-motorised treadmill ergometry. *International Journal of Sports Medicine*, 27, 900–904.

Hultman, E., Sjöholm, H., Jäderholm-Ek, I. and Krynicki, J. (1983). Evaluation of methods for electrical stimulation of human skeletal muscle in situ. *Pflugers Archives*, 398, 139–141.

Iotti, S., Frassineti, C., Alderighi, L., Sabatini, A., Vacca, A. and Barbiroli, B. (2000) In vivo (31)P-MRS assessment of cytosolic [Mg(2+)] in the human skeletal muscle in different metabolic conditions. *Magnetic Resonance Imaging*, 18, 607–614.

Jubrias, S.A., Esselman, P.C., Price, L.B., Cress, M.E. and Conley, K.E. (2001). Large energetic adaptations of elderly muscle to resistance and endurance training. *Journal of Applied Physiology*, 90, 1663–1670.

Keenan, K.G., Farina, D., Maluf, K.S., Merletti, R. and Enoka, R.M. (2005). Influence of amplitude cancellation on the simulated surface electromyogram. *Journal of Applied Physiology*, 98, 120–131.

Kemp, G.J., Roussel, M., Bendahan, D., Le Fur, Y. and Cozzone, P.J. (2001). Interrelations of ATP synthesis and proton handling in ischaemically exercising human forearm muscle studied by ^{31}P magnetic resonance spectroscopy. *Journal of Physiology*, 535, 901–928.

Kemp, G.J., Taylor, D.J., Thompson, C.H., Hands, L.J., Rajagopalan, B., Styles, P. and Radda, G.K. (1993). Quantitative analysis by ^{31}P magnetic resonance spectroscopy of abnormal mitochondrial oxidation in skeletal muscle during recovery from exercise. *NMR in Biomedicine*, 6, 302–310.

Kent-Braun, J.A. and Le Blanc, R. (1996). Quantitation of central activation failure during maximal voluntary contractions in humans. *Muscle and Nerve*, 19, 861–869.

Kent-Braun, J.A. and Ng, A.V. (2000). Skeletal muscle oxidative capacity in young and older women and men. *Journal of Applied Physiology*, 89, 1072–1078.

Kent-Braun, J.A., McCully, K.K and Chance, B. (1990). Metabolic effects of training in humans: a ^{31}P-MRS study. *Journal of Applied Physiology*, 69, 1165–1170.

Kohrt, W.M., Malley, M.T., Coggan, A.R., Spina, R.J., Ogawa, T., Ehsani, A.A., Bourey, R.E., Martin, W.H. and Holloszy, J.O. (1991). Effects of gender, age, and fitness level on response of $\dot{V}O_2$max to training in 60–71 yr olds. *Journal of Applied Physiology*, 71, 2004–2011.

Kriketos, A.D., Pan, D.A., Lillioja, S., Cooney, G.J., Baur, L.A., Milner, M.R., Sutton, J.R., Jenkins, A.B., Bogardus, C. and Storlien, L.H. (1996). Interrelationships between muscle morphology, insulin action, and adiposity. *American Journal of Physiology*, 270, R1332–1339.

Kufel, T.J., Pineda, L.A. and Mador, M.J. (2002). Comparison of potentiated and unpotentiated twitches as an index of muscle fatigue. *Muscle and Nerve*, 25, 438–444.

Lakomy, H.K.A. (1984). An ergometer for measuring the power generated during sprinting. *Journal of Physiology*, 354, 33.

Larson-Meyer, D.E., Newcomer, B.R., Hunter, G.R., Hetherington, H.P. and Weinsier, R.L. (2000a). ^{31}P MRS measurement of mitochondrial function in skeletal muscle: reliability, force-level sensitivity and relation to whole body maximal oxygen uptake. *NMR in Biomedicine*, 13, 14–27.

Larson-Meyer D.E., Newcomer B.R., Hunter G.R., McLean J.E., Hetherington H.P. and Weinsier R.L. (2000b). Effect of weight reduction, obesity predisposition, and aerobic fitness on skeletal muscle mitochondrial function. *American Journal of Physiology, Endocrinology and Metabolism*, 278, E153–161.

Layec, G., Bringard, A., Vilmen, C., Micallef, J.P., Le Fur, Y., Perrey, S., Cozzone, P.J. and Bendahan, D. (2008). Accurate work-rate measurements during in vivo MRS studies of exercising human quadriceps. *Magma*, 21, 227–235.

Lodi, R., Muntoni, F., Taylor, J., Kumar, S., Sewry, C.A., Blamire, A., Styles, P. and Taylor, D.J. (1997). Correlative MR imaging and ^{31}P-MR spectroscopy study in sarcoglycan deficient limb girdle muscular dystrophy. *Neuromuscular Disorders*, 7, 505–511.

Lodi, R., Rajagopalan, B., Bradley, J.L., Taylor, D.J., Crilley, J.G., Hart, P.E., Blamire, A.M., Manners, D., Styles, P., Schapira, A.H. and Cooper, J.M. (2002). Mitochondrial dysfunction in Friedreich's ataxia: from pathogenesis to treatment perspectives. *Free Radical Research*, 36, 461–466.

McCartney, N., Heigenhauser, G.J., Sargeant, A.J. and Jones, N.L. (1983). A constant-velocity cycle ergometer for the study of dynamic muscle function. *Journal of Applied Physiology*, 55, 212–217.

McCreary, C.R., Chilibeck, P.D., Marsh, G.D., Paterson, D.H., Cunningham, D.A. and Thompson R.T. (1996). Kinetics of pulmonary oxygen uptake and muscle phosphates during moderate-intensity calf exercise. *Journal of Applied Physiology*, 81, 1331–1338.

McCully, K.K., Kent, J.A. and Chance, B. (1988). Application of ^{31}P magnetic resonance spectroscopy to the study of athletic performance. *Sports Medicine*, 5, 312–321.

Maffiuletti, N.A., Bizzini, M., Desbrosses, K., Babault, N. and Munzinger, U. (2007). Reliability of knee extension and flexion measurements using the Con-Trex isokinetic dynamometer. *Clinical Physiology and Functional Imaging*, 27, 346–353.

Margaria, R., Aghemo, P. and Rovelli, E. (1966). Measurement of muscular power (anaerobic) in man. *Journal of Applied Physiology*, 21, 1662–1664.

Marsh, G.D., Paterson, D.H., Thompson, R.T. and Driedger, A.A. (1991). Coincident thresholds in intracellular phosphorylation potential and pH during progressive exercise. *Journal of Applied Physiology*, 71, 1076–1081.

Mattei, J.P., Kozak-Ribbens, G., Roussel, M., Le Fur, Y., Cozzone, P.J. and Bendahan, D. (2002). New parameters reducing the interindividual variability of metabolic changes during muscle contraction in humans. A ^{31}P MRS study with physiological and clinical implications. *Biochimica et Biophysica Acta*, 1554, 129–136.

Merton, P.A. (1954). Voluntary strength and fatigue. *Journal of Physiology*, 123, 553–564.

Meyer, R.A., Prior, B.M., Siles, R.I. and Wiseman, R.W. (2001) Contraction increases the T(2) of muscle in fresh water but not in marine invertebrates. *NMR in Biomedicine*, 14, 199–203.

Miller, R.G., Boska, M.D., Moussavi, R.S., Carson, P.J. and Weiner, M.W. (1988) ^{31}P nuclear magnetic resonance studies of high energy phosphates and pH in human muscle fatigue. Comparison of aerobic and anaerobic exercise. *Journal of Clinical Investigation*, 81, 1190–1196.

Miller, R.G., Carson, P.J., Moussavi, R.S., Green, A., Baker, A., Boska, M.D. and Weiner, M.W. (1995). Factors which influence alterations of phosphates and pH in exercising human skeletal muscle: measurement error, reproducibility, and effects of fasting, carbohydrate loading, and metabolic acidosis. *Muscle and Nerve*, 18, 60–67.

Miller, R.G., Giannini, D., Milner-Brown, H.S., Layzer, R.B., Koretsky, A.P., Hooper, D. and Weiner, M.W. (1987). Effects of fatiguing exercise on high-energy phosphates, force, and EMG: evidence for three phases of recovery. *Muscle and Nerve*, 10, 810–821.

Millet, G.Y. and Lepers, R. (2004). Alterations of neuromuscular function after prolonged running, cycling and skiing exercises. *Sports Medicine*, 34, 105–116.

Millet, G.Y., Lepers, R., Maffiuletti, N.A., Babault, N., Martin, V. and Lattier, G. (2002). Alterations of neuromuscular function after an ultramarathon. *Journal of Applied Physiology*, 92, 486–492.

Nygren, A.T. and Kaijser, L. (2002). Water exchange induced by unilateral exercise in active and inactive skeletal muscles. *Journal of Applied Physiology*, 93, 1716–1722.

O'Brien, T.D., Reeves, N.D., Baltzopoulos, V., Jones, D.A. and Maganaris, C.N. (2008). Assessment of voluntary muscle activation using magnetic stimulation. *European Journal of Applied Physiology*, 104, 49–55.

Park, J.H., Vansant, J.P., Kumar, N.G., Gibbs, S.J., Curvin, M.S., Price, R.R., Partain, C.L. and James Jr, A.E. (1990). Dermatomyositis: correlative MR imaging and P-31 MR spectroscopy for quantitative characterization of inflammatory disease. *Radiology*, 177, 473–479.

Paton, C.D. and Hopkins, W.G. (2001). Tests of cycling performance. *Sports Medicine*, 31, 489–496.

Pierrot-Deseilligny, E. and Mazevet, D. (2000). The monosynaptic reflex: a tool to investigate motor control in humans. Interest and limits. *Clinical Neurophysiology*, 30, 67–80.

Pincivero, D.M., Gear, W.S. and Sterner, R.L. (2001). Assessment of the reliability of high-intensity quadriceps femoris muscle fatigue. *Medicine and Science in Sports and Exercise*, 33, 334–338.

Place, N., Maffiuletti, N.A., Martin, A. and Lepers, R. (2007). Assessment of the reliability of central and peripheral fatigue after sustained maximal voluntary contraction of the quadriceps muscle. *Muscle and Nerve*, 35, 486–495.

Place, N., Yamada, T., Bruton, J.D. and Westerblad, H. (2008). Interpolated twitches in fatiguing single mouse muscle fibres: implications for the assessment of central fatigue. *Journal of Physiology*, 86, 2799–2805.

Polkey, M.I., Kyroussis, D., Hamnegard, C.H., Mills, G.H., Green, M. and Moxham, J. (1996). Quadriceps strength and fatigue assessed by magnetic stimulation of the femoral nerve in man. *Muscle and Nerve*, 19, 549–555.

Quistorff, B., Nielsen, S., Thomsen, C., Jensen, K.E. and Henriksen, O. (1990). A simple calf muscle ergometer for use in a standard whole-body MR scanner. *Magnetic Resonance in Medicine*, 13, 444–449.

Racinais, S., Girard, O., Micallef, J.P. and Perrey, S. (2007). Failed excitability of spinal motoneurons induced by prolonged running exercise. *Journal of Neurophysiology*, 97, 596–603.

Raymer, G.H., Marsh, G.D., Kowalchuk, J.M. and Thompson, R.T. (2004). Metabolic effects of induced alkalosis during progressive forearm exercise to fatigue. *Journal of Applied Physiology*, 96, 2050–2056.

Rodenburg, J.B., de Boer, R.W., Jeneson, J.A., van Echteld, C.J. and Bar, P.R. (1994). ^{31}P-MRS and simultaneous quantification of dynamic human quadriceps exercise in a whole body MR scanner. *Journal of Applied Physiology*, 77, 1021–1029.

Russ, D.W. and Kent-Braun, J.A. (2004). Is skeletal muscle oxidative capacity decreased in old age? *Sports Medicine*, 34, 221–229.

Ryschon, T.W., Fowler, M.D., Arai, A.A., Wysong, R.E., Leighton, S.B., Clem Sr, T.R. and Balaban, R.S. (1995). A multimode dynamometer for in vivo MRS studies of human skeletal muscle. *Journal of Applied Physiology*, 79, 2139–2147.

Saab, G., Thompson, R.T. and Marsh, G.D. (2000). Effects of exercise on muscle transverse relaxation determined by MR imaging and in vivo relaxometry. *Journal of Applied Physiology*, 88, 226–233.

Smith, S.A., Montain, S.J., Matott, R.P., Zientara, G.P., Jolesz, F.A. and Fielding, R.A. (1998). Creatine supplementation and age influence muscle metabolism during exercise. *Journal of Applied Physiology*, 85, 1349–1356.

Tarnopolsky, M.A. and Parise, G. (1999). Direct measurement of high-energy phosphate compounds in patients with neuromuscular disease. *Muscle and Nerve*, 22, 1228–1233.

Taylor, D.J. (2000). Clinical utility of muscle MR spectroscopy. *Seminars in Musculoskeletal Radiology*, 4, 481–502.

Thomas, C.K., Woods, J.J. and Bigland-Ritchie, B. (1989). Impulse propagation and muscle activation in long maximal voluntary contractions. *Journal of Applied Physiology*, 67, 1835–1842.

Thorstensson, A. and Karlsson, J. (1976). Fatiguability and fibre composition of human skeletal muscle. *Acta Physiologica Scandinavica*, 98, 318–322.

Todd, G., Taylor, J.L. and Gandevia, S.C. (2003). Measurement of voluntary activation of fresh and fatigued human muscles using transcranial magnetic stimulation. *Journal of Physiology*, 551, 661–671.

Tong, R.J., Bell, W., Ball, G. and Winter, E.M. (2001). Reliability of power output measurements during repeated treadmill sprinting in rugby players. *Journal of Sports Sciences*, 19, 289–297.

Treuth, M.S., Butte, N.F. and Herrick, R. (2001). Skeletal muscle energetics assessed by (^{31})P-NMR in prepubertal girls with a familial predisposition to obesity. *International Journal of Obesity and Related Metabolic Disorders*, 25, 1300–1308.

Trimble, M.H. and Harp, S.S. (1998). Postexercise potentiation of the H-reflex in humans. *Medicine and Science in Sports and Exercise*, 30, 933–941.

Vergès, S., Maffiuletti, N.A., Kerherve, H., Decorte, N., Wuyam, B. and Millet, G.Y. (2009). Comparison of electrical and magnetic stimulations to assess quadriceps muscle function. *Journal of Applied Physiology*, 106, 701–710.

Vøllestad, N.K. (1997). Measurement of human muscle fatigue. *Journal of Neuroscience Methods*, 74, 219–227.

Wade, A.J., Marbut, M.M. and Round, J.M. (1990). Muscle fibre type and aetiology of obesity. *Lancet*, 335, 805–808.

Ward, K.M., Rajan, S.S., Wysong, M., Radulovic, D. and Clauw, D.J. (1996). Phosphorus nuclear magnetic resonance spectroscopy: in vivo magnesium measurements in the skeletal muscle of normal subjects. *Magnetic Resonance in Medicine*, 36, 475–480.

Weidman, E.R., Charles, H.C., Negro-Vilar, R., Sullivan, M.J. and MacFall, J.R. (1991). Muscle activity localization with ^{31}P spectroscopy and calculated T2-weighted 1H images. *Investigative Radiology*, 26, 309–316.

Whipp, B.J., Rossiter, H.B., Ward, S.A., Avery, D., Doyle, V.L., Howe, F.A. and Griffiths, J.R. (1999). Simultaneous determination of muscle ^{31}P and O_2 uptake kinetics during whole body NMR spectroscopy. *Journal of Applied Physiology*, 86, 742–747.

Wilson, G.J. and Murphy, A.J. (1996). The use of isometric tests of muscular function in athletic assessment. *Sports Medicine*, 22, 19–37.

CHAPTER 3

CELLULAR MECHANISMS OF SKELETAL MUSCLE FATIGUE

Håkan Westerblad and David G. Allen

OBJECTIVES

The objectives of this chapter are:

- To define fatigue and understand when it occurs in normal and diseased subjects.
- To identify the sites at which muscle fatigue can occur and understand how to distinguish these sites experimentally.
- To understand the metabolic changes which occur in muscle when ATP consumption exceeds resynthesis.
- To describe how the action potential triggers Ca^{2+} release in muscle and how the various processes involved are affected by intense muscle activity.
- To assess how various metabolic changes affect the myofibrillar performance.

INTRODUCTION

High-intensity exercise leads to a decline in muscle performance, known as fatigue. This decline in performance is manifested in the muscles as decreased force production and slowed contractions. In this chapter we will discuss changes within the skeletal muscle cells that contribute to fatigue.

Figure 3.1 illustrates the activation pathway in skeletal muscle cells. The muscle cells are activated by the α-motoneurons. At the neuromuscular endplate, acetylcholine (ACh) is released from the end terminals of the motoneurons. ACh binds to receptors on the surface membrane of the muscle cells, which triggers the opening of channels that allow passage of both Na^+ and K^+. This leads to a localised depolarisation of the surface membrane, which activates nearby voltage-dependent Na^+ channels and an action potential is generated. The safety margin in the neuromuscular transmission is large, and this process is not limiting during intense exercise under normal conditions. However, in a disease called myasthenia gravis there are antibodies which act against the ACh receptors. This results in impaired neuromuscular transmission, meaning patients with myasthenia suffer from early fatigue development.

The neuromuscular junctions are located at the middle of the muscle cells. The action potential is conducted along the surface membrane of the muscle cells and also into the transverse tubular (t-tubular) system. This system consists of invaginations of the surface membrane that form a fine network of tubules that surround each myofibril. Thus, the t-tubules contain extracellular fluid. They also contain voltage-activated Na^+ and K^+ channels and hence they can conduct action potentials.

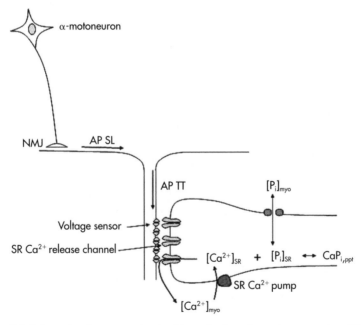

Figure 3.1 Schematic of structures involved in activation of muscle contraction. NMJ: neuromuscular junction; AP SL: action potential in the sarcolemma; AP TT: action potential in the t-tubule; voltage sensor or dihydropyridine receptor; SR Ca^{2+} release channel or ryanodine receptor

Source: Modified from Allen *et al.*, 2008a.

There are voltage sensors in the t-tubular membrane known as dihydro-pyridine receptors (DHPRs). In fact, these are L-type voltage-activated Ca^{2+} channels, which are too slow to pass any significant amount of Ca^{2+} in response to the fast action potentials. Thus, only their voltage-sensing part is utilised. The DHPRs are located in close connection to the Ca^{2+} channels of the sarcoplasmic reticulum (SR), the ryanodine receptors (RyRs). When activated by an action potential, the DHPRs interact mechanically with the RyRs, which results in the opening of the latter Ca^{2+} channels (Dulhunty, 2006). SR is the intracellular Ca^{2+} storage space, and when the RyRs open, Ca^{2+} is released from the SR into the cytosol (myoplasm).

Skeletal muscle cells contain thousands of myofibrils, which are in direct contact with the myoplasm. The major components of the myofibrils are actin and myosin filaments, which are organised in long series of sarcomeres. The functional stage of the actin filament is controlled by two regulatory proteins, troponin and tropomyosin. When Ca^{2+} is released from the SR, and hence increases in the myoplasm, Ca^{2+} binds to troponin. This leads to a conformational change of troponin that is transmitted to tropomyosin, which moves away from the myosin-binding sites of actin. The myosin heads (the cross-bridges) can then bind to actin, so cross-bridge cycling starts and the muscle cell contracts.

Ca^{2+} is continuously pumped back into the SR by the SR Ca^{2+} pumps (SERCAs). When the activation of α-motoneurons ceases and SR Ca^{2+} release is stopped, the Ca^{2+} is rapidly pumped into the SR, and the free Ca^{2+} concentration in the myoplasm ($[Ca^{2+}]_i$) decreases. Ca^{2+} then dissociates from troponin, the myosin heads can no longer bind to actin and the muscle cell relaxes.

All the individual events described above, from action potential propagation to SR Ca^{2+} pumping, are affected by intense activation and hence could contribute to the impaired muscle function in fatigue. In this chapter we will first describe different methods used to study fatigue at the muscle cell level. We will then describe changes in muscle metabolism and function during fatiguing stimulation and the marked differences between muscle cells of different types in this respect. Finally, the largest section of this chapter will discuss changes that occur in different cellular processes during intensive muscle activity and their importance for muscle fatigue.

METHODS USED TO STUDY FATIGUE AT THE CELLULAR LEVEL

Fatigue is easily recognisable in humans, and the phenomenon has long generated interest in the underlying mechanisms and how to minimise it. Both in escaping from a predator and in winning a race, the speed and magnitude of the development of fatigue is one important factor. A turning point in the mechanistic understanding of fatigue was the ability to stimulate the motor nerve to a muscle and thereby potentially distinguish whether fatigue of the muscle was central (i.e. in the brain and spinal cord) or peripheral (i.e. in the neuromuscular junction or muscle) (Merton, 1954). While central fatigue is common, particularly with long-lasting, low-intensity activities (Gandevia, 2001), it is generally accepted that muscle is the major site of weakness in many situations, and this is the focus of this chapter.

Another important breakthrough in the study of human fatigue includes the development of the muscle biopsy technique, which allows sampling of muscle for metabolic levels, histology, protein expression, muscle fibre typing, etc. (Bergström et al., 1967). Subsequently, nuclear magnetic resonance (NMR) has transformed our ability to measure phosphorus metabolites (adenosine triphosphate (ATP), phosphocreatine (PCr) and inorganic phosphate (P_i)) and pH (Dawson et al., 1978). The NMR technique is developing rapidly to include measurements of other compounds (see Chapter 2).

It is generally recognised that some muscles fatigue rapidly (e.g. the elbow flexor biceps brachii), while other muscles fatigue very slowly (e.g. the postural muscles of the legs). A turning point in our understanding of these issues came with a study in which single motor units were stimulated to exhaustion (Burke et al., 1973). The activated muscle fibres could then be identified because they showed glycogen depletion. This study showed that fast-twitch fibres fatigued within minutes, whereas slow-twitch fibres were essentially unfatigable.

Much of the development of ideas about the mechanism of fatigue has depended on the use of isolated muscles. Whole muscles are convenient and simple but suffer from the serious disadvantage that the absence of blood supply allows the development of an anoxic core and leads to the accumulation of K^+, lactate, H^+ and CO_2 in the centre of the muscle. For these reasons the dissection of small bundles of fibres or single fibres is a popular alternative. In a single fibre, the problems of fibre type mixtures are eliminated, and it is possible to measure changes in force and metabolites or ions continuously from a single cell.

A parallel approach is the use of skinned fibres, which involves the removal of the surface membrane, either mechanically or by dissolving with detergents. Skinned fibres have the advantage that the composition of solutions bathing the contractile proteins and SR can be specified completely, so the effect of any particular metabolic change can be investigated in the absence of the multitude of changes that occur during fatigue. A disadvantage with skinned fibre preparations is that soluble intracellular molecules will diffuse into the solution bathing the cell and hence their concentration will be substantially decreased in the vicinity of the myofibrils and SR. This means that important cellular constituents (soluble enzymes, buffers, etc.) may no longer be present at sufficient concentrations, which can have a large impact on the results.

CHANGES IN METABOLISM DURING FATIGUE

Energy consumption in skeletal muscle fibres increases dramatically during high-intensity exercise. In fact, it can substantially exceed the aerobic capacity of the cells, especially in fast-twitch muscle fibres. A large fraction of the ATP consumed is then produced by anaerobic metabolism, which includes lactate formation and breakdown of PCr. Fatigue occurs more rapidly during intense exercise where anaerobic metabolism is required compared to exercises of lower intensity where aerobic metabolism is dominating.

Glycogen is the storage form of carbohydrates in skeletal muscle, and when energy is required the enzyme phosphorylase catalyses the sequential removal of glycosyl residues from the glycogen molecule. The glycosyl residues are then metabolised in a sequence of reactions ending with the formation of pyruvate. Pyruvate can enter the mitochondria, where it is oxidised to CO_2 and H_2O, and in this way 39 ATP are formed from each glycosyl residue. During intense exercise, the rate of pyruvate formation by glycogen breakdown exceeds the rate of its oxidation by the citric acid cycle in the mitochondria; at the same time the formation of the reduced form of nicotinamide adenine nucleotide (NADH) exceeds the rate at which it can be oxidised by the mitochondrial respiratory chain. This forces pyruvate to be reduced to lactate, and only three ATP are then formed from each glycosyl residue. Anaerobic glycogenolysis results in the accumulation of lactate ions which are accompanied by H^+. During vigorous exercise, the lactate concentration can reach 30 mM or higher and pH can decrease to ~6.5 (Fitts, 1994). Glycogen is rapidly depleted during intense exercise with a large component of anaerobic glycogen breakdown, which only yields three ATP per glycosyl residue, and more slowly during aerobic exercise, where 39 ATP per glycosyl residue are obtained. There is a correlation between glycogen depletion and exhaustion (Bergström *et al.*, 1967), but the exact mechanism(s) by which this occurs is not fully understood.

Contracting skeletal muscle cells can consume ATP, producing ADP and P_i, much faster than it is regenerated. Still, the change in ATP concentration is limited mainly because of the creatine kinase reaction ($PCr + ADP \leftrightarrow Cr + ATP$), but also because of the adenylate kinase reaction ($2ADP \leftrightarrow AMP + ATP$). These reactions are close to equilibrium and therefore the net consumption of ATP leads to relatively stereotyped changes in the concentrations of ATP, ADP, P_i, PCr, Cr and AMP. During net consumption of ATP, [ATP] is initially unchanged and the net effect is a decline in [PCr] and increases in [Cr] and [P_i], which can reached concentrations as high as ~30 mM. Subsequently, when [PCr] has fallen to less than ~10 mM, [ATP] starts to fall and [ADP] rises substantially (from a few µM to 100–300 µM). When [ADP] is increased to such levels, the [AMP] also becomes significant and can be degraded by AMP deaminase to NH_3 and inosine monophosphate (IMP) (Fitts, 1994). It is worth noting that while changes in [ADP] and [AMP] are mainly transient and probably also spatially restricted, the increase in [IMP] is stable and can increase to ~5 mM during intense exercise. Thus, a substantial increase in [IMP] in fatigue indicates that [ADP] and [AMP] have been significantly increased during the exercise period, although these increases are difficult to measure directly.

Exercise is accompanied by an increased production of reactive oxygen and nitrogen species (ROS/RNS) (Allen *et al.*, 2008b; Reid, 2008). The most important ROS are superoxide ($O_2^{\bullet-}$), hydrogen peroxide (H_2O_2), and the hydroxyl radical ($OH^{\bullet}$). Superoxide is produced in mitochondria as a by-product of oxidative phosphorylation. Superoxide can also be produced by various enzymes, including NADPH oxidase, xanthine oxidase, and lipo- and cyclo-oxygenases. Superoxide is moderately reactive and is rapidly broken down by superoxide dismutase to hydrogen peroxide, which easily diffuses through cell membranes and can affect various cellular processes. Hydrogen peroxide can be broken down by catalase ($2 H_2O_2 => 2 H_2O + O_2$) or by glutathione peroxidase, which produces water while converting reduced glutathione (GSH) to the oxidised form (GSSG). Hydrogen peroxide may also be converted to the very reactive hydroxyl radical by free transition metals, such as Fe^{2+}. Hydroxyl radicals bind to – and can damage – proteins, DNA and cell membranes. Thus, the important ROS form a cascade, which makes it difficult to define if a change in muscle function is associated with increased ROS production as such, or by a particular species of ROS. Furthermore, there is a complex interaction between ROS and RNS (Ferreira and Reid, 2008). The parental RNS is nitric oxide ($NO^{\bullet}$), which is mainly produced by nitric oxide synthases. Nitric oxide can nitrosylate proteins, and in this way alter their function. In addition, nitric oxide can react with superoxide to form peroxynitrite ($ONOO^-$), which is a highly reactive oxidant.

CHANGES IN FORCE PRODUCTION DURING FATIGUE

Force production is submaximal during most types of exercise, and a decreased ability to generate force is then not necessarily reflected in impaired performance. Experiments can be designed so that maximal contractions are produced at regular intervals during submaximal exercise, and it is in this way the time-course of changes in force-generating capacity can be studied (see Chapter 1). Experiments with electrically stimulated muscles or muscle cells often employ stimulation schemes with maximal contractions. For instance, a fatiguing protocol with short tetanic contractions repeated at regular intervals is often used. An attraction with this type of stimulation is that changes in force production can be studied in great detail and correlated with associated changes in, for example, SR Ca^{2+} release or metabolism.

During fatigue induced by repeated short tetanic contractions, a characteristic pattern is observed in fast-twitch (type II) muscle fibres: initially there is fast decline of tetanic force by 10–20%; then follows a period of relatively constant tetanic force; and finally there is rapid decline in force (Figure 3.2a). The force decline in early fatigue is caused by impaired myofibrillar function, whereas the decline in late fatigue is mainly due to decreased SR Ca^{2+} release (Allen et al., 2008b); these issues will be discussed in detail in the following sections.

In contrast to fast-twitch fibres, slow-twitch fibres show little change in force production during most types of repeated tetanic stimulation (Figure 3.2b). In other words, slow-twitch fibres are generally much more fatigue-resistant than fast-twitch fibres. This can be explained by a larger capacity

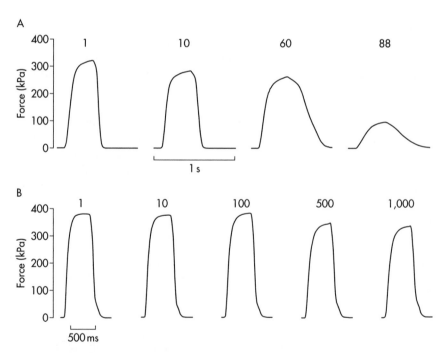

Figure 3.2 Tetanic force records obtained during fatiguing stimulation of a fast-twitch (A) and a slow-twitch (B) mouse muscle fibre. Note the severe changes in force production in the fast-twitch fibre and the very limited changes in the slow-twitch fibre. The number of fatiguing tetani are shown above the records. Stimulation protocol: 350 ms, 70 Hz tetani given at 2.5 s intervals (A) and 500 ms, 70 Hz tetani given at 2 s intervals (B)

Source: Adapted from Dahlstedt et al., 2000 (A) and Bruton et al., 2003 (B).

to use aerobic metabolism in slow-twitch fibres than in fast-twitch fibres. Moreover, the rate of both cross-bridge cycling and SR Ca^{2+} pumping, and hence the rate of ATP consumption, is lower in slow-twitch than in fast-twitch fibres.

CHANGES IN $[Ca^{2+}]_i$ DURING FATIGUE

Force production in a muscle fibre depends on the action potential triggering release of Ca^{2+} from the SR. As indicated above and in Figure 3.1, this depends on the coordinated action of the action potential in the surface membrane and the t-tubules, the activation of DHPRs and RyRs and the storage of Ca^{2+} in the SR, which requires activity of the SERCAs. In this section we focus on those sites that have been implicated in the fatigue process.

The concept that failure of SR Ca^{2+} release contributes to fatigue was first demonstrated by the use of caffeine and high extracellular K^+, agents capable of facilitating SR Ca^{2+} release, which were found to partially overcome the reduced force of fatigue (Eberstein and Sandow, 1963). Later it became possible to measure $[Ca^{2+}]_i$ in muscle fibres during fatigue (Allen *et*

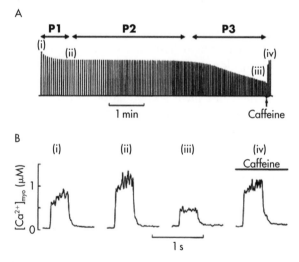

Figure 3.3 $[Ca^{2+}]_i$ transients at various stages of fatigue. (A) Force from a single mouse muscle fibre fatigued by repeated tetani. Phase 1 (P1) is a rapid reduction of force over 1 min to 80–90% maximum force. Phase 2 (P2) is a period of stable force production. Phase 3 (P3) is a more rapid decline of force when the muscle fibre is becoming exhausted. In this experiment 10 mM caffeine was suddenly applied when force was reduced to about 50%. (B) Representative $[Ca^{2+}]_i$ signals from tetani at the times indicated on panel A. Note that P3 is accompanied by a substantial decline of $[Ca^{2+}]_i$. Both the force and $[Ca^{2+}]_i$ were considerable increased when caffeine was applied

Source: Modified from Westerblad and Allen, 1991.

al., 1989; Westerblad and Allen, 1991). Figure 3.3 shows the force decline in an isolated fast-twitch muscle fibre and representative $[Ca^{2+}]_i$ records. Note that $[Ca^{2+}]_i$ first increases and then declines. The functional importance of this decline in SR Ca^{2+} release in fatigued muscle has been established in several ways. The application of caffeine to a fatigued muscle cell overcomes much of the late decline of force and is associated with a substantial increase in tetanic $[Ca^{2+}]_i$ (Figure 3.3b(iv)). This provides good evidence that the impairment of SR Ca^{2+} release is the cause of the final force decrease in fatigue. Possible mechanisms underlying the failure of SR Ca^{2+} release during fatigue are considered in the next five sub-sections.

Failure of action potential propagation

Normal tetanic contraction of a muscle fibre requires action potentials to propagate rapidly along the length of a muscle fibre, and then into and throughout the t-tubular system. A key feature of the t-tubules is that they enclose a small volume of extracellular space (~1% of muscle volume), which is in diffusion equilibrium with the extended extracellular space with a time-course of a few seconds. During a normal action potential, Na^+ enters and K^+ leaves the fibre, so that there is temporarily a depletion of the extracellular Na^+ and an accumulation of K^+ (Sejersted and Sjøgaard, 2000). These changes are much larger in the restricted t-tubular space, although they would only last for a few seconds after a single action potential. Obviously these extracellular ionic changes will be larger when action potentials are repeated at high frequencies. The changes are further augmented by restricted blood flow, which occurs in maximal tetani because the circulation is collapsed by the high pressure developed inside muscles. It is thought that the consequent accumulation of extracellular K^+ is functionally most important (Sejersted and Sjøgaard, 2000). This is because it will lead to depolarisation of the membrane potential and, if this is sufficiently great, cause inactivation of Na^+ channels and a reduction in the action potential amplitude. Once the action potential amplitude is reduced below a critical level, the activation of the voltage sensors is reduced, and hence SR Ca^{2+} release via the RyRs is decreased, resulting in force reduction.

There is currently debate on whether impaired action potential activation of the voltage sensors is an important mechanism of fatigue in humans. During repeated short tetani this mechanism is probably unimportant, because there is insufficient time for K^+ to accumulate in a short tetanus and it will diffuse out of the t-tubules during the time between tetani. On the other hand, if an isolated muscle fibre is continuously stimulated at a relatively high frequency, force declines quite rapidly (high-frequency fatigue). In this situation there is reduced SR Ca^{2+} release in the centre of the fibre, suggesting that K^+ is accumulating in the t-tubules of the fibre and that the above sequence of changes is occurring (Westerblad *et al.*, 1990; Duty and Allen, 1994). This interpretation is strongly supported by the observation that if the stimulation frequency is reduced for just a few seconds, the failure of SR Ca^{2+} release in the centre of the fibre recovers, and force also shows some recovery.

A key question is whether action potential failure occurs during normal muscle activity in intact animals. As noted above, such a mechanism would be expected to be of importance in continuous, maximal tetani because this is the situation where action potential frequency is highest. However, it is known that during continuous tetani, the rate of action potential triggered by central motor areas rapidly falls from a high initial level to a much lower frequency (Bigland-Ritchie *et al.*, 1983), which would reduce K^+ accumulation. Furthermore, a number of other mechanisms have been described which would be expected to reduce the risk of action potential failure. For instance, the accumulation of H^+ during intense activity causes a reduction in membrane chloride ion (Cl^-) conductance (Pedersen *et al.*, 2004; Pedersen *et al.*, 2005). This helps the action potentials to continue to propagate both along the sarcolemma and within the t-tubular system, despite substantial increases in extracellular K^+ concentration and the accompanying membrane depolarisation.

One important way to investigate whether action potential failure occurs is to record the EMG from muscles during activity. Unfortunately, in EMG records it is difficult to distinguish between the contribution of changes in action potential frequency, amplitude, duration and conduction velocity. A better approach might be to record the contractile response to a single, maximal electrical stimulation of the motor nerve, but again there is some debate about the interpretation of these results (Place *et al.*, 2008). At any rate, many groups have concluded that changes in the action potentials do not seem to be an important cause of fatigue during continuous maximal tetani *in vivo*.

In conclusion, while external high-frequency stimulation can cause t-tubular action potential failure and fatigue, the risk of action potential failure *in vivo* is decreased by a gradual reduction of the α-motoneuron firing frequency. Moreover, acidosis may also help prevent the occurrence of t-tubular failure, by decreasing the Cl^- conductance.

Metabolic effects on the SR Ca^{2+} release channels

It is possible to study the SR Ca^{2+} RyRs either by isolating and incorporating them in lipid bilayers or by measuring Ca^{2+} release (or the resulting force production) in skinned muscle fibres in which the t-tubules still make functional connections with the SR. Both of these approaches show that SR Ca^{2+} release is sensitive to metabolite changes. Specifically, substantial falls in [ATP] and rises in $[Mg^{2+}]_i$ both reduce SR Ca^{2+} release (Blazev and Lamb, 1999). Whenever [ATP] falls, $[Mg^{2+}]_i$ rises, because under normal conditions ATP exists as MgATP, and none of the breakdown products of

ATP bind Mg^{2+} with great affinity. Furthermore, reduced [ATP] and increased $[Mg^{2+}]_i$, when tested together produce larger changes than either alone and so mutually reinforce each other (Allen et al., 2008b). Thus, the critical question is whether the fall of [ATP] and the rise of $[Mg^{2+}]_i$ during fatigue are large enough to contribute to the decrease in SR Ca^{2+} release.

Traditionally, it has been thought that the [ATP] in muscle cells does not fall to very low levels during normal exercise. This seems reasonable because breakdown of PCr occurs rapidly and regenerates the ATP that has been used by the muscle. However, normal methods of measuring [ATP] determine the average change in whole muscles, which are composed of fibre types with different metabolic profiles and recruitment patterns. Recent studies have instead measured [ATP] in individual fibres from human muscle immediately after a 25 s maximal cycling bout (Karatzaferi et al., 2001). In fast-twitch fibres, which are recruited in this type of exercise, [PCr] had fallen to low levels so that the buffering of ATP by PCr would be reduced. [ATP] in these fibres had fallen from its resting level of 5–6 mM to around 1 mM, and might be even lower in regions of the cells where ATP consumption is highest (e.g. in the middle of a myofibril). Another site in muscle with high ATP consumption and restricted diffusion is the region between the t-tubule and the SR, so this may be another region in which metabolic changes are greater than in the bulk myoplasm.

These findings suggest that metabolic changes may contribute to the decline of SR Ca^{2+} release in fatigue. However, it is very hard to be quantitative about this issue because it is not known how large the metabolic changes are at the critical site (the myoplasmic face of the RyR) and because the measured SR Ca^{2+} release depends on many factors (e.g. SR Ca^{2+} content and myoplasmic Ca^{2+} buffering). During fatigue induced by intermittent tetani, $[Mg^{2+}]_i$ starts to increase at the time when tetanic $[Ca^{2+}]_i$ starts to fall, suggesting a causal relationship (Westerblad and Allen, 1992). In the absence of creatine kinase activity, changes in [ATP] and $[Mg^{2+}]_i$ during contraction should be larger, and hence SR Ca^{2+} release more affected. In accordance with this assumption, SR Ca^{2+} release decreases rapidly at the onset of high-intensity stimulation of muscle fibres from mice lacking creatine kinase (Dahlstedt et al., 2000; Dahlstedt et al., 2003). However, these creatine kinase-deficient fibres display smaller changes in tetanic $[Ca^{2+}]_i$ than control fibres during prolonged stimulation at lower intensity.

In conclusion, direct inhibition of SR Ca^{2+} release channels by reduced [ATP] and increased $[Mg^{2+}]_i$ appears to be of greatest importance at the onset of high-intensity fatiguing stimulation, whereas a role in later stages of fatigue remains uncertain.

Ca^{2+}-P_i precipitation in the SR

A recent proposal is that Ca^{2+}-P_i might precipitate within the SR during fatigue, leading to a reduced amount of free Ca^{2+} available for release

(Allen and Westerblad, 2001). The underlying theory is that during fatigue, P_i first accumulates in the myoplasm due to breakdown of PCr. Some P_i are then transported into the SR. Where the Ca^{2+}-P_i solubility product is exceeded, precipitation occurs, and the releasable pool of Ca^{2+} is reduced. The initial evidence for this theory came from skinned fibres with intact SR in which raised (myoplasmic) $[P_i]$ caused a fall in SR Ca^{2+} release, presumably due to Ca^{2+}-P_i precipitation (Fryer et al., 1995).

Recent experiments have provided various lines of support for the Ca^{2+}-P_i precipitation mechanism in fatigue. First, when P_i was microinjected into normal fibres it caused a large fall in SR Ca^{2+} release (Westerblad and Allen, 1996). Second, the amount of Ca^{2+} in the SR that could be released by rapid caffeine application was found to be reduced in fatigued muscles (Westerblad and Allen, 1991; Kabbara and Allen, 1999). Third, measurement of the free $[Ca^{2+}]$ in the sarcoplasmic reticulum using a low-affinity Ca^{2+} indicator (fluo-5N) showed that it declined throughout a period of fatiguing stimulation and then returned to normal during recovery (Kabbara and Allen, 2001). Fourth, the decline of tetanic $[Ca^{2+}]_i$ during fatiguing stimulation was markedly delayed in fibres where the creatine kinase reaction was inhibited (Dahlstedt et al., 2000). In the absence of creatine kinase, $[P_i]$ does not rise significantly, so Ca^{2+}-P_i precipitation will not occur. Fifth, a necessary feature of this theory is some pathway for P_i to enter the SR. Anion channels in the SR that conduct P_i have been discovered (Laver et al., 2001). These channels are sensitive to [ATP] and open when [ATP] falls, so this may mean that these channel open relatively late in fatigue and helps to explain why the fall in SR Ca^{2+} release occurs in late fatigue, whereas the rise in $[P_i]$ would occur much earlier.

In conclusion, several lines of evidence show that Ca^{2+}-P_i precipitation in the SR can be a major cause of reduced tetanic $[Ca^{2+}]_i$ in the late stages of fatigue.

Failure of SR Ca^{2+} release due to glycogen depletion

The development of the muscle biopsy technique allowed correlations to be made between exhaustion and muscle metabolites. Studies of this sort in humans showed that in prolonged endurance activities, muscle fatigue becomes pronounced at about the time when muscle glycogen levels fall to low levels (Bergström et al., 1967). Studies on single mouse fibres have also shown that glycogen content fell to ~25% during fatigue caused by repeated tetani, and this coincided with reduced $[Ca^{2+}]_i$ transients (Chin and Allen, 1997). When a muscle fibre was allowed to recover in the absence of glucose, glycogen did not recover and there was limited recovery of tetanic force and $[Ca^{2+}]_i$. When re-stimulated with repeated tetani in the absence of glucose so that glycogen could not recover, the fibre fatigued much more rapidly. Thus, there is a correlation between the level of glycogen and the decrease of tetanic $[Ca^{2+}]_i$ during fatigue.

Electron microscopy of human muscle biopsies shows that in fatigued samples, glycogen particles were preferentially depleted in the region of the t-tubular–SR junction (Friden *et al.*, 1989). Furthermore, more rapidly fatigable fibres showed greater depletion than fatigue-resistant fibres. These results suggest that glycogen depletion in the region of the t-tubular–SR junction may contribute to the failure of SR Ca^{2+} release during fatigue. In studies of skinned fibres, SR Ca^{2+} release in response to depolarisation of the t-tubules correlated closely with the muscle glycogen content (Stephenson *et al.*, 1999). In these skinned fibre experiments, ATP and PCr were present in the bathing solutions so the absence of glycogen is probably not being signalled by decreased [ATP] in this case. Instead, these experiments suggest that the glycogen performed a structural, rather than a metabolic role.

In conclusion, depletion of glycogen during prolonged, exhausting exercise may contribute to fatigue by causing decreased SR Ca^{2+} release.

Reduction of SR Ca^{2+} release after fatiguing stimulation

After intense or prolonged exercise in humans, muscles may show a component of fatigue that lasts many hours, or even days. Characteristically, the force produced at high frequencies of stimulation recovers quickly to normal levels, whereas the force produced at low stimulus frequencies shows prolonged decline (Edwards *et al.*, 1977). This phenomenon was originally named "low-frequency fatigue", but since this term is now frequently misused (e.g. to describe fatigue induced by stimulation at low frequencies), we propose the less ambiguous name "prolonged low-frequency force depression" (PLFFD) (Allen *et al.*, 2008b). This phenomenon is thought to be responsible for the feeling of weakness that is experienced for a few days after any very intense period of exercise. Presumably the brain increases the firing frequency in order to achieve the expected force and interprets this as a feeling of weakness even though the maximum force of the muscle is unchanged. In experiments on isolated muscle, PLFFD has been attributed to decreased SR Ca^{2+} release (Westerblad *et al.*, 1993). However, more recent studies have shown that reduced myofibrillar Ca^{2+} sensitivity can also lead to the same phenomenon (Bruton *et al.*, 2008) (see later section).

One proposed mechanism for the prolonged reduction in SR Ca^{2+} release is that it is a consequence of the period of elevated $[Ca^{2+}]_i$ which accompanies fatiguing stimulation. Such a mechanism has been described in skinned fibres in which elevating intracellular $[Ca^{2+}]$ to very high levels ($>10\,\mu M$) for a few seconds, or to more moderate levels (2–$10\,\mu M$) for several minutes, causes long-term disruption of voltage-sensor-induced activation of the SR Ca^{2+} release channels (Lamb *et al.*, 1995; Verburg *et al.*, 2006). The involvement of Ca^{2+}-dependent proteases (e.g. calpains) in this process has been suggested, since the deleterious changes can sometimes be slowed or prevented by leupeptin, a calpain inhibitor.

Recent data indicate that the development of PLFFD is related to increased ROS/RNS during fatiguing stimulation. The increased ROS/RNS may then cause the force depression either by inhibiting the SR Ca^{2+} release or by decreasing the myofibrillar Ca^{2+} sensitivity (Bruton *et al.*, 2008).

In conclusion, PLFFD can be caused by Ca^{2+}-dependent and/or ROS/RNS-induced damage to the SR Ca^{2+} release mechanism.

CHANGES IN MYOFIBRILLAR FUNCTION DURING FATIGUE

Fatigue is associated with several changes in myofibrillar function. The cross-bridge force production decreases during fatigue and myofibrillar Ca^{2+} sensitivity is reduced, both of which tend to decrease force production. Furthermore, the rate of cross-bridge cycling declines, which decreases the shortening velocity and possibly also the rate of relaxation. In this section, we will first discuss mechanisms underlying the decreased cross-bridge force production and the reduced myofibrillar Ca^{2+} sensitivity.

Figure 3.4 illustrates how changes in myofibrillar function affects the force–$[Ca^{2+}]_i$ relationship. The decrease in cross-bridge force production is seen as a general reduction, which in relative terms is equally large at all $[Ca^{2+}]_i$. It is most easily assessed in experiments where saturating $[Ca^{2+}]_i$ can be achieved (Figure 3.4a). The reduced myofibrillar Ca^{2+} sensitivity causes a right-ward shift of the force–$[Ca^{2+}]_i$ relationship, which means that it has a large effect on force production at low $[Ca^{2+}]_i$, whereas it has no effect at saturating $[Ca^{2+}]_i$ (Figure 3.4b). A standard way to quantify the myofibrillar Ca^{2+} sensitivity is to measure the $[Ca^{2+}]_i$ required to produce

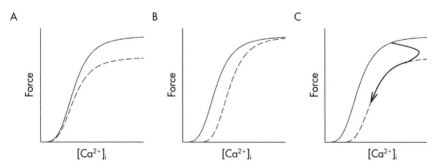

Figure 3.4 Schematic force–$[Ca^{2+}]_i$ relationships illustrating changes in myofibrillar function in fatigue. (A) Decreased cross-bridge force production (dashed line) results in the same relative force reduction at all $[Ca^{2+}]_i$. (B) Reduced myofibrillar Ca^{2+} sensitivity (dashed line) shifts the force–$[Ca^{2+}]_i$ relationship to the right. (C) The combined effect of decreased cross-bridge force and reduced myofibrillar Ca^{2+} sensitivity (dashed line). The arrow illustrates the typical pattern of changes during fatigue induced by repeated tetanic contractions in a fast-twitch fibre

50% of the maximum force (Ca_{50}), and a reduced sensitivity is then manifested as an increase in Ca_{50}.

The arrow in Figure 3.4c shows the typical time-course of changes in $[Ca^{2+}]_i$ and force during fatigue induced by repeated tetanic stimulation in a fast-twitch muscle fibre (Westerblad and Allen, 1991). The early decrease in force production is associated with increased tetanic $[Ca^{2+}]_i$, and hence the dominant factor behind the decrease is a decreased ability of the cross-bridges to generate force. As fatigue progresses, tetanic $[Ca^{2+}]_i$ decreases and reaches the steep part of the force–$[Ca^{2+}]_i$ relationship. Tetanic force then declines more rapidly due to the combined effect of the decreased $[Ca^{2+}]_i$ and the reduced myofibrillar Ca^{2+} sensitivity. Possible mechanisms underlying the decreased cross-bridge force production and the reduced myofibrillar Ca^{2+} sensitivity during fatigue are considered in the next three sub-sections.

Acidosis and myofibrillar function

Anaerobic breakdown of glycogen can cause large increases in the intracellular concentrations of H^+ and lactate ions, and intracellular pH can decrease to ~6.5 during intense exercise. Earlier studies indicated that low pH decreased force production by the contractile apparatus. However, recent studies performed under more physiological conditions show that the effects of low pH on myofibrillar function are small. For instance, experiments with skinned fibres have shown that while maximum force production and maximum shortening velocity are reduced when pH is reduced at low temperatures (<15°C), these parameters are little-affected by decreased pH at temperatures closer to normal body temperature (>30°C) (Pate et al., 1995). Similar results have also been obtained in isolated muscle fibres and whole muscles (Westerblad et al., 2002). Moreover, the rate of fatigue development was not increased in experiments where acidosis was exaggerated in isolated muscles or muscle cells during fatigue (Bruton et al., 1998; Zhang et al., 2006). Thus, acidosis has little effect on myofibrillar function and fatigue development.

Measurement of blood lactate is still a useful aid for assessing exercise performance, although the accumulation of lactate and hydrogen ions as such is not a major causative factor in fatigue. This is because a rise in blood lactate is easy to measure. Moreover, a major rise in blood lactate indicates that muscles heavily depend on anaerobic metabolism, which clearly is related to pronounced fatigue. Thus, there is often a good correlation between fatigue development and blood lactate, although there is no causative relationship.

P_i and myofibrillar function

During periods of high-intensity exercise, PCr breaks down to Cr and P_i via the creatine kinase reaction, and in this way [ATP] remains almost constant as long as [PCr] is not decreased to very low levels. Creatine has no important effect on myofibrillar function. Increased $[P_i]$, on the other hand, has an important inhibitory effect on both cross-bridge force production and myofibrillar Ca^{2+} sensitivity. According to current models of cross-bridge cycling, P_i is released in the transition from low-force, weakly attached cross-bridge states to high-force, strongly attached cross-bridge states (Takagi et al., 2004). This means that increased $[P_i]$ inhibits the transition to high-force cross-bridge states. Fewer cross-bridges will then be in high-force states as $[P_i]$ increases during fatigue and hence cross-bridge production decreases. Accordingly, experiments on skinned fibres consistently show decreased maximum force production in the presence of elevated $[P_i]$ (Pate and Cooke, 1989; Millar and Homsher, 1990). Similar to the situation with decreased pH (see above), the P_i-induced inhibition of cross-bridge force production becomes smaller with increasing temperature (Coupland et al., 2001; Debold et al., 2004).

Genetically engineered mice that completely lack creatine kinase in their skeletal muscles fatigue without any major increase in $[P_i]$. Fibres of these creatine kinase-deficient muscles do not display the 10–20% reduction of maximum force observed at the onset of fatigue induced by repeated tetanic stimulation (Dahlstedt et al., 2000). Thus, this early decrease in force during fatigue can be ascribed to increased $[P_i]$. Further support for a coupling between $[P_i]$ and force production in intact muscle cells comes from experiments where $[P_i]$ was reduced below the normal resting level during the recovery period after a series of contractions. This subnormal $[P_i]$ was associated with increased tetanic force production (Phillips et al., 1993; Bruton et al., 1997).

The force–$[Ca^{2+}]_i$ relationship is determined by the complex interaction between Ca^{2+}-induced activation of the thin (actin) filament and cross-bridge attachment to this filament. Skinned fibre experiments have shown that increased $[P_i]$ shifts the force–$[Ca^{2+}]_i$ relationship towards higher $[Ca^{2+}]_i$. That is, it decreases the myofibrillar Ca^{2+} sensitivity (Millar and Homsher, 1990; Martyn and Gordon, 1992). Thus, the fatigue-induced increase in $[P_i]$ may decrease force production by reducing both cross-bridge force production and myofibrillar Ca^{2+} sensitivity.

ROS/RNS and myofibrillar function

The production of ROS/RNS may increase during exercise due to the increased energy turnover and elevated $[Ca^{2+}]_i$ (Ferreira and Reid, 2008). Furthermore, the muscle temperature increases during exercise and this also speeds up ROS/RNS production (Allen et al., 2008b). Some studies have shown improved muscle performance when fatigue was induced in the presence of antioxidants (e.g. N-acetylcysteine), which supports a causative role of ROS/RNS in fatigue. However, the magnitude of the improvement induced by antioxidants is quite variable. This variability seems to depend on the stimulation protocol. For example, when human tibialis anterior muscles were fatigued by electrical stimulation using intermittent tetani at either 40 Hz or 10 Hz, a positive effect of N-acetylcysteine was only observed with 10 Hz stimulation (Reid *et al.*, 1994). The effect of ROS/RNS is temperature dependent, and the effect becomes larger with increasing temperature.

The mechanisms by which ROS/RNS affect force production are not fully understood. A recent study on single fibres of a mouse toe muscle fatigued by repeated tetani at 37°C showed a rapid force decrease (Moopanar and Allen, 2005; Moopanar and Allen, 2006). This rapid fatigue development was due to decreased myofibrillar Ca^{2+} sensitivity and was prevented by pre-application of ROS scavengers and reversed by a reducing agent. In accordance, experiments on intact muscle fibres show that myofibrillar Ca^{2+} sensitivity is very responsive to application of ROS (Andrade *et al.*, 2001). This finding contrasts with earlier studies on skinned fibres where the Ca^{2+} sensitivity was affected little by application of ROS. However, more recent skinned fibre studies have shown that the effect requires the presence of glutathione (Lamb and Posterino, 2003). Thus, when ROS/RNS affect myofibrillar function during fatigue, the most sensitive parameter is the myofibrillar Ca^{2+} sensitivity and the effect involves glutathione.

CHANGES IN SHORTENING VELOCITY AND POWER OUTPUT DURING FATIGUE

Most types of human locomotion are driven by mechanical work produced by skeletal muscles, and hence the speed of locomotion depends on the muscles' mechanical power output. Mechanical power equals shortening velocity × force, and both these parameters may decrease during fatigue. Consequently, the power output can decrease to very low levels during fatigue (Figure 3.5). The mechanisms underlying fatigue-induced changes in isometric force production are not the same as those affecting the shortening speed. Therefore, decreased force production may dominate in some situations, whereas decreased shortening velocity can be of greater importance in other cases. Furthermore, decreased force production has a larger impact on power output during movements requiring high forces, and decreased shortening velocity becomes more important as the speed of movement increases. Fatigue-induced changes in shortening velocity are displayed as decreased maximal shortening velocity (i.e. velocity at zero load) and/or altered curvature of the force–velocity relationship. During fatigue the changes in these parameters develop with different time-courses, indicating different underlying mechanisms (Jones *et al.*, 2006).

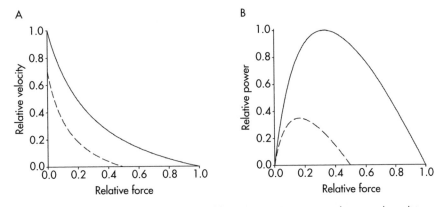

Figure 3.5 Schematic force–velocity (A) and force–power (B) curves under control conditions and in fatigue (dashed line). Note the marked decrease in relative power in fatigue; in this example isometric force was decreased to 50% and maximum shortening velocity to 70% in fatigue, and this reduced the maximum power to 35% of the control

Acidosis was classically a major cause of the fatigue-induced decrease in shortening speed. However, while acidosis decreases shortening velocity at low temperatures, it has little effect at mammalian body temperature (cf. the effects on isometric force discussed above) (Westerblad *et al.*, 2002). Increased [P_i] inhibits force production under isometric conditions, but has limited effects on shortening velocity. Reducing [ATP] to very low levels (~0.5 mM) decreases the shortening speed, but such low levels will not be reached in most types of fatigue. Reduced [Ca^{2+}]$_i$ is an important cause of decreased isometric force in fatigue, but has little effect on the maximal shortening velocity (Allen *et al.*, 2008b).

Skinned fibre experiments have shown that ADP has a major inhibitory effect on maximal shortening velocity (Cooke and Pate, 1985). The free [ADP] in the myoplasm cannot be measured with standard methods. Transient increases in [ADP] have been suggested to occur during contractions, especially when the PCr store becomes depleted (Sahlin, 1992). Substantial increases in IMP and a corresponding decrease in the total adenine nucleotide concentration have been observed in fatigued muscles (Jansson *et al.*, 1987; Sahlin *et al.*, 1990). These changes support the occurrence of transient, large increases in [ADP] during fatigue, because the formation of IMP is driven by an increase in [ADP], and the production would be limited if [ADP] remained low throughout fatigue. Thus, transient increases in [ADP] during contractions appear to be an important cause of the decreased shortening speed in fatigued muscles.

Studies have shown both an increased and a decreased curvature of the force–velocity relationship in fatigued muscles (Allen *et al.*, 2008b). The mechanisms underlying the changes in curvature as well as the basis for the opposite results are uncertain, but the temperature at which the experiments were performed seems to be an important factor. For instance, application of 30 mM P_i, which has little effect on the maximum shortening velocity, increased the curvature in both fast- and slow-twitch fibres when studied at 30°C, but had no effect at 15°C (Debold et al., 2004). A decreased curvature of the force–velocity relationship counteracts the decline in power output during fatigue, whereas an increased curvature has the opposite effect.

In conclusion, fatigued muscles show decreased maximal shortening velocity and altered curvature of the force–velocity relationship. The slowed shortening, together with the reduction in isometric force, decreases the power output of fatigued muscles.

CHANGES IN RELAXATION DURING FATIGUE

Fatigued skeletal muscles generally show a slowing of relaxation. This slowing can limit performance during dynamic exercise where rapidly alternating movements are performed (Allen et al., 1995). Conversely, slowed relaxation can be beneficial during a prolonged isometric contraction where it increases fusion of the force output at lower stimulation frequencies, thus minimising the force decline when the motoneuron firing rate decreases (Jones et al., 1979).

Relaxation of skeletal muscle cells involves: (1) SR Ca^{2+} release stops; (2) ATP-driven pumps transport Ca^{2+} back into the SR (myoplasmic buffers, such as parvalbumin, may contribute to the decrease in $[Ca^{2+}]_i$ in muscles that contain such buffers, provided they are not already saturated with Ca^{2+}); (3) Ca^{2+} dissociates from troponin; and (4) cross-bridge cycling ceases. All these steps can potentially be slowed in fatigue, hence cause slowing of relaxation. It is methodologically difficult to measure changes in the rate of each of these four steps during fatiguing stimulation. Moreover, it is doubtful whether one step can be accurately measured without interference from the other steps. For simplicity, the relaxation process can be divided into a Ca^{2+} component (steps 1 and 2) and a cross-bridge component (steps 3 and 4). The relative importance of the Ca^{2+} component and the cross-bridge component to the observed slowing of relaxation depends on many factors, such as the species and muscle studied, as well as the temperature and stimulation protocol. For instance, in a study on fatigue induced by repeated tetani, a large Ca^{2+} component of the slowing of relaxation was observed in frog muscle fibres, whereas the slowing in mouse muscle fibres was due to the cross-bridge component (Westerblad et al., 1997).

Several metabolic changes that occur during fatigue may contribute to the slowing of relaxation. Acidosis causes a marked slowing of relaxation at low temperatures, and the effect gets smaller as the temperature is increased. Nevertheless, acidification induces some slowing also at physiological temperatures, both in mouse and human muscle (Bruton et al., 1998; Cady et al., 1989). Changes in [ADP] and [P_i] directly affect both cross-bridge cycling and SR Ca^{2+} pumping, and changes in these metabolites are therefore likely to be involved in the fatigue-induced slowing of relaxation. Adenylate kinase limits the accumulation of ADP during fatigue, and muscles deficient of this enzyme show more slowing of relaxation during fatigue than control muscles (Hancock et al., 2005). Creatine kinase-deficient muscle fibres fatigue without any major increase in [P_i] and they do not display the normal fatigue-induced slowing of relaxation (Dahlstedt et al., 2000). Thus, increases in [H^+], [ADP] and [P_i] are likely metabolic factors causing slowing of relaxation in fatigued muscles.

CAUSES OF EXERCISE LIMITATION

So far we have concentrated mainly on studies carried out in isolated muscles, since mechanisms are most easily studied in such preparations. To what extent can these types of studies be applied to humans? Figure 3.6 is a schematic diagram in which the force decline observed in single fibres when maximally activated is shown as a dotted line. When maximal repeated tetani are carried out in a small human muscle, such as the adductor pollicis, curves not unlike the dotted line are obtained. The rate of fatigue is often quite similar for voluntary contractions and electrical stimulation, which indicates that under these conditions most fatigue is in the muscle, and any central component is quite small. If, instead, a substantial but submaximal contraction is carried out in a small muscle group (A), then the performance can continue unchanged for a period before task failure (TF) occurs. Examples would be pull-ups (lifting the body weight using the biceps muscle) where task failure is often close to the performance of the muscle when stimulated directly, so that most fatigue in this situation is in the muscle, but a small and variable component is central. However, more typical of many human activities (e.g. running, swimming and bicycle riding) would be line B, in which only a small fraction of the maximal force is used. These activities involve a substantial fraction of the body muscle mass so there are large increases in cardiac output and respiration to support the oxidative needs of the muscles. In this situation task failure, often called exhaustion (E), can be quite sudden and, if maximal activation of the muscle can be initiated, the muscles are still capable of producing high forces, at least for a short period (Beelen and Sargeant, 1991). This type of exhaustion can theoretically have two quite different causes. One is that receptors in the muscles sense, for example, raised K^+ and/or reduced extracellular pH and, via neuronal feedback, this inhibits motoneuron firing. Alternatively, the subject may simply stop the activity

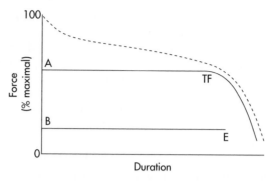

Figure 3.6 Schematic to illustrate various patterns of muscle fatigue. Dotted line shows fatigue of the pattern seen during repeated maximal tetani (cf. Figure 3.3). (A) illustrates the pattern of fatigue expected if force is substantial but submaximal. The activity can continue until the decline in maximal force is equivalent to the force required for (A). Thereafter there is task failure (TF) and the achievable force falls with a similar pattern to maximal force. (B) illustrates an activity in which only a small fraction of the maximum force is used, but which involves a large muscle mass so that there will be large stress on circulation and respiration. Often the subject ceases the activity at (E) (exhaustion), where the muscles still can produce greater force, at least for a short period

because the perception of muscle discomfort or breathlessness is so great that there is voluntary disengagement from the task (Marcora *et al.*, 2008). An important and interesting challenge is to define the details and relative importance of these pathways, which are probably the key to many kinds of task failure in humans.

CONCLUSION

At the muscle cell level, fatigue is manifested as decreased force production and slowed contractions. The reduced force production is due to (1) decreased SR Ca^{2+} release; (ii) reduced cross-bridge force production; and (iii) decreased myofibrillar Ca^{2+} sensitivity. The slowing involves both decreased shortening speed and slowed relaxation. Many metabolic changes in fatigued muscle cells (e.g. increased $[P_i]$ and ROS/RNS) can contribute to the contractile impairment.

FIVE KEY PAPERS THAT SHAPED THE TOPIC AREA

Study 1. Merton, P.A. (1954). Voluntary strength and fatigue. *Journal of Physiology*, 123, 553–564.

A turning point in the mechanistic understanding of fatigue was the ability to stimulate the motor nerve to a muscle and thereby potentially distinguish whether fatigue of the muscle was central (i.e. in the brain and spinal cord) or peripheral (i.e. in the neuromuscular junction or muscle). Merton showed that the ulnar nerve could be stimulated through the skin, causing contraction of the adductor pollicis muscle, and one key finding was that the maximum voluntary contraction and the maximum stimulated contraction were nearly identical. When the adductor pollicis muscle was fatigued by a continuous maximal voluntary contraction, Merton found that the fatigue was almost entirely in the muscle since maximal nerve stimulation was unable to increase the force. Subsequently, it has become apparent that for larger muscle groups and for lower intensity contractions central fatigue can be a substantial contributor to fatigue though peripheral (muscle) fatigue also contributes.

Study 2. Bergström, J., Hermansen, L., Hultman, E. and Saltin, B. (1967). Diet, muscle glycogen and physical performance. *Acta Physiologica Scandinavica*, 71, 140–150.

Another important breakthrough in the study of fatigue mechanisms in humans was the development of the muscle biopsy technique, which allows sampling of muscle for metabolic levels, histology, protein expression, muscle fibre typing, etc. Of particular importance was the discovery that loss of glycogen correlated with a decline in performance after several hours of moderate-intensity exercise. It was also recognised that dietary methods that boosted muscle glycogen, specifically a carbohydrate-rich diet, improved performance in this type of activity. This study provided the scientific basis for "carbo-loading" diets and the muscle biopsy technique has been used in numerous studies of fatigue mechanisms.

Study 3. Burke, R.E., Levine, D.N., Trairis, P. and Zajac, F.E. (1973). Physiological types and histochemical profiles in muscle motor units of the cat gastrocnemius. *Journal of Physiology*, 234, 723–748.

It is generally recognised that some muscles fatigue rapidly while other muscles fatigue very slowly. In this animal study by Burke *et al.*, single motor units were stimulated to exhaustion. The results showed that fast-twitch fibres fatigued within minutes, whereas slow-twitch fibres were essentially unfatiguable. This study provided the basis for the development of ideas about recruitment and the fatigability of different fibre types.

Study 4. Westerblad, H. and Allen, D.G. (1991). Changes of myoplasmic calcium concentration during fatigue in single mouse muscle fibers. *Journal of General Physiology*, 98, 615–635.

Much of our knowledge about cellular mechanisms of fatigue comes from studies on isolated muscle preparations. Whole muscles are convenient and simple to use but suffer from the serious disadvantage that the absence of blood supply allows the development of an anoxic core and leads to the accumulation of K^+, lactate, H^+ and CO_2 in the centre of the muscle. For these reasons, the dissection of small bundles of fibres or single fibres is a popular alternative. In a single fibre, the problems of fibre type mixtures are eliminated, and it is possible to measure changes in force and metabolites or ions continuously. This paper by Westerblad and Allen provided some of the earliest evidence that Ca^{2+} release declined in the later phase of fatigue and that this contributed to the reduction in force. The study also allowed the measurement of the relation between force and Ca^{2+}, which is critical for the understanding of the effects of metabolites on the contractile proteins.

Study 5. Pate, E., Bhimani, M., Franks-Skiba, K. and Cooke, R. (1995). Reduced effect of pH on skinned rabbit psoas muscle mechanics at high temperatures: implications for fatigue. *Journal of Physiology*, 486, 689–694.

Another approach to studies of fatigue is the use of skinned fibres. Skinned fibres have the advantage that the composition of solutions bathing the contractile proteins and SR can be specified completely, so the effect of any particular metabolic change can be investigated in the absence of the multitude of changes that occur during fatigue. This study by Pate *et al.* was particularly important as it was the first to show that while acidosis inhibits force production at low temperatures, it has very little effect at mammalian body temperature. This observation was the key to overturning the lactic acid theory of fatigue in mammals.

GLOSSARY OF TERMS

ACh	acetylcholine
ADP	adenosine diphosphate
AMP	adenosine monophosphate
ATP	adenosine triphosphate
$[Ca^{2+}]_i$	free myoplasmic Ca^{2+} concentration
Cr	creatine
DHPR	dihydropyridine receptor
EMG	electromyogram
IMP	inosine monophosphate
$[Mg^{2+}]_i$	free myoplasmic Mg^{2+} concentration
NMR	nuclear magnetic resonance
Pcr	phosphocreatine
P_i	inorganic phosphate
PLFFD	prolonged low-frequency force depression
RNS	reactive nitrogen species

ROS   reactive oxygen species
RyR   ryanodine receptor
SERCA   sarcoplasmic reticulum
t-tubules   transverse tubules

REFERENCES

Allen, D.G. and Westerblad, H. (2001). Role of phosphate and calcium stores in muscle fatigue. *Journal of Physiology*, 536, 657–665.

Allen, D.G., Lamb, G.D. and Westerblad, H. (2008a). Impaired calcium release during fatigue. *Journal of Applied Physiology*, 104, 296–305.

Allen, D.G., Lamb, G.D., and Westerblad, H. (2008b). Skeletal muscle fatigue: cellular mechanisms. *Physiological Review*, 88, 287–332.

Allen, D.G., Lee, J.A. and Westerblad, H. (1989). Intracellular calcium and tension during fatigue in isolated single muscle fibres from *Xenopus laevis*. *Journal of Physiology*, 415, 433–458.

Andrade, F.H., Reid, M.B. and Westerblad, H. (2001). Contractile response of skeletal muscle to low peroxide concentrations: myofibrillar calcium sensitivity as a likely target for redox-modulation. *Federation of American Societies for Experimental Biology Journal*, 15, 309–311.

Beelen, A. and Sargeant, A.J. (1991). Effect of fatigue on maximal power output at different contraction velocities in humans. *Journal of Applied Physiology*, 71, 2332–2337.

Bergström, J., Hermansen, L., Hultman, E. and Saltin, B. (1967). Diet, muscle glycogen and physical performance. *Acta Physiologica Scandinavica*, 71, 140–150.

Bigland-Ritchie, B., Johansson, R., Lippold, O.C., Smith, S. and Woods, J.J. (1983). Changes in motoneurone firing rates during sustained maximal voluntary contractions. *Journal of Physiology*, 340, 335–346.

Blazev, R. and Lamb, G.D. (1999). Low [ATP] and elevated [Mg^{2+}] reduce depolarization-induced Ca^{2+} release in rat skinned skeletal muscle fibres. *Journal of Physiology*, 520, 203–215.

Bruton, J.D., Lännergren, J. and Westerblad, H. (1998). Effects of CO_2-induced acidification on the fatigue resistance of single mouse muscle fibers at 28°C. *Journal of Applied Physiology*, 85, 478–483.

Bruton, J.D., Place, N., Yamada, T., Silva, J.P., Andrade, F.H., Dahlstedt, A.J., Zhang, S.J., Katz, A., Larsson, N.G. and Westerblad, H. (2008). Reactive oxygen species and fatigue-induced prolonged low-frequency force depression in skeletal muscle fibres of rats, mice and SOD2 overexpressing mice. *Journal of Physiology*, 586, 175–184.

Bruton, J.D., Tavi, P., Aydin, J., Westerblad, H. and Lännergren, J. (2003). Mitochondrial and myoplasmic Ca^{2+} in single fibres from mouse limb muscles during repeated tetanic contractions. *Journal of Physiology*, 551, 179–190.

Bruton, J.D., Wretman, C., Katz, A. and Westerblad, H. (1997). Increased tetanic force and reduced myoplasmic [P_i] following a brief series of tetani in mouse soleus muscle. *American Journal of Physiology*, 272, C870–874.

Burke, R.E., Levine, D.N., Tsairis, P. and Zajac III, F.E. (1973). Physiological types and histochemical profiles in motor units of the cat gastrocnemius. *Journal of Physiology*, 234, 723–748.

Cady, E.B., Elshove, H., Jones, D.A. and Moll, A. (1989). The metabolic causes of slow relaxation in fatigued human skeletal muscle. *Journal of Physiology*, 418, 327–337.

Chin, E.R. and Allen, D.G. (1997). Effects of reduced muscle glycogen concentration on force, Ca^{2+} release and contractile protein function in intact mouse skeletal muscle. *Journal of Physiology*, 498, 17–29.

Cooke, R. and Pate, E. (1985). The effects of ADP and phosphate on the contraction of muscle fibers. *Biophysical Journal*, 48, 789–798.

Coupland, M.E., Puchert, E. and Ranatunga, K.W. (2001). Temperature dependence of active tension in mammalian (rabbit psoas) muscle fibres: effect of inorganic phosphate. *Journal of Physiology*, 536, 879–891.

Dahlstedt, A.J., Katz, A., Tavi, P. and Westerblad, H. (2003). Creatine kinase injection restores contractile function in creatine-kinase-deficient mouse skeletal muscle fibres. *Journal of Physiology*, 547, 395–403.

Dahlstedt, A.J., Katz, A., Wieringa, B. and Westerblad, H. (2000). Is creatine kinase responsible for fatigue? Studies of skeletal muscle deficient of creatine kinase. *Federation of American Societies for Experimental Biology Journal*, 14, 982–990.

Dawson, M.J., Gadian, D.G. and Wilkie, D.R. (1978). Muscular fatigue investigated by phosphorus nuclear magnetic resonance. *Nature*, 274, 861–866.

Debold, E.P., Dave, H. and Fitts, R.H. (2004). Fiber type and temperature dependence of inorganic phosphate: implications for fatigue. *American Journal of Physiology: Cell Physiology*, 287, C673–681.

Dulhunty, A.F. (2006). Excitation–contraction coupling from the 1950s into the new millennium. *Clinical and Experimental Pharmacology and Physiology*, 33, 763–772.

Duty, S. and Allen, D.G. (1994). The distribution of intracellular calcium concentration in isolated single fibres of mouse skeletal muscle during fatiguing stimulation. *Pflügers Archives*, 427, 102–109.

Eberstein, A. and Sandow, A. (1963). Fatigue mechanisms in muscle fibres. In E. Gutman and P. Hnik (eds), *The Effect of Use and Disuse on Neuromuscular Functions*. Amsterdam: Elsevier, pp. 515–526.

Edwards, R.H., Hill, D.K., Jones, D.A. and Merton, P.A. (1977). Fatigue of long duration in human skeletal muscle after exercise. *Journal of Physiology*, 272, 769–778.

Ferreira, L.F. and Reid, M.B. (2008). Muscle-derived ROS and thiol regulation in muscle fatigue. *Journal of Applied Physiology*, 104, 853–860.

Fitts, R.H. (1994). Cellular mechanisms of muscle fatigue. *Physiological Review*, 74, 49–94.

Friden, J., Seger, J. and Ekblom, B. (1989). Topographical localization of muscle glycogen: an ultrahistochemical study in the human vastus lateralis. *Acta Physiologica Scandinavica*, 135, 381–391.

Fryer, M.W., Owen, V.J., Lamb, G.D. and Stephenson, D.G. (1995). Effects of creatine phosphate and P_i on Ca^{2+} movements and tension development in rat skinned skeletal muscle fibres. *Journal of Physiology*, 482, 123–140.

Gandevia, S.C. (2001). Spinal and supraspinal factors in human muscle fatigue. *Physiological Review*, 81, 1725–1789.

Hancock, C.R., Janssen, E. and Terjung, R.L. (2005). Skeletal muscle contractile performance and ADP accumulation in adenylate kinase-deficient mice. *American Journal of Physiology: Cell Physiology*, 288, C1287–1297.

Jansson, E., Dudley, G.A., Norman, B. and Tesch, P. (1987). ATP and IMP in single human muscle fibres after high intensity exercise. *Clinical Physiology*, 7, 337–345.

Jones, D.A., Bigland-Ritchie, B. and Edwards, R.H. (1979). Excitation frequency and muscle fatigue: mechanical responses during voluntary and stimulated contractions. *Experimental Neurology*, 64, 401–413.

Jones, D.A., De Ruiter, J. and de Haan, A. (2006). Change in contractile properties of human muscle in relationship to the loss of power and slowing of relaxation seen with fatigue. *Journal of Physiology*, 526, 671–681.

Kabbara, A.A. and Allen, D.G. (1999). The role of calcium stores in fatigue of isolated single muscle fibres from the cane toad. *Journal of Physiology*, 519, 169–176.

Kabbara, A.A. and Allen, D.G. (2001). The use of fluo-5N to measure sarcoplasmic reticulum calcium in single muscle fibres of the cane toad. *Journal of Physiology*, 534, 87–97.

Karatzaferi, C., de Haan, A., Ferguson, R.A., van Mechelen, W. and Sargeant, A.J. (2001). Phosphocreatine and ATP content in human single muscle fibres before and after maximum dynamic exercise. *Pflugers Archives*, 442, 467–474.

Lamb, G.D. and Posterino, G.S. (2003). Effects of oxidation and reduction on contractile function in skeletal muscle fibres of the rat. *Journal of Physiology*, 546, 149–163.

Lamb, G.D., Junankar, P.R. and Stephenson, D.G. (1995). Raised intracellular [Ca^{2+}] abolishes excitation–contraction coupling in skeletal muscle fibres of rat and toad. *Journal of Physiology*, 489, 349–362.

Laver, D.R., Lenz, G.K. and Dulhunty, A.F. (2001). Phosphate ion channels in sarcoplasmic reticulum of rabbit skeletal muscle. *Journal of Physiology*, 535, 715–728.

Marcora, S.M., Bosio, A. and de Morree, H.M. (2008). Locomotor muscle fatigue increases cardiorespiratory responses and reduces performance during intense cycling exercise independently from metabolic stress. *American Journal of Physiology: Regulatory, Integrative and Comparative Physiology*, 294, R874–883.

Martyn, D.A. and Gordon, A.M. (1992). Force and stiffness in glycerinated rabbit psoas fibers. Effects of calcium and elevated phosphate. *Journal of General Physiology*, 99, 795–816.

Merton, P.A. (1954). Voluntary strength and fatigue. *Journal of Physiology*, 123, 553–564.

Millar, N.C. and Homsher, E. (1990). The effect of phosphate and calcium on force generation in glycerinated rabbit skeletal muscle fibers. A steady-state and transient kinetic study. *Journal of Biological Chemistry*, 265, 20234–20240.

Moopanar, T.R. and Allen, D.G. (2005). Reactive oxygen species reduce myofibrillar Ca^{2+} sensitivity in fatiguing mouse skeletal muscle at 37 degrees C. *Journal of Physiology*, 564, 189–199.

Moopanar, T.R. and Allen, D.G. (2006). The activity-induced reduction of myofibrillar Ca^{2+} sensitivity in mouse skeletal muscle is reversed by dithiothreitol. *Journal of Physiology*, 571, 191–200.

Pate, E. and Cooke, R. (1989). Addition of phosphate to active muscle fibers probes actomyosin states within the powerstroke. *Pflügers Archives*, 414, 73–81.

Pate, E., Bhimani, M., Franks-Skiba, K. and Cooke, R. (1995). Reduced effect of pH on skinned rabbit psoas muscle mechanics at high temperatures: implications for fatigue. *Journal of Physiology*, 486, 689–694.

Pedersen, T.H., de Paoli, F. and Nielsen, O.B. (2005). Increased excitability of acidified skeletal muscle: role of chloride conductance. *Journal of General Physiology*, 125, 237–246.

Pedersen, T.H., Nielsen, O.B., Lamb, G.D. and Stephenson, D.G. (2004). Intracellular acidosis enhances the excitability of working muscle. *Science*, 305, 1144–1147.

Phillips, S.K., Wiseman, R.W., Woledge, R.C. and Kushmerick, M.J. (1993). The effect of metabolic fuel on force production and resting inorganic phosphate levels in mouse skeletal muscle. *Journal of Physiology*, 462, 135–146.

Place, N., Yamada, T., Bruton, J.D. and Westerblad, H. (2008). Interpolated twitches in fatiguing single mouse muscle fibres: implications for the assessment of central fatigue. *Journal of Physiology*, 586, 2799–2805.

Reid, M.B. (2008). Free radicals and muscle fatigue: Of ROS, canaries, and the IOC. *Free Radical Biology and Medicine*, 44, 169–179.

Reid, M.B., Stokic, D.S., Koch, S.M., Khawli, F.A. and Leis, A.A. (1994). N-acetylcysteine inhibits muscle fatigue in humans. *Journal of Clinical Investigation*, 94, 2468–2474.

Sahlin, K. (1992). Metabolic factors in fatigue. *Sports Medicine*, 13, 99–107.

Sahlin, K., Katz, A. and Broberg, S. (1990). Tricarboxylic acid cycle intermediates in human muscle during prolonged exercise. *American Journal of Physiology*, 259, C834–841.

Sejersted, O.M. and Sjøgaard, G. (2000). Dynamics and consequences of potassium shifts in skeletal muscle and heart during exercise. *Physiological Review*, 80, 1411–1481.

Stephenson, D.G., Nguyen, L.T. and Stephenson, G.M. (1999). Glycogen content and excitation–contraction coupling in mechanically skinned muscle fibres of the cane toad. *Journal of Physiology*, 519, 177–187.

Takagi, Y., Shuman, H. and Goldman, Y.E. (2004). Coupling between phosphate release and force generation in muscle actomyosin. *Philosophical transactions of the Royal Society of London. Series B, Biological Sciences*, 359, 1913–1920.

Verburg, E., Dutka, T.L. and Lamb, G.D. (2006). Long-lasting muscle fatigue: partial disruption of excitation–contraction coupling by elevated cytosolic Ca^{2+} concentration during contractions. *American Journal of Physiology: Cell Physiology*, 290, C1199–1208.

Westerblad, H. and Allen, D.G. (1991). Changes of myoplasmic calcium concentration during fatigue in single mouse muscle fibers. *Journal of General Physiology*, 98, 615–635.

Westerblad, H. and Allen, D.G. (1992). Myoplasmic free Mg^{2+} concentration during repetitive stimulation of single fibres from mouse skeletal muscle. *Journal of Physiology*, 453, 413–434.

Westerblad, H. and Allen, D.G. (1996). The effects of intracellular injections of phosphate on intracellular calcium and force in single fibres of mouse skeletal muscle. *Pflügers Archives*, 431, 964–970.

Westerblad, H., Allen, D.G. and Lännergren, J. (2002). Muscle fatigue: lactic acid or inorganic phosphate the major cause? *News in Physiological Sciences*, 17, 17–21.

Westerblad, H., Duty, S. and Allen, D.G. (1993). Intracellular calcium concentration during low-frequency fatigue in isolated single fibers of mouse skeletal muscle. *Journal of Applied Physiology*, 75, 382–388.

Westerblad, H., Lännergren, J. and Allen, D.G. (1997). Slowed relaxation in fatigued skeletal muscle fibers of *Xenopus* and mouse. Contribution of $[Ca^{2+}]_i$ and cross-bridges. *Journal of General Physiology*, 109, 385–399.

Westerblad, H., Lee, J.A., Lamb, A.G., Bolsover, S.R. and Allen, D.G. (1990). Spatial gradients of intracellular calcium in skeletal muscle during fatigue. *Pflügers Archives*, 415, 734–740.

Zhang, S.J., Bruton, J.D., Katz, A. and Westerblad, H. (2006). Limited oxygen diffusion accelerates fatigue development in mouse skeletal muscle. *Journal of Physiology*, 572, 551–559.

PART II

FATIGUE AND EXTRANEOUS FACTORS

CHAPTER 4

MUSCLE FATIGUE IN CHILDREN

Sébastien Ratel, Pascale Duché and Craig A. Williams

OBJECTIVES

The aim of this chapter is to provide an overview on:

- the capacity of children to resist fatigue during aerobic and anaerobic exercises as compared with adults;
- the potential factors explaining the differences in fatigue resistance between children and adults according to the type of exercise;
- the practical implications that emerge from the studies conducted on muscle fatigue in children.

INTRODUCTION

Over the last decades, exercise-induced muscle fatigue in children has received much more attention. One reason for this growing interest is the increasing involvement of children and youth in high-level sports. Today's prepubertal children are often exposed to training regimens that are considered as highly demanding, even compared to adult standards. In some sports, such as female gymnastics, children and adolescents excel and reach world standards. In other sports, such as athletics or swimming, children do not reach their peak performance levels before the second decade of life, but their specialised training might start as early as the first decade. Knowledge of muscle performance and physiological demand during exhaustive exercise in children is therefore of fundamental importance for coaches and practitioners in paediatric research. The main objective of this chapter will be to provide the reader an overview of the effects of growth and maturation on muscle fatigue during exercise. The approach of this chapter will focus on the investigation of muscle fatigue in children during prolonged sustained exercise and short-term high-intensity exercise.

FATIGUE DURING PROLONGED EXERCISE

Prolonged exercise is defined here as any exercise that last 30 min or more and is continuous in nature. The work rates are submaximal (less than peak $\dot{V}O_2$) and oxidative processes predominantly meet the ATP supply. Although research on physical activity has characterised children as spontaneously choosing multiple-sprint activities that usually last no longer than 10 s and are typically interspersed with varying levels of low-to-moderate activity (e.g. football, basketball, hockey), many children participate in long-distance events (e.g. running, cycling, swimming, cross-country skiing). Furthermore, other young athletes, as part of their training regimens, also practise prolonged activities. A frequent question usually asked by coaches is whether, compared with adults, children have a greater endurance, i.e. an ability to exercise longer at a given absolute intensity? Or if children display a greater resistance to fatigue during prolonged exercise? We shall try to answer this question in the following section.

Age-related differences

One approach to evaluate fatigue during prolonged exercise is to measure the endurance time, i.e. the time an individual can maintain a given exercise speed or at a percentage of maximal aerobic speed (%MAS). In running, the endurance time expressed as a function of %MAS decreases much faster in children as compared with adults (Léger, 1996). As shown in Figure 4.1, the exercise duration at any speed and in %MAS is shorter in younger individuals. As a result, children's exercise performance in long-duration athletics events improves with age. This is illustrated in Figure 4.2, considering world-best performances in a marathon from five-year-old children to 40-year-old adults.

As outlined by numerous authors, endurance time is conditioned by the highest volume of O_2 that can be consumed by the body per time unit (peak $\dot{V}O_2$), the true endurance (endurance time at any percentage of peak $\dot{V}O_2$) and the amount of O_2 required to move at a given speed (running economy) (Léger, 1996). Conventional comparisons of peak $\dot{V}O_2$ per unit of body mass between prepubertal children, pubertal children and adults revealed no significant difference in males, but a significant decline in

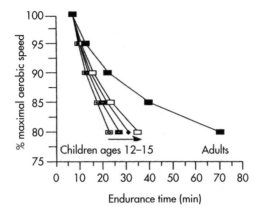

Figure 4.1 Endurance time expressed in percentage of maximal aerobic speed (%MAS) for running in children and adults

Source: Reprinted from Léger, 1996.

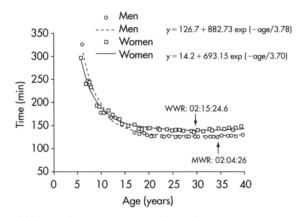

Figure 4.2 World-best performances in marathon with increasing age. WWR: women's world record, MWR: men's world record

females (Welsman *et al.*, 1996; Armstrong, 2006). However, when data are allometrically scaled, a progressive increase in peak $\dot{V}O_2$ can be observed both in males and females, thereby indicating a lower aerobic fitness in the younger children (Welsman *et al.*, 1996; Armstrong and Welsman, 2001; Armstrong, 2006). Furthermore, although the true endurance (as percentage of peak $\dot{V}O_2$) stays constant with age, running economy increases during growth (Léger, 1996). When expressed per kilogram of body mass, the O_2 cost of locomotion was found to be higher in children (Robinson, 1938; Astrand, 1952; Davies, 1980). For instance, Astrand (1952) showed at various submaximal speeds on a motorised treadmill that O_2 consumption ($ml \cdot min^{-1} \cdot kg^{-1}$ body mass) decreased from five-year-old to 18-year-old girls and boys. At a constant speed of

$12\,km\cdot h^{-1}$, nine-year-old boys exhibit a greater O_2 uptake of about $7\,ml\cdot min^{-1}\cdot kg^{-1}$ body mass than 18-year-old men. These results were supported by longitudinal changes of the O_2 cost of running in males and females aged between 13 and 27 years (Ariens *et al.*, 1997). A significant decrease in O_2 uptake ($ml\cdot min^{-1}\cdot kg^{-1}$ body mass) was obtained with increasing age at a constant speed of $8\,km\cdot h^{-1}$ and various treadmill slopes (0%, 2.5% and 5.0%). Males showed significantly higher $\dot{V}O_2$ values than females at all ages measured and for all three slopes, indicating females to be more economical than males. Gender-related differences in submaximal $\dot{V}O_2$ and running economy appeared larger when allometric analysis was used to describe the possible effect of body mass. Consistent with this latter finding, a three-year longitudinal study investigating of the O_2 cost of running at $8\,km\cdot h^{-1}$ found girls to be more economical than boys between the ages of 11–13 years following the use of allometric modelling procedures to partition out the influence of body mass and body fatness (Welsman and Armstrong, 2000). However, in contrast to the study by Ariens *et al.* (1997), these authors found the O_2 cost of running at $8\,km\cdot h^{-1}$ to be independent of age and maturity, the latter assessed using Tanner's indices of sexual maturity, after scaling for body mass and body fatness.

Despite the ambiguity as to whether the O_2 cost of submaximal running (i.e. exercise economy) increases or remains stable during growth and maturation, one implication of these results is that when normalised to peak O_2 uptake (per unit of body mass), the O_2 required for a given submaximal speed represents a greater percentage in young children as compared with adolescents (Bar-Or, 1982; Bar-Or and Rowland, 2004) (see Figure 4.3). The smaller "metabolic reserve" (so named by Bar-Or, 1982) in children associated with a peak $\dot{V}O_2$ (in $ml\cdot kg^{-1}\cdot min^{-1}$) and true endurance (at any percentage of peak $\dot{V}O_2$) that stay constant with age may explain their lesser ability to run as long and as intensively as adults over long-distance

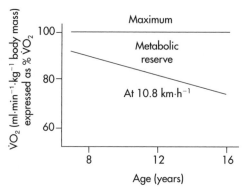

Figure 4.3 Metabolic reserve as a function of age

Source: Redrawn from Bar-Or and Rowland, 2004.

events, even though they can maintain a slow speed for long periods (Roberts, 2007) (see Figure 4.2).

However, such evidence does not seem to be applied in non-weight-bearing activities such as swimming and cycling. By considering that the energy cost of swimming is dependent on body length (BL) and not on body mass, Kjendlie et al. (2004) indicated that at a swim velocity of $1\,m\cdot s^{-1}$, O_2 consumption when normalised to BL was similar in children and adults. The same results were obtained more recently by Ratel and Poujade (2009) at the same relative swimming velocity. In the same way, previous studies have indicated that the mechanical efficiency of cycling, defined as the ratio between external mechanical work produced by the muscle and the chemical energy utilised during the contraction, was comparable in prepubertal and postpubertal subjects, either for submaximal absolute workload or similar relative intensity expressed in percentage of peak $\dot{V}O_2$ (Rowland et al., 1990). In addition, Rowland et al. (2008) have recently showed that endurance performance and thermoregulatory and cardiovascular responses were similar during steady-load cycling until exhaustion at an intensity equivalent to approximately 65% peak $\dot{V}O_2$ in prepubertal boys and adult men in either hot or normal thermal environments (hot: men $30.5\pm8.8\,min$, boys $29.3\pm6.2\,min$; normal: men $42.9\pm11.8\,min$, boys $41.4\pm6.3\,min$). It should be noted however, that studies examining the O_2 cost of cycling using the "gain" concept ($\dot{V}O_2/$ power output) which is akin to delta efficiency, have shown children to be less efficient than older children or adults during 6–9 min of cycling exercise above and below the ventilatory threshold (Armon et al., 1991; Fawkner and Armstrong, 2004).

Therefore, in contrast to locomotion, the energy cost should not be considered as a limiting factor of endurance in non-weight-bearing activities in children. However, we need to be cautious on this suggestion since few data regarding child–adult differences of endurance over long-distance swimming and cycling events are available, and given the ambiguity present for cycling data.

Explanatory factors

Numerous mechanisms may be suggested to explain the greater O_2 cost of locomotion and the lower endurance performance during prolonged exercise in children. The most likely factors are changes in body mass, stride frequency, ventilatory cost, coactivation of antagonist muscles and locomotion style.

Higher stride frequency

Because of their shorter legs, young children have shorter step length and higher step rate than adults at a given walking and running speed

(Rowland *et al.*, 1987; Frost *et al.*, 2002; Ratel *et al.*, 2006b). According to Rowland *et al.* (1988), greater running economy is associated with superior treadmill endurance performance, and stride frequency may influence submaximal energy expenditure in children. However, multiple regression analysis including kinematic and electromyographic data from three maturational groups of children showed that stride frequency did not appear as a good indicator of the metabolic cost of locomotion in 7–16-year-old girls and boys (Frost *et al.*, 2002). In the same way, Davies (1980) and Rowland *et al.* (1987) showed that the stride frequency was unrelated to running economy in active and athletic children. Furthermore, the metabolic cost expressed per stride has been shown to be similar between boys and men, except at speeds higher than 8 km·h^{-1} (Unnithan and Eston, 1990; Ebbeling *et al.*, 1992; Frost *et al.*, 2002). As a result, stride frequency does not seem to be the best predictor of running economy and should not explain the lesser ability of children to compete in long-distance events as intensively as adults.

Higher ventilatory cost

Ventilatory efficiency can be measured from the ventilatory equivalent, i.e. the ventilation (VE) to oxygen uptake ($\dot{V}O_2$) ratio (VE/$\dot{V}O_2$). A high ratio reflects a high ventilatory cost and low efficiency. The longitudinal study from Rowland and Cunningham (1997) showed that the VE/$\dot{V}O_2$ ratio fell progressively between the ages of nine and 13 years at submaximal exercise levels in both girls and boys, but declined only in the boys with maximal steady-state treadmill walking testing. The decline in submaximal and maximal VE/$\dot{V}O_2$ with age has also been described in cross-sectional studies (Astrand, 1952; Andersen *et al.*, 1974). The major implication of high VE/$\dot{V}O_2$ is a greater O_2 cost of breathing in younger children. This may contribute to their relatively high metabolic demands during submaximal running and their lesser ability to perform exercise at the same absolute intensity for as long as older subjects.

Higher coactivation

By definition, coactivation, also known as co-contraction, is a result of simultaneous activation of agonist and antagonist muscle groups during voluntary contractions (Psek and Cafarelli, 1993). While coactivation of antagonist muscles is important for joint stability, especially at the ankle joint during locomotion, excessive coactivation may cause an inefficient movement, high metabolic cost and excessive muscle fatigue. Thigh and leg coactivation and $\dot{V}O_2$ were measured in three different age groups of children aged 7–8, 10–12 and 15–16 years across a total of six speeds (Frost *et al.*, 2002). It has been found that during walking and running at the same absolute speed, $\dot{V}O_2$ when expressed per kilogram of body mass was significantly higher in the youngest age group. In the same way, the coactivation values of thigh and leg were found to be higher in the youngest children. According to Frost *et al.* (2002), muscle coactivation would be an important predictor of net $\dot{V}O_2$ and efficiency (after age) during locomotion and would account for a larger explained variance in the youngest children than in the oldest (Frost *et al.*, 2002). The implication is that young children (7–8 years old) may not yet have the required neuromotor control to optimally synchronise the action between muscle groups. The unnecessary activation of antagonist muscles would result in a "waste" of energy and thereby in a greater O_2 cost during locomotion in children.

"Non optimal" body mass

Davies (1980) studied the effects of external loading on the metabolic cost of running in nine athletic boys, five athletic girls and nine active boys aged 11–13 years. The author showed that external loading equivalent to 5% of body mass reduced $\dot{V}O_2$, particularly at the highest speeds. It was suggested for young active and athletic children, due to their relatively light body masses, that the required frequency of leg movement was not optimally matched to the force necessary to produce the most economic conversion of aerobic energy into mechanical work.

Locomotion style

Another factor that may increase the O_2 cost in children is the instability of locomotion during growth. It has been shown that stride-to-stride variations in gait cycle duration are significantly larger in healthy three- and four-year-old children compared with six- and seven-year-old children, and in six- and seven-year-old children compared with children aged 11–14 (Hausdorff *et al.*, 1999). Furthermore, within-session variability of muscle coordination pattern during gait measured by surface electromyography (EMG) was found to be approximately twice as high in healthy 6.5-year-old children as compared with adults (Granata *et al.*, 2005). This high variability in the temporal organisation of stride dynamics could result from biomechanical and neural properties that are known to mature only in older children (Hausdorff *et al.*, 1999). Inconsistent stride-to-stride walking pattern may therefore be considered as another important cause of the high metabolic cost of locomotion in children. This increased variability may result in a "waste" of energy and hence contribute to the lesser ability of children to compete in long-distance events as intensively as adults.

Other factors

Other factors include substrate utilisation. Previous investigations have shown that prepubertal children utilise proportionally more fat and less carbohydrate as compared with adults during exercise performed at the same relative intensity (Martinez and Haymes, 1992; Timmons *et al.*, 2003, 2007). The preferential use of fat as fuel during submaximal exercise in young children may be explained by higher intramuscular triglyceride availability, an enhanced oxidative metabolism and an immature glycolytic system (Eriksson *et al.*, 1971; Bell *et al.*, 1980; Haralambie, 1982; Berg *et al.*, 1986). Given that fat oxidation provides much less ATP per time unit than endogenous carbohydrate oxidation, and considering lower intramuscular glycogen stores in children (Eriksson, 1980), children will be physiologically limited in running as intensively as adults in long-distance events. However, while endogenous carbohydrate utilisation during exercise is lower, the relative oxidation of exogenous carbohydrate is considerably higher in boys than in men (Timmons *et al.*, 2003). The greater reliance on exogenous carbohydrate in boys may therefore be considered as a way to preserve endogenous fuels and delay muscle fatigue in children during long-distance events.

Furthermore, physiological and anatomical characteristics of children (i.e. high ratio of body surface area to mass, diminished sweating capacity, low cardiac output) were considered to impair thermoregulatory responses and physical performance in long-distance events, especially in the heat. However, recent investigations have failed to indicate child–adult differences in heat dispersal when exercise was appropriately performed at the same relative intensity (Rowland, 2008). For instance, Rowland *et al.* (2008) have showed no differences in exercise tolerance, thermoregulatory adaptation, or cardiovascular response to submaximal cycling exercise until exhaustion in the heat (31°C) between prepubertal boys and adult men. Therefore, these findings imply that the regulation of thermal balance during prolonged exercise in the heat should not be considered as a limiting factor in children.

Summary

There do not seem to be any underlying physiological factors that limit the ability of children to perform prolonged continuous exercise. However, apart from weight-bearing activities, economy of locomotion is significantly lower in children than in adults. Hence, children fatigue faster than adults during prolonged exercise performed at various absolute or relative walking and running speeds. The various mechanisms that have been suggested are a "non-optimal" body mass, higher stride frequency, higher ventilatory cost, higher coactivation of antagonist muscles and lesser muscle coordination, leading to a mechanically "wasteful" locomotion style. According to Frost *et al.* (2002), it is possible to explain up to 77% of the variance in net $\dot{V}O_2$ and 62% of the variance in efficiency in children using a combination of mechanical power, coactivation, stride length and age.

FATIGUE DURING HIGH-INTENSITY EXERCISE

In this section, high-intensity exercise will be defined as any exercise that exceeds the ATP turnover rate of the maximal power of the oxidative metabolism, such as maximal or "all-out" cycling or running sprints. Muscle fatigue appears more precipitously, particularly during this type of exercise, and is more evident during multiple-sprint activities such as team sports (soccer, basketball, rugby). A question often asked by coaches is whether high-intensity intermittent training regimens should be advised for young children? Or, as compared to adults, what are the possibilities of children to repeat successive high-intensity exercises interspersed by short recovery intervals?

Age-related differences

In contrast to prolonged moderate-intensity exercise, children appear to fatigue less than adults during successive bouts of high-intensity exercise. Sports coaches and other personnel have often commented on this observation also. However, data showing these differences are very limited. In 1993, Hebestreit et al. published the first study examining the time-course of power output during two successive Wingate anaerobic tests (30 s "all-out" cycle sprints) in young boys and men. The authors showed that the boys decreased less of their cycling peak power during the first sprint and were able to duplicate their peak power after a 2 min recovery period, whereas adults required at least 10 min (Hebestreit et al., 1993). Similar age-related differences were also reported in our laboratory within a series of 10 s bouts of cycling (Ratel et al., 2002a, 2004b) (Figure 4.4).

Using 30 s recovery periods, we showed that young boys were able to maintain their cycling peak power during ten repeated 10 s sprints, whereas adolescent boys and men decreased their peak power output by 18.5% and 28.5%, respectively (Ratel et al., 2002a). In this last study, 5 min recovery periods were necessary in adolescent boys and men to sustain their cycling peak power from the first to the tenth sprint. Comparable findings were also reported during consecutive maximal isokinetic knee extensions by comparing the time-course of peak torque in young boys, adolescent boys and men (Zafeiridis et al., 2005; Paraschos et al., 2007). Similar results were also obtained within a series of 10 s bouts of running (Ratel et al., 2006b). However, young boys and men experienced greater declines in work rate during running compared with cycling (Ratel et al., 2004b) (Figure 4.4).

Although rarely studied, age-related differences were confirmed in females by comparing the power output profiles during a series of three 15 s Wingate test-like cycling bouts, interspersed by 45 s active recovery intervals in young girls and women (Chia, 2001). For more details on apparent age differences in the rate of fatigue development during high-

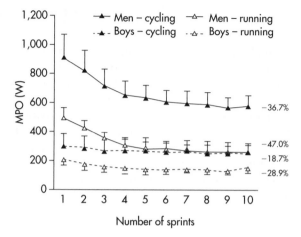

Figure 4.4 Mean power output profile during ten repeated 10s cycling and running sprints separated by 15s recovery intervals in boys and men

Source: Reprinted with permission from Ratel *et al.*, 2004a.

intensity exercise, we advise the readers to read the review papers written by Falk and Dotan (2006) and Ratel *et al.* (2006a).

In summary, the ability to repeat high levels of power output during intermittent high-intensity activities depends upon age and probably maturation level. Prepubertal children have lower but more consistent power output in repetitious sprint activities. Therefore, during high-intensity intermittent training regimens, young children may cope better with shorter recovery intervals than are commonly used by adults (Bar-Or, 1995; Ratel *et al.*, 2004a).

Explanatory factors

The exact causative factors explaining the greater resistance to fatigue during high-intensity intermittent exercise in children have not been clearly discriminated, but several hypotheses related to quantitative and qualitative muscle changes during growth have been postulated. These explanatory factors could be linked to smaller body dimensions, changes in the muscle structure and metabolism during growth and lower neuromuscular activation, resulting in limited energy disturbance in the exercising muscle and the enhanced recovery rates between exercise bouts in children (Falk and Dotan, 2006; Ratel *et al.*, 2006a).

Smaller body dimensions

Because of their smaller muscle mass, children are characterised by an inferior maximal power-generating capacity. This smaller muscle mass results in limited power and work capacity, and may contribute to faster recovery from physical exertion, and thereby limit the development of muscle fatigue during repeated bouts of high-intensity exercise. In this respect, similar recovery rates from fatigue after two bouts of maximal cycling were found in boys and men when, on a time basis, the same amount of relative work done was considered (Williams, 2007).

Furthermore, the significant changes in body mass, body length and proportionality that occur during growth also involve large changes in the dynamic characteristics of the locomotor system. The significant increase in body mass and body length means a greater mass moment of inertia and as a consequence a greater difficulty in setting the mass in motion or stopping it once it is moving. The practical implication of the consequences of increased muscle mass and segment length during growth results in the initiation and cessation of motion being more effortful and fatiguing in older subjects (Jensen and Bothner, 1998).

Lower neuromuscular activation

Neuromuscular activation could be considered as another factor explaining the greater resistance to fatigue in children. It has been shown that adults who have a poor ability to achieve full voluntary activation of plantar flexors and knee extensor motor units before fatigue ensues develop less muscle fatigue during consecutive maximal and submaximal isometric contractions (Yamada et al., 2002; Nordlund et al., 2004). Considering that children are limited in their ability to recruit and use the totality of available motor units (Belanger and McComas, 1989), they cannot perform a maximal effort as intensively as adults. This difference limits the magnitude of force and power production and thereby affects the extent of the fatigue in children, especially during repeated bouts of high-intensity exercise. In this regard, Paraschos et al. (2007) have shown during an isokinetic fatigue task that the peak torque of knee extensors and EMG activity are reduced significantly less in prepubertal boys compared with men, thereby indicating a lower decrease of the agonist drive in children. However, in this study, the EMG activity of vastii in children during the fatigue test was, on average, above the maximum EMG recorded during non-fatiguing maximal contraction conditions. This argues for a lower voluntary activation level for the children in a non-fatigue state, and might explain why children are less susceptible to fatigue during sustained supramaximal contraction conditions. The authors have also shown that the antagonist activation level of biceps femoris is always higher throughout the fatigue protocol in children. Therefore, this higher coactivation level does not substantially contribute to the torque reduction during fatigue in children (Paraschos et al., 2007).

Higher percentage of type I fibres

It is generally accepted that the rate of fatigue development during exercise is more evident in muscles composed primarily of fibres with low oxidative capacity (i.e. fast-twitch fibres or type II fibres) compared with those of a more oxidative type (i.e. slow-twitch fibres or type I fibres) (Carvalho *et al.*, 1996; Howlett and Hogan, 2007). Although rarely investigated in children for ethical reasons, muscle biopsy measurements showed a higher percentage of type I fibres in prepubertal children as compared with young adults (Lexell *et al.*, 1992). The proportion of type I fibres would decrease from around 65% at the age of five years, to 50% at the age of 20 (Lexell *et al.*, 1992). In addition, in the only longitudinal study investigating the relationship between age and percentage of type I fibres, Glenmark *et al.* (1992) conducted biopsy samples on 55 males and 28 females at age 16 and again at age 27. The percentage of type I fibres tended to decrease with age, but only for the males. The decrease in proportion of type I fibres would most likely be caused by a transformation of type I fibres to type II fibres (Lexell *et al.*, 1992).

Besides the limited use of muscle biopsies in children, 31-phosphorus magnetic resonance spectroscopy (^{31}P MRS) could be considered as a potent tool for non-invasively estimating muscle fibre composition (Ratel *et al.*, 2008; Ratel and Williams, 2008). It has been shown in animal muscle that the phosphocreatine (PCr) to inorganic phosphate (P_i) ratio at rest is clearly lower in the type I fibres than in the type II fibres (Meyer *et al.*, 1985). In the same way, Clark *et al.* (1988) showed *in vivo* that chronic stimulation of dog muscles is linked to muscle fibre type conversion (from type II to type I fibres) and to a corresponding significant PCr/P_i decrease at rest. This could be related to a lower P_i content and a higher PCr concentration in type II fibres compared to type I fibres. The few data available on the resting PCr/P_i ratio in children are, however, conflicting. A number of studies have demonstrated a significantly lower ratio of P_i/PCr in children compared to their adult counterparts, thereby illustrating a lower PCr content and possibly a higher proportion of type I muscle fibres (Eriksson, 1980; Taylor *et al.*, 1997; Ratel *et al.*, 2008). In contrast, a recent study by Barker *et al.* (2008) found no age- or sex-related difference in resting quadriceps muscle P_i/PCr in 9–10-year-old children and young adults, despite men having a higher resting PCr content that the boys (46.1 mM versus 35.9 mM respectively) and no differences between the women and girls (40.1 mM versus 37.2 mM, respectively). Moreover, a recent study has shown that the within-subject variation of measured resting P_i/PCr in children has a coefficient of variation of ~35% when using the non-invasive technique of ^{31}P MRS (Barker *et al.*, 2006), indicating that any inferences based on this variable should be made with caution.

Type I muscle fibres are also associated with a lower breakdown of muscle PCr, a fall in pH and a rise in P_i during muscle contractions

(Mizuno et al., 1994), and faster kinetics of $\dot{V}O_2$ at the onset of exercise (Jones et al., 2005). In this sense it is pertinent to note that children are characterised by faster kinetics of $\dot{V}O_2$ at the cycling and treadmill exercise (see Barker and Armstrong (in press) for a recent review) and require a lower breakdown of muscle PCr, fall in pH and rise in muscle P_i when compared to adults during high-intensity exercise (Zanconato et al., 1993; Kuno et al., 1995), which may reflect a preferential recruitment of type I fibres in young people.

Therefore, on the basis of these results, the greater resistance to fatigue in children could be explained by higher type I fibre content. However, this should be viewed with caution given that, according to Elder and Kakulas (1993), less than 10% of fibres would be immature in the prepubertal period, and the normal type II-to-type I fibres transformation during development could be completed within the first two postnatal years (Oertel, 1988). These conflicting findings could be related to the action of the gross muscle studied, the micro sample of the studied muscle, the limited number of samples investigated, biopsy sample errors and the training status of individuals. Further studies using ^{31}P MRS and muscle biopsies are therefore required to investigate the morphological properties of different muscle groups in both children and adults, hence developing the analysis and implication of muscle fatigue.

Oxidative and glycolytic profiles

There is some evidence suggesting that muscle by-products produced by glycolytic metabolism, such as lactate and H^+ ions, are closely related to the decline of force or power output during fatiguing muscle contractions (Fitts, 2008). The reduced accumulation and the fast removal of lactate and H^+ ions from the muscle compartment are therefore of fundamental importance for preserving muscle function and delaying muscle fatigue during continuous and intermittent supramaximal exercises (Thomas et al., 2004, 2005; Messonnier et al., 2006). It has been suggested that prepubertal children experience a lower level of glycolytic activity than older children and adults during maximal exercise (Eriksson et al., 1971, 1973; Eriksson, 1980). This suggestion was based on lower glycolytic anaerobic enzyme activity (i.e. phosphofructokinase and lactate dehydrogenase), resulting in a lesser muscle lactate accumulation in children (Eriksson et al., 1971, 1973; Berg et al., 1986; Eriksson, 1980; Kaczor et al., 2005). Furthermore, some ^{31}P MRS studies have reported lower intramuscular H^+ concentration values following an incremental exercise to exhaustion in young children (Zanconato et al., 1993; Kuno et al., 1995; Taylor et al., 1997). However, others have showed no significant difference in end-of-exercise muscle pH values after submaximal and supramaximal exercises in children, adolescents and adults (Petersen et al., 1999; Ratel et al., 2008). In addition, no age-related significant difference was reported in glycolytic anaerobic enzyme activity (Haralambie, 1982). These conflicting results could be related to differences in sample sizes, subject maturity, training status of individuals and the muscle group investigated. Furthermore, the buffering capacity of H^+ ions in skeletal muscle is yet unknown in children, thereby limiting the value of measuring muscle pH for comparing glycolytic activity between children and adults. Hence, further work needs to be conducted before definitive conclusions can be drawn about the effects of age and maturation on glycolytic metabolism. The lesser accumulation of glycolytic metabolites in children might be because of an enhanced oxidative enzyme activity, which points to the possibility of children being able to oxidise pyruvate and free fatty acids at a higher rate than adolescents or adults (Haralambie, 1982; Berg et al., 1986; Kaczor et al., 2005).

It has also been shown after a 30 s Wingate anaerobic test that the lactate disappearance rate towards pre-exercise levels is faster in children (Dotan et al., 2003; Beneke et al., 2005). This would be due to a lesser muscle production and faster diffusion of lactate into the blood compartment, resulting in significantly shorter peak blood lactate concentration lag-time values in children as compared with adults (5.0 min versus 7.6 min, respectively). As suggested by Falk and Dotan (2006), the blood lactate's earlier peaking in children after high-intensity exercise would be explained by their smaller muscle fibre diameter, resulting in a shorter mean muscle–blood diffusion distance and a higher capillary density. Besides, the removal rate of lactate out of the blood compartment remains yet debatable (Dotan et al., 2003; Beneke et al., 2005). Recently, we have shown that the proton efflux rate following moderate-intensity exercise is significantly higher in boys as compared to men (Ratel et al., 2008). In the same way, we showed a faster removal rate of H^+ ions into blood in children because they are able to ventilate relatively more than adults to exhale the carbon dioxide more quickly and decrease the partial pressure of carbon dioxide to a lower level (Ratel et al., 2002b). On the basis of these findings, one can therefore suggest that the lesser accumulation and the faster removal of lactate and H^+ ions into and from the muscle compartment in children preserves longer muscle function and hence delays the development of muscle fatigue during repeated bouts of high-intensity exercise.

Depletion and recovery rates of PCr

Currently, it is accepted that in adults the restoration of short-term muscle power following exercise is dependent on PCr resynthesis, a purely oxidative process (Bogdanis et al., 1996). Individuals with a high oxidative capacity (high capillary network and oxidative enzyme activity and greater proportion of type I fibres) can recover their initial PCr stores faster, thereby limiting the fatigue phenomenon during repeated bouts of high-intensity exercise, which could be linked to a shortfall in aerobic ATP synthesis (Bogdanis *et al.*, 1996). Using ^{31}P MRS, we have recently shown that PCr depletion (as percentage of initial value) during moderate-intensity exercise is significantly lower in children than in adults, and post-exercise PCr recovery indices (i.e. the constant rate of PCr recovery from a single exponential model and the maximum rate of aerobic ATP production) are about two-fold higher in children, thereby illustrating a greater mitochondrial oxidative capacity (Taylor *et al.*, 1997; Ratel *et al.*, 2008) (Figure 4.5). On the basis of these results, the greater fatigue resistance in children during intermittent-type fatigue protocols could be explained by a lesser PCr depletion in exercising muscle and a faster ability to recover their initial PCr stores (via a higher muscle oxidative capacity) following each strenuous exercise bout. However, in the light of a recent ^{31}P MRS study that has shown no age- or sex-related differences in the recovery kinetics of muscle PCr or a theoretical estimation of the maximal rate of oxidative ATP flux following a bout of moderate-intensity quadriceps exercise in 9–10-year-old children and young adults (Barker *et al.*, 2008), this conclusion warrants further investigation.

Summary

Young children fatigue less than adults during repeated short bouts of high-intensity exercise interspersed with limited rest periods. In other words, children have lower but more consistent power output in repetitious sprint activities. The factors underlying the greater ability of children

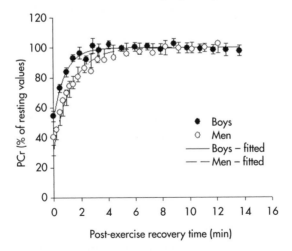

Figure 4.5 PCr time-dependent changes during the post-exercise recovery period from a finger flexor exercise and the corresponding line-fitting in boys (●) and men (○). Post-exercise recovery of PCr was fitted to a single exponential curve. PCr data are expressed as percentage of the resting values. Results are means with error bars representing SD

Source: Reprinted with permission from Ratel *et al.*, 2008.

to resist fatigue are not understood fully, but potential mechanisms related to metabolic changes during or between exercise bouts have been postulated. In this respect, the greater resistance to fatigue in children could be linked to limited energy changes in exercising muscle and/or enhanced recovery rates between exercise bouts (Falk and Dotan, 2006; Ratel *et al.*, 2006a).

CASE STUDY EXAMPLE: PROBLEM AND RESOLUTION

In exercise protocols involving repeated short bouts of high-intensity exercise interspersed with limited rest periods, fatigue is manifested as a progressive decline in power output. However, during the first few bouts of intermittent work, fatigue can often be masked by a potentiation effect, i.e. an increase in power output, which remains largely unexplained (Glaister, 2005). The possible explanatory factors of this phenomenon could be related to an insufficient warm-up, leading to low muscle temperature, subjects inexperienced in testing procedures and underlying motivational aspects. When researchers are led to investigate muscle fatigue, they have to ensure the willing cooperation of the individuals, especially in children during intermittent high-intensity exercise. It is therefore necessary to familiarise the subjects with the procedures used during the subsequent sessions. In this respect, the peak-performance data measured during the habituation sessions should be highly reproducible, and then compared to those scores obtained during the first bout of maximal intermittent work at each subsequent protocol session. Furthermore, the subjects should be continuously encouraged throughout the tests in order to ensure maximal effort.

Other technical points should be considered to analyse muscle fatigue during intermittent short-burst activities in children. It has been recently shown that the resistive force employed to optimise power output (i.e. optimal force or Fopt) resulted in lower fatigue (i.e. lower decline in peak power output) compared with a load corresponding to 50% Fopt within and between repeated short cycle sprints in children (Bogdanis *et al.*, 2007). A greater relative contribution of fatigue-sensitive type II fibres when pedalling against the lighter load (50% Fopt) may explain this result, since the largest part of pedalling rates is well above the optimum (or optimal velocity). Even though in our laboratory, we have clearly shown significant differences in power output profiles during repeated bouts of sprint cycling exercise between children and adults (Ratel *et al.*, 2002a, 2004b), the resistive load should be taken into account in future studies to ensure a valid comparison of muscle fatigue between individuals.

CONCLUSION

Because of their lower metabolic reserve, children have a lower ability than adults to perform long-distance running at any given absolute speed, even though they can exercise at a slow speed for a long time. Hence, they fatigue earlier than adults when the same running speed is fixed. In contrast, while children have lower peak power outputs, they have more consistent power outputs during repeated bouts of high-intensity exercise than adults, thereby indicating that they are more resistant to fatigue during high-intensity intermittent activities.

FIVE KEY PAPERS THAT SHAPED THE TOPIC AREA

Study 1. Hebestreit, H., Mimura, K. and Bar-Or, O. (1993). Recovery of muscle power after high-intensity short-term exercise: comparing boys and men. *Journal of Applied Physiology*, 74, 2875–2880.

This study was the first to investigate muscle fatigue during repeated high-intensity short-term exercises in children and adults. On three different occasions, eight prepubertal boys (9–12 years old) and eight young men (19–23 years old) performed two consecutive 30 s all-out cycling tasks (Wingate anaerobic test (WAnT)), separated by a 1, 2 or 10 min recovery interval. In the boys, mean power reached $89.9 \pm 3.6\%$ of the initial value after 1 min of recovery, $96.4 \pm 2.3\%$ after 2 min, and $103.5 \pm 1.3\%$ after 10 min. For the men, the values were 71.2 ± 2.6, 77.1 ± 2.4 and $94.0 \pm 1.3\%$, respectively (boys versus men, $P < 0.0001$). Relative to the external work performed during the first WAnT, the boys had a higher net oxygen uptake (exercise–resting) during the test than the men (83.8 ± 18.4 versus $57.8 \pm 6.1 \text{ ml} \cdot \text{kJ}^{-1}$). After the WAnT, the net CO_2 output and the respiratory exchange ratio were lower in the boys, and they recovered faster. The authors concluded that boys fatigued less and recovered faster than men from high-intensity short-term exercise. They attributed this finding to a lower reliance on glycolysis during the WAnT in the boys, leading to less acidosis.

Study 2. Ratel, S., Bedu, M., Hennegrave, A., Doré, E. and Duché, P. (2002). Effects of age and recovery duration on peak power output during repeated cycling sprints. *International Journal of Sports Medicine*, 23, 397–402.

This study aimed to further understand the effects of age and maturation on the ability to repeat power activities as a function of recovery intervals. In this study, 11 prepubescent boys (9.6 ± 0.7 years old), nine pubescent boys (15.0 ± 0.7 years old) and ten men (20.4 ± 0.8 years old) performed ten consecutive 10s cycling sprints separated by either 30s, 1 min, or 5 min passive recovery intervals. Arterialised capillary blood samples were collected at rest and during the sprint exercises to measure the time-course of lactate concentrations. It has been shown with 30 s recovery intervals that the prepubescent boys are able to maintain their cycling peak power, whereas the adolescent boys and the men decrease their peak power output by 18.5% and 28.5%, respectively. Longer recovery periods of 5 min were necessary in the adolescent boys and men to sustain their cycling peak power from the first to the tenth sprint. For each recovery interval, the increase in blood lactate concentration over the ten sprints was significantly lower in the prepubescent boys compared with the pubescent boys and the men. The authors concluded that the prepubescent boys fatigued less than the pubescent boys and the men during repeated short-term cycling sprints, and that the maturation status could account for these differences. The authors suggested that the faster recovery of peak power output in the prepubescent boys was due to their lower muscle glycolytic activity and their higher muscle oxidative capacity, allowing a faster resynthesis in PCr.

Study 3. Eriksson, B.O. (1980) Muscle metabolism in children: a review. *Acta Physiologica Scandinavica*, 283: 20–28.

This review article summarises the series of innovative biopsy studies of exercise metabolism carried out by Eriksson and his colleagues in the 1970s in boys aged 11–16 years. Eriksson and his colleagues. observed that resting ATP stores in the quadriceps muscle were invariant with age. On the contrary, children's PCr concentration and muscle glycogen concentration at rest were lower than values that had previously been recorded in adults. Over the series of exercise sessions, muscle glycogen stores gradually decreased and the utilisation rate was directly related to age group, being three-fold greater in the oldest compared to the youngest boys. Furthermore, Eriksson and his colleagues described levels of PFK and SDH activity in the *vastus lateralis* muscle of five 11-year-old boys which were respectively 50% lower and 20% higher than had previously been reported in adults. Also, these authors reported muscle lactate concentration following a peak $\dot{V}O_2$ test as 8.8, 10.7, 11.3 and 15.3 $mmol \cdot kg^{-1}$ wet weight for boys aged 11.6, 12.6, 13.5 and 15.5 years, respectively, suggesting an age dependency for lactate production. Therefore, the children's metabolic profile was described as more oxidative than glycolytic, which could be one of the factors accounting for their greater resistance to fatigue during repeated bouts of high-intensity exercise.

Study 4. Astrand, P.O. (1952) *Experimental Studies of Physical Working Capacity in Relation to Sex and Age*. Copenhagen: Ejnar Munksgaard.

This study by Astrand is one of the first studies conducted on the metabolic cost of walking and running during growth. Astrand showed that on a treadmill the submaximal O_2 uptake during walking and running was higher in children at various speeds when expressed per kilogram of body mass. For instance, it has been reported that at $10\,km\cdot h^{-1}$ that the five-year-old child displays an $8\,ml\cdot kg^{-1}\cdot min^{-1}$ (20%) greater metabolic cost than the 17-year-old adolescent. This study is therefore one of the first studies showing the greater metabolic cost of locomotion and the lesser resistance to fatigue during submaximal prolonged walking and running at any given speed in children as compared with older individuals.

Study 5. Frost, G., Bar-Or, O., Dowling, J. and Dyson, K. (2002). Explaining differences in the metabolic cost and efficiency of treadmill locomotion in children. *Journal of Sports Sciences*, 20, 451–461.

Incomplete explanations for the age-related differences in the metabolic cost of locomotion motivated this study. A multidisciplinary method was used to examine metabolic, kinematic and EMG data from three maturational groups of children. Thirty children aged 7–8 (n = 10), 10–12 (n = 10) and 15–16 (n = 10) years completed 4 min bouts of submaximal treadmill exercise at six speeds (two walking and four running) assigned in random order. Metabolic (net $\dot{V}O_2$), kinematic (total body mechanical power, energy transfer rates, stride rate) and EMG (coactivation of agonist and antagonist muscles in thigh and leg segments) data were collected. Multiple regression analysis was performed with net $\dot{V}O_2$ or efficiency as the dependent variable and mechanical power, thigh and leg coactivation, stride rate and age as independent variables. The results showed that it was possible to explain up to 77% of the age-related variance in net $\dot{V}O_2$ and 62% of the variation in efficiency using combinations of these variables. Age was the best single predictor of both $\dot{V}O_2$ and efficiency. Coactivation, possibly used to enhance joint stability, was an important component of the observed age-related differences, although mechanical power was not.

GLOSSARY OF TERMS

EMG	electromyography
PCr	phosphocreatine
P_i	inorganic phosphate
$\dot{V}O_2$	oxygen uptake
WAnT	Wingate anaerobic test

REFERENCES

Andersen, K.L., Seliger, V., Rutenfranz, J. and Messel S. (1974). Physical performance capacity of children in Norway. III. Respiratory responses to graded exercise loadings: population parameters in a rural community. *European Journal of Applied Physiology and Occupational Physiology*, 33, 265–274.

Ariëns, G.A., van Mechelen, W., Kemper, H.C. and Twisk J.W. (1997). The longitudinal development of running economy in males and females aged between 13 and 27 years: the Amsterdam Growth and Health Study. *European Journal of Applied Physiology and Occupational Physiology*, 76, 214–220.

Armon, Y., Cooper, D.M., Flores, R., Zanconato, S. and Barstow, T.J. (1991). Oxygen uptake dynamics during high-intensity exercise in children and adults. *Journal of Applied Physiology*, 70, 841–848.

Armstrong, N. (2006). Aerobic fitness of children and adolescents. *Journal of Pediatrics*, 82, 406–408.

Armstrong, N. and Welsman, J. (2001). Peak oxygen uptake in relation to growth and maturation. *European Journal of Applied Physiology*, 85, 546–551.

Astrand, P.O. (1952). *Experimental Studies of Physical Working Capacity in Relation to Sex and Age*. Copenhagen: Ejnar Munksgaard.

Barker, A.R. and Armstrong, N. (in press). Oxygen uptake kinetic in children and adolescents: a review. *Pediatric Exercise Science*.

Barker, A.R., Welsman, J.R., Fulford, J., Welford, D. and Armstrong, N. (2008). Muscle phosphocreatine kinetics in children and adults at the onset and offset of moderate-intensity exercise. *Journal of Applied Physiology*, 105, 446–456.

Barker, A., Welsman, J.R., Welford, D., Fulford, J., Williams, C. and Armstrong, N. (2006). Reliability of ^{31}P-magnetic resonance spectroscopy during an exhaustive incremental exercise test in children. *European Journal of Applied Physiology*, 98, 556–565.

Bar-Or, O. (1982). Physiologische gesetzmassigkeiten sportlicher aktivitat beim kind. In H. Howald and E. Han (eds), *Kinder im Leistungssport*. Basel: Birkhauser, pp. 18–30.

Bar-Or, O. (1995). The young athlete: some physiological considerations. *Journal of Sports Sciences*, 13, S31–33.

Bar-Or, O. and Rowland, T.W. (2004). *Pediatric Exercise Medicine: From Physiologic Principles to Health Care Application*. Champaign: Human Kinetics.

Belanger, A.Y. and McComas, A.J. (1989). Contractile properties of human skeletal muscle in childhood and adolescence. *European Journal of Applied Physiology and Occupational Physiology*, 58, 563–567.

Bell, R.D., MacDougall, J.D., Billeter, R. and Howald, H. (1980). Muscle fiber types and morphometric analysis of skeletal msucle in six-year-old children. *Medicine and Science in Sports and Exercise*, 12, 28–31.

Beneke, R., Hütler, M., Jung, M. and Leithäuser, R.M. (2005). Modeling the blood lactate kinetics at maximal short-term exercise conditions in children, adolescents, and adults. *Journal of Applied Physiology*, 99, 499–504.

Berg, A., Kim, S.S. and Keul, J. (1986). Skeletal muscle enzyme activities in healthy young subjects. *International Journal of Sports Medicine*, 7, 236–239.

Bogdanis, G.C., Nevill, M.E., Boobis, L.H. and Lakomy, H.K. (1996). Contribution of phosphocreatine and aerobic metabolism to energy supply during repeated sprint exercise. *Journal of Applied Physiology*, 80, 876–884.

Bogdanis, G.C., Papaspyrou, A., Theos, A. and Maridaki, M. (2007). Influence of resistive load on power output and fatigue during intermittent sprint cycling exercise in children. *European Journal of Applied Physiology*, 101, 313–320.

Carvalho, A.J., McKee, N.H. and Green, H.J. (1996). Metabolic and contractile responses of fast- and slow-twitch rat skeletal muscles to ischemia. *Canadian Journal of Physiology and Pharmacology*, 74, 1333–1341.

Chia, M.Y.H. (2001). Recovery of Wingate anaerobic test power following prior sprints of a short duration: a comparison between girls and women. In T. Rowland (ed.), *Pediatric Exercise Science*, Champaign: Human Kinetics, p. 273.

Clark III, B.J., Acker, M.A., McCully, K., Subramanian, H.V., Hammond, R.L., Salmons, S., Chance, B. and Stephenson, L.W. (1988). In vivo ^{31}P-NMR spectroscopy of chronically stimulated canine skeletal muscle. *American Journal of Physiology*, 254, C258–266.

Davies, C.T. (1980). Metabolic cost of exercise and physical performance in children with some observations on external loading. *European Journal of Applied Physiology and Occupational Physiology*, 45, 95–102.

Dotan, R., Ohana, S., Bediz, C. and Falk, B. (2003). Blood lactate disappearance dynamics in boys and men following exercise of similar and dissimilar peak-lactate concentrations. *Journal of Pediatric Endocrinology and Metabolism*, 16, 419–429.

Ebbeling, C.J., Hamill, J., Freedson, P.S. and Rowland, T. (1992). An examination of efficiency during walking in children and adults. *Pediatric Exercise Science*, 4, 36–49.

Elder, G.C. and Kakulas, B.A. (1993). Histochemical and contractile property changes during human muscle development. *Muscle and Nerve*, 16, 1246–1253.

Eriksson, B.O. (1980). Muscle metabolism in children: a review. *Acta Paediatrica Scandinavica*, 283: S20–28.

Eriksson, B.O., Gollnick, P.D. and Saltin, B. (1973). Muscle metabolism and enzyme activities after training in boys 11–13 years old. *Acta Physiologica Scandinavica*, 87, 485–497.

Eriksson, B.O., Karlsson, J. and Saltin, B. (1971). Muscle metabolites during exercise in pubertal boys. *Acta Paediatrica Scandinavica*, 217, S154–157.

Falk, B. and Dotan, R. (2006). Child–adult differences in the recovery from high-intensity exercise. *Exercise and Sport Science Review*, 34, 107–112.

Fawkner, S.G. and Armstrong, N. (2004). Longitudinal changes in the kinetic response to heavy-intensity exercise in children. *Journal of Applied Physiology*, 97, 460–466.

Fitts, R.H. (2008). The cross-bridge cycle and skeletal muscle fatigue. *Journal of Applied Physiology*, 14, 551–558.

Frost, G., Bar-Or, O., Dowling, J. and Dyson, K. (2002). Explaining differences in the metabolic cost and efficiency of treadmill locomotion in children. *Journal of Sports Sciences*, 20, 451–461.

Glaister, M. (2005). Multiple sprint work: physiological responses, mechanisms of fatigue and the influence of aerobic fitness. *Sports Medicine*, 35, 757–777.

Glenmark, B., Hedberg, G. and Jansson, E. (1992). Changes in muscle fibre type from adolescence to adulthood in women and men. *Acta Physiologica Scandinavica*, 146, 251–259.

Granata, K.P., Padua, D.A. and Abel, M.F. (2005). Repeatability of surface EMG during gait in children. *Gait Posture*, 22, 346–350.

Haralambie, G. (1982). Enzyme activities in skeletal muscle of 13–15 years old adolescents. *Bulletin Européen de Physiopathologie Respiratoire*, 18, 65–74.

Hausdorff, J.M., Zemany, L., Peng, C. and Goldberger A.L. (1999). Maturation of gait dynamics: stride-to-stride variability and its temporal organization in children. *Journal of Applied Physiology*, 86, 1040–1047.

Hebestreit, H., Mimura, K.I. and Bar-Or, O. (1993). Recovery of muscle power after high-intensity short-term exercise: comparing boys and men. *Journal of Applied Physiology*, 74, 2875–2880.

Howlett, R.A. and Hogan, M.C. (2007). Effect of hypoxia on fatigue development in rat muscle composed of different fibre types. *Experimental Physiology*, 92, 887–894.

Jensen, J.L. and Bothner, K.E. (1998). Infant motor development: the biomechanics of change. In E. Van Praagh (ed.), *Pediatric Anaerobic Performance*. Champaign: Human Kinetics, pp. 23–43.

Jones, A.M., Pringle, J.S.M. and Carter, H. (2005). Influence of muscle fibre type and motor unit recruitment on $\dot{V}O_2$ kinetics. In A.M. Jones and D.C. Poole (eds), *Oxygen Uptake Kinetics in Sport, Exercise and Medicine*. London and New York: Routledge, pp. 261–293.

Kaczor, J.J., Ziolkowski, W., Popinigis, J. and Tarnopolsky, M.A. (2005). Anaerobic and aerobic enzyme activities in human skeletal muscle from children and adults. *Pediatric Research*, 57, 331–335.

Kjendlie, P.L., Ingjer, F., Madsen, Ø., Stallman, R.K. and Stray-Gundersen, J. (2004). Differences in the energy cost between children and adults during front crawl swimming. *European Journal of Applied Physiology*, 91, 473–480.

Kuno, S., Takahashi, H., Fujimoto, K., Akima, H., Miyamaru, M., Nemoto, I., Itai, Y. and Katsuta, S. (1995). Muscle metabolism during exercise using phosphorus-31 nuclear magnetic resonance spectroscopy in adolescents. *European Journal of Applied Physiology and Occupational Physiology*, 70, 301–304.

Léger, L. (1996). Aerobic performance. In D. Docherty (ed.), *Measurement in Pediatric Exercise Science*. Champaign: Human Kinetics, pp. 183–223.

Lexell, J., Sjöström, M., Nordlund, A.S. and Taylor, C.C. (1992). Growth and development of human muscle: a quantitative morphological study of whole vastus lateralis from childhood to adult age. *Muscle and Nerve*, 15, 404–409.

Martinez, L.R. and Haymes, E.M. (1992). Substrate utilization during treadmill running in prepubertal girls and women. *Medicine and Science in Sports and Exercise*, 24, 975–983.

Messonnier, L., Denis, C., Feasson, L. and Lacour, J.R. (2006). An elevated sarcolemmal lactate (and proton) transport capacity is an advantage during muscle activity in healthy humans. *Journal of Applied Physiology*. Online, available at: http://jap.physiology.org/cgi/reprint/00807.2006v1.pdf.

Meyer, R.A., Brown, T.R. and Kushmerick, M.J. (1985). Phosphorus nuclear magnetic resonance of fast- and slow-twitch muscle. *American Journal of Physiology*, 248, C279–287.

Mizuno, M., Secher, N.H. and Quistorff, B. (1994). 31P-NMR spectroscopy, rsEMG, and histochemical fiber types of human wrist flexor muscles. *Journal of Applied Physiology*, 76, 531–538.

Nordlund, M.M., Thorstensson, A. and Cresswell, A.G. (2004). Central and peripheral contributions to fatigue in relation to level of activation during repeated maximal voluntary isometric plantar flexions. *Journal of Applied Physiology*, 96, 218–225.

Oertel, G. (1988). Morphometric analysis of normal skeletal muscles in infancy, childhood and adolescence: an autopsy study. *Journal of Neurological Science*, 88, 303–313.

Paraschos, I., Hassani, A., Bassa, E., Hatzikotoulas, K., Patikas, D. and Kotzamanidis, C. (2007). Fatigue differences between adults and prepubertal males. *International Journal of Sports Medicine*, 28, 958–963.

Petersen, S.R., Gaul, C.A., Stanton, M.M. and Hanstock, C.C. (1999). Skeletal muscle metabolism during short-term, high-intensity exercise in prepubertal and pubertal girls. *Journal of Applied Physiology*, 87, 2151–2156.

Psek, J.A. and Cafarelli, E. (1993). Behaviour of coactive muscle during fatigue. *Journal of Applied Physiology*, 74, 170–175.

Ratel, S. and Poujade, B. (2009) Comparative analysis of the energy cost during front crawl swimming in children and adults. *European Journal of Applied Physiology*, 105, 543–549

Ratel, S. and Williams, C.A. (2008). Children's musculoskeletal system: new research perspectives. In N.P. Beaulieu (ed.), *Physical Activity and Children: New Research*. New York: Nova Science Publishers, pp. 117–135.

Ratel, S., Bedu, M., Hennegrave, A., Doré, E. and Duché, P. (2002a). Effects of age and recovery duration on peak power output during repeated cycling sprints. *International Journal of Sports Medicine*, 23, 397–402.

Ratel, S., Duché, P., Hennegrave, A., Van Praagh, E. and Bedu, M. (2002b). Acid-base balance during repeated cycling sprints in boys and men. *Journal of Applied Physiology*, 92, 479–485.

Ratel, S., Duché, P. and Williams, C.A. (2006a). Muscle fatigue during high-intensity exercise in children. *Sports Medicine*, 36, 1031–1065.

Ratel, S., Lazaar, N., Doré, E., Baquet, G., Williams, C.A., Berthoin, S., Van Praagh, E., Bedu, M. and Duché, P. (2004a). High-intensity intermittent activities at school: controversies and facts. *Journal of Sports Medicine and Physical Fitness*, 44, 272–280.

Ratel, S., Tonson, A., Le Fur, Y., Cozzone, P.J. and Bendahan, D. (2008). Comparative analysis of skeletal muscle oxidative capacity in children and adults: a ^{31}P-MRS study. *Applied Physiology, Nutrition and Metabolism*, 33, 720–727.

Ratel, S., Williams, C.A., Oliver, J. and Armstrong, N. (2004b). Effects of age and mode of exercise on power output profiles during repeated sprints. *European Journal of Applied Physiology*, 92, 204–210.

Ratel, S., Williams, C.A., Oliver, J. and Armstrong, N. (2006b). Effects of age and recovery duration on performance during multiple treadmill sprints. *International Journal of Sports Medicine*, 27, 1–8.

Roberts, W.O. (2007). Can children and adolescents run marathons? *Sports Medicine*, 37, 299–301.

Robinson, S. (1938). Experimental studies of physical fitness in relation to age. *Zeitschrift für angewandte Physiologie, einschliesslich Arbeitsphysiologie*, 10, 251–323.

Rowland, T.W. (2008). Thermoregulation during exercise in the heat in children: old concepts revisited. *Journal of Applied Physiology*, 15, 718–724.

Rowland, T.W. and Cunningham, L.N. (1997). Development of ventilatory responses to exercise in normal white children: a longitudinal study. *Chest*, 111, 327–332.

Rowland, T.W., Auchinachie, J.A., Keenan, T.J. and Green, G.M. (1987). Physiologic responses to treadmill running in adult and prepubertal males. *International Journal of Sports Medicine*, 8, 292–297.

Rowland, T.W., Auchinachie, J.A., Keenan, T.J. and Green, G.M. (1988). Submaximal aerobic running economy and treadmill performance in prepubertal boys. *International Journal of Sports Medicine*, 9, 201–204.

Rowland, T., Hagenbuch, S., Pober, D. and Garrison, A. (2008). Exercise tolerance and thermoregulatory responses during cycling in boys and men. *Medicine and Science in Sports and Exercise*, 40, 282–287.

Rowland, T.W., Staab, J.S., Unnithan, V.B., Rambusch, J.M. and Siconolfi, S.F. (1990). Mechanical efficiency during cycling in prepubertal and adult males. *International Journal of Sports Medicine*, 11, 452–455.

Taylor, D.J., Kemp, G.J., Thompson, C.H. and Radda, G.K. (1997). Ageing: effects on oxidative function of skeletal muscle in vivo. *Molecular and Cellular Biochemistry*, 174, 321–324.

Thomas, C., Perrey, S., Lambert, K., Hugon, G., Mornet, D. and Mercier, J. (2005). Monocarboxylate transporters, blood lactate removal after supramaximal exercise, and fatigue indexes in humans. *Journal of Applied Physiology*, 98, 804–809.

Thomas, C., Sirvent, P., Perrey, S., Raynaud, E. and Mercier, J. (2004). Relationships between maximal muscle oxidative capacity and blood lactate removal after supramaximal exercise and fatigue indexes in humans. *Journal of Applied Physiology*, 97, 2132–2138.

Timmons, B.W., Bar-Or, O. and Riddell, M.C. (2003). Oxidation rate of exogenous carbohydrate during exercise is higher in boys than in men. *Journal of Applied Physiology*, 94, 278–284.

Timmons, B.W., Bar-Or, O. and Riddell, M.C. (2007). Energy substrate utilization during prolonged exercise with and without carbohydrate intake in preadolescent and adolescent girls. *Journal of Applied Physiology*, 103, 995–1000.

Unnithan, V.B. and Eston, R.G. (1990). Stride frequency and submaximal treadmill running economy in adults and children. *Pediatric Exercise Science*, 2, 149–155.

Welsman, J.R. and Armstrong, N. (2000). Longitudinal changes in submaximal oxygen uptake in 11- to 13-year-olds. *Journal of Sports Sciences*, 18, 183–189.

Welsman, J.R., Armstrong, N., Nevill, A.M., Winter, E.M. and Kirby, B.J. (1996). Scaling peak VO_2 for differences in body size. *Medicine and Science in Sports and Exercise*, 28, 259–265.

Williams, C.A. (2007). Recovery from fatigue after two bouts of maximal 20-s cycling in boys and men. *Journal of Sports Sciences*, 25, 283.

Yamada, H., Kaneko, K. and Masuda, T. (2002). Effects of voluntary activation on neuromuscular endurance analyzed by surface electromyography. *Perceptual Motor Skills*, 95, 613–619.

Zafeiridis, A., Dalamitros, A., Dipla, K., Manou, V., Galanis, N. and Kellis, S. (2005). Recovery during high-intensity intermittent anaerobic exercise in boys, teens, and men. *Medicine and Science in Sports and Exercise*, 37, 505–512.

Zanconato, S., Buchthal, S., Barstow, T.J. and Cooper, D.M. (1993). ^{31}P-magnetic resonance spectroscopy of leg muscle metabolism during exercise in children and adults. *Journal of Applied Physiology*, 74, 2214–2218.

CHAPTER 5

MUSCLE FATIGUE IN ELDERLY PEOPLE

Jane A. Kent-Braun, Damien M. Callahan, Stephen A. Foulis and Linda H. Chung

OBJECTIVES

The objectives for this chapter are to:

- discuss the changes in the ageing neuromuscular system that affect a person's ability to produce muscle force and power;
- describe the current state of the literature with regard to muscle fatigue in older persons, highlighting the impact of study design on the outcomes of this research;
- provide an understanding of the metabolic basis of muscle fatigue in humans, and how this plays a role in the fatigue resistance of the elderly in some situations;
- present and discuss a new hypothesis about how neural and metabolic factors may combine to increase muscle fatigue resistance in healthy older adults, as well as how these may change as limitations in physical function develop.

INTRODUCTION

While it is acknowledged that the term "ageing" broadly incorporates changes across the entire lifespan – including those that occur during development – we will restrict the use of the term here to mean senescence. Due to improvements in nutrition and medicine, the life expectancy for people in many countries has increased markedly in the past century. As a result, the average age of our population is increasing, bringing with it a critical need for understanding the basic biology of ageing and its effects on various physiological systems. With this knowledge, effective approaches for encouraging healthy ageing can be developed.

Fatigue can mean many things, particularly in older adults. For example, fatigue can be characterised as the lassitude or sense of exhaustion that develops over the course of a day. Muscle fatigue is a type of fatigue that is both quantifiable and relevant to adequate physical functioning. A further distinction can be made between muscle fatigue (the decline in maximal force-producing capability during contractions) and muscle endurance (the time for which target force can be maintained).

Studies of human skeletal muscle fatigue or endurance typically involve a single muscle group. As a result, the effects of systemic changes, such as the ability of the cardiovascular system to provide adequate blood flow and oxygen to the working tissue, are minimised. While this approach presents something of a constraint in terms of understanding the factors that limit vigorous whole-body activity, it does allow the contributions of age-related changes in neuromuscular function to be clarified in the absence of confounding influences.

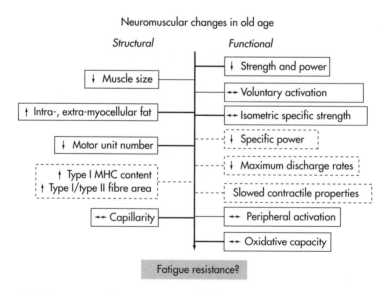

Figure 5.1 Alterations in the ageing neuromuscular system. A number of neuromuscular changes that occur in the elderly could play a role in their ability to resist fatigue under various conditions. Dotted lines indicate those variables most likely to be important in explaining age-related differences in fatigue resistance

The ability to maintain force or power production, and thereby resist muscle fatigue, is a primary task for the human neuromuscular system. While old age generally brings with it a decline in physiological function, it may be that certain changes within the ageing neuromuscular system can prove beneficial to the maintenance of force production during contractile activity. The major changes that occur to the neuromuscular system in old age are summarised in Table 5.1, and several key components are illustrated in Figure 5.1. Primary among these changes is the loss of muscle mass that occurs beyond ~60 years of age. For this reason, in studies of muscle fatigue it is necessary to express the decline of force or power during muscular activity relative to each person's baseline strength. In this way, comparisons can be drawn across individuals with different force-generating capacity, and thus weakness can be distinguished from fatigability. In this chapter, relevant information about changes in baseline function in old age will be incorporated into the discussion of muscle fatigue, as appropriate.

MUSCLE FATIGUE IN OLDER ADULTS

A growing body of literature has focused on identifying whether the ability to maintain force production, and therefore resist fatigue, is altered in old age. Upper and lower extremity muscle groups have been studied under a variety of conditions in a wide range of individuals. It has become apparent that factors such as contraction mode and intensity may impact the comparisons of fatigue between young and older adults.

Effect of contraction mode in studies of fatigue

Although studies of muscle fatigue in the elderly indicate a range of results, a review of the literature reveals a pattern of fatigue resistance that suggests a dependence on contraction mode in the fatigue of young compared with older men and women (Tables 5.2 and 5.3). Studies using isometric endurance tasks consistently report shorter endurance times in young compared to older adults, regardless of the muscle group examined and the intensity of contraction (Aniansson et al., 1978; Bilodeau et al., 2001a; Bilodeau et al., 2001b; Bazzucchi et al., 2005; Hunter et al., 2005a; Hunter et al., 2005b; Mademli and Arampatzis, 2008b; Yoon et al., 2008). When the contraction protocol involves isometric maximal voluntary contractions (MVC), the results also generally indicate greater fatigue in young than old (Chan et al., 2000; Ditor and Hicks, 2000; Kent-Braun et al., 2002; Rubinstein and Kamen, 2005; Chung et al., 2007). However, this consistency of results is lost for those studies in which fatigue is elicited by dynamic muscle contractions. In these cases, fatigue in older adults may be similar

Table 5.1 Age-related neuromuscular changes that may affect muscle function

System	Variable (technique)	Old relative to young	Reference
Neural			
Motor unit behaviour	Maximal discharge rate	→	(Kamen et al., 1995; Connelly et al., 1999; Rubinstein and Kamen, 2005)
	Submaximal discharge rate	↕ ↕→ ↕	(Roos et al., 1999) (Kamen et al., 1995; Roos et al., 1999)
Motor control	Central activation — CAR	↕	(Connelly et al., 1999; McNeil et al., 2005)
	Interpolated doublet	↕	(Kent-Braun and Ng, 1999; Kent-Braun et al., 2002; Lanza et al., 2004; Chung et al., 2007; Russ et al., 2008)
	Interpolated twitch	→	(Allman and Rice, 2004; Connelly et al., 1999; Klass et al., 2005; Klein et al., 2001; Roos et al., 1999; De Serres and Enoka, 1998)
	Antagonist coactivation	←→ ↕←→ ↕	(Thelen et al., 1996; Klein et al., 2001) (Thelen et al., 1996) (McNeil et al., 2007) (Thelen et al., 1996)
Peripheral activation	Pre-motor time		(Klass et al., 2005; Lanza et al., 2004; Chung et al., 2007;
	M wave amplitude	↕	Ng and Kent-Braun, 1999; Kent-Braun et al., 2002; Lanza et al., 2004; Russ et al., 2008)
Morphological			
Muscle structure	Pennation angle	↕	(Binzoni et al., 2001; Kubo et al., 2003; Mademli and Arampatzis, 2008b)
	Fascicle length	→ ↕ →	(Kubo et al., 2003; Narici et al., 2003; Morse et al., 2005) (Kubo et al., 2003; Morse et al., 2005) (Mademli and Arampatzis, 2008b; Narici et al., 2003; Thom et al., 2005)
Fibre number (proportional)	Type I	→ ← ↕	(McNeil et al., 2005) (Jakobsson et al., 1990) (Lexell et al., 1988; Coggan et al., 1992; McCully et al., 1993)
	Type II	→ ↕	(Jakobsson et al., 1990) (Coggan et al., 1992; McCully et al., 1993)

Measure		Change	References
Fibre area (proportional)	Type I	←→	(Larsson et al., 1979)
	Type II	→	(Larsson et al., 1979; Essen-Gustavsson and Borges, 1986; Lexell et al., 1988; Proctor et al., 1995)
MHC content	Type I	↕→	(D'Antona et al., 2003)
	Type IIa	←	(D'Antona et al., 2003)
	type IIx	→	(D'Antona et al., 2003)
Capillarity	Capillary-to-fibre ratio	↕→→→	(Coggan et al., 1992)
			(Coggan et al., 1990; Jakobsson et al., 1990)
Mitochondrial content	mtDNA concentration		(Welle et al., 2003; Short et al., 2005)
	mtProtein concentration		(Short et al., 2005)
Contractile			
Contractile characteristics (stimulated)	Rate of force development (Tetanic)	→→	(Kent-Braun and Ng, 1999; Kent-Braun et al., 2002)
	(Doublet)	↕←	(Klass et al., 2005)
	(Twitch)		(Klass et al., 2005)
	Contraction (twitch) duration	←	(Vandervoort and McComas, 1986; Connelly et al., 1999; Lanza et al., 2004; Klass et al., 2005; Roos et al., 1999; van Schaik et al., 1994)
	Time-to-peak tension (twitch)	↕↕↕→	(Davies et al., 1983; Connelly et al., 1999; Roos et al., 1999; McNeil et al., 2005; McNeil et al., 2007)
			(Ng and Kent-Braun, 1999)
	Rate of force relaxation (Twitch)		(Klass et al., 2005)
	(Tetanus)	↕←	(Lanza et al., 2004; Russ et al., 2008)
			(Kent-Braun et al., 2002; Allman and Rice, 2004; Chung et al., 2007)
	Half-time of force relaxation		(Klass et al., 2005;McNeil et al., 2007)
			(Davies et al., 1983;Vandervoort and McComas, 1986; Connelly et al., 1999; Roos et al., 1999; Allman and Rice, 2004; Chung et al., 2007; Lanza et al., 2004; McNeil et al., 2005; McNeil et al., 2007; Russ, et al., 2008)
Contractile characteristics (voluntary)	Force–frequency curve	Leftward shift ↕→	(Allman and Rice, 2004)
	Rate of force development	↕	(Kent-Braun and Ng, 1999)
			(McNeil et al., 2007)
	Torque–angle relationship	Downward shift	(Lanza et al., 2003)
	Torque–velocity relationship	Downward shift	(Lanza et al., 2003)
	Specific strength	↕	(Kent-Braun and Ng, 1999; McNeil et al., 2007; Vandervoort and McComas, 1986)
	Specific power	→→	(Klein et al., 2001; McNeil et al., 2007)
			(Thom et al., 2005; McNeil et al., 2007)

continued

Table 5.1 continued

System	Variable (technique)		Old relative to young	Reference
Vascular				
Blood flow	Vascular conductance, rest		↔→ →	(Jasperse et al., 1994; Proctor et al., 2005) (Proctor et al., 2005)
	Vascular conductance, contractions		↔→ → →	(Jasperse et al., 1994; Lawrenson et al., 2003) (Magnusson et al., 1994) (Proctor et al., 2005)
	Vascular conductance, reactive hyperaemia		↔ →	(Jasperse et al., 1994; Proctor et al., 2005)
	Flow-mediated dilation		→	(Parker et al., 2006)
Energetics				
Metabolic characteristics (in vitro)	Aerobic enzyme activity	CS	→	(Coggan et al., 1992; McCully et al., 1993; Pastoris et al., 2000; Short et al., 2005)
		SDH	→	(Coggan et al., 1992)
		β-HADH	→	(Coggan et al., 1992)
		HK	→	(Pastoris et al., 2000)
	Anaerobic enzyme activity	LDH	←	(Pastoris et al., 2000)
Muscle energetics (in vivo)	Metabolic economy		↔	(Lanza et al., 2007)
	Glycolytic function		↔	(Lanza et al., 2007)
	Oxidative capacity		↔	(Chilibeck et al., 1998; Kutsuzawa et al., 2001; Kent-Braun et al., 2002; Lanza et al., 2005; Lanza et al., 2007; McCully et al., 1993; Taylor et al., 1997)
			→	

Notes

For each variable, results are summarised by arrows indicating higher (↑), lower (↓) or similar (↔) capacity in older relative to young adults.
CAR, central activation ratio; MHC, myosin heavy chain; CS, citrate synthase; SDH, succinate dehydrogenase; β-HADH, beta-hydroxyacyl-CoA dehydrogenase; HK, hexokinase; LDH, lactate dehydrogenase.

Table 5.2 Intermittent contraction protocols: results from cross-sectional studies in which muscle fatigue was measured during intermittent contraction protocols in young (Y) and older (O) volunteers.

Muscle group	Subject sex; ages (n)	Contraction intensity (% MVC)	Endurance time of old relative to young (Δ)	Reference
FF	M; 25–70 (~120)	100	↑ (123%)	(Aniansson et al., 1978)
EF	M, F; 26, 70 (19)	35	↑ (264%)	(Bilodeau et al., 2001b)
EF	M, F; 25, 76 (21)	100	↑ (169%)	(Bilodeau et al., 2001a)
EF	M; 28, 71 (12)	30	↔ (103%)	(Bazzucchi et al., 2005)
		50	↑ (174%)	
		80	↑ (183%)	
EF	M, F; 22, 72 (45)	20	↑ (158%)	(Hunter et al., 2005b)
EF	M; 21, 71 (16)	20	↑ (173%)	(Hunter et al., 2005a)
PF	M; 30, 65 (26)	40	↑ (163%)	(Mademli and Arampatzis, 2008b)
EF	M, F; 21, 70 (30)	20	↑ (204%)	(Yoon et al., 2008)
		80	↔ (133%)	

Notes
Fatigue was quantified as the relative decline in muscle force-generating capacity during contractions. Results are summarised in the table by arrows indicating greater fatigue in young than old (↑), no significant difference in fatigue between young and old (↔), or less fatigue in young than old (↓). Although in some studies sex-based differences in fatigue were reported, this table reflects only differences observed by age.
EF, elbow flexors; PF, plantar flexors.

Table 5.3 Sustained contraction protocols: summary of cross-sectional studies in which muscle fatigue and endurance were examined during sustained, voluntary, isometric contractions

Contraction mode	Muscle group	Subject sex, ages (n) relative to young	Fatigue in old	Reference
Isometric	AP	M; 20–91 (70)	↓	(Narici et al., 1991)
	FF, ThA, DF, PF	M; 22–72 (153)	↕↑→→	(Bemben et al., 1996)
	AP	M, F; 24, 70 (48)	↕↑	(Ditor and Hicks, 2000)
	ThA	M, F; 30, 70 (19)	↕↑→→	(Chan et al., 2000)
	EF	M; 24, 84 (14)	↕↑	(Allman and Rice, 2001)
	KE	M, F; 23, 72 (37)	↕↑→→	(Stackhouse et al., 2001)
	DF	M, F; 33, 75 (41)	↕↑	(Kent-Braun et al., 2002)
	DF	F; 20, 73 (19)	↕←	(Rubinstein and Kamen, 2005)
Isotonic	KE	F; 25, 48, 68 (23)	↕↑	(Hakkinen, 1995)
	KE	M, F; 27, 64 (52)	↕←	(Petrella et al., 2005)
	DF	M; 26, 64, 84 (32)	↕↑	(McNeil and Rice, 2007)
Isokinetic Concentric	KE	M; 25, 40, 70 (~120); M; active 30, 64; sedentary 29, 65; F; active 27, 65; sedentary 25, 65 (40)	↑	(Aniansson et al., 1978)
Eccentric, concentric	DF	M, F; 28, 73 (38)	↕↑	(Laforest et al., 1990)
Isometric, isokinetic	KE	M,F; 31, 77 (32)	↕↑←	(Lindstrom et al., 1997)
	DF	M; 26–61 (50)	↕↑→	(Baudry et al., 2007)
	DF	M, F; 26, 72 (18)	↕↑	(Larsson and Karlsson, 1978)
Isometric, stimulated	AP	M, F; 33–67 (23)	↕↑ ↑(OM)	(Lanza et al., 2004)
Isometric, stimulated	PF	M, F; 22, 69 (51)	↕↑← (Nf)	(Lennmarken et al., 1985)
Isometric, stimulated	DF	M; 26, 67 (18)	↕↑ (Cf)	(Davies et al., 1986)
Isometric, stimulated	KE	M; 27, 78 (18)	→	(Cupido et al., 1992)
Norm'd f, constant f			↕↑→→→	(Allman and Rice, 2004)
Isometric (free-flow)	DF	M; 26, 72 (24)	→→	(Chung et al., 2007)
Isometric (occluded)				
Isometric (free-flow)	DF	M; 27,70 (38)	→	(Lanza et al., 2007)
Isometric (occluded)				

Notes
Contraction intensity refers to the target force during the protocol, expressed as a percentage of MVC.
Results are summarised in this table by arrows indicating longer endurance time (↑) for old than young, or no difference between groups (↔). Arrows reflecting greater endurance time for the older adults is also provided, expressed as a percentage of the endurance time for the young adults (based on mean values).
AP, adductor policis; FF, finger flexors; ThA, thumb abductors; DF, dorsiflexors; EF, elbow flexors; PF, plantar flexors; KE, knee extensors.

(Laforest *et al.*, 1990; Hakkinen, 1995; Lindstrom *et al.*, 1997), less (Lanza *et al.*, 2004) or greater (Aniansson *et al.*, 1978; Baudry *et al.*, 2007; McNeil and Rice, 2007; Petrella *et al.*, 2005) than fatigue in young adults. Some of this inconsistency in results may be the result of differences in the characteristics of individuals included in each study, as well as the specific contraction protocols used to induce fatigue.

A number of early studies designed to understand age-related changes in muscle fatigue used electrically stimulated contractions, which circumvent the central nervous system (CNS) and focus on changes in the periphery (Table 5.3). Two studies in the mid-1980s reported greater fatigue in older men than young men during stimulated contraction protocols (Lennmarken *et al.*, 1985; Davies *et al.*, 1986). Subsequent work by Cupido *et al.* (1992) and Allman and Rice (2004) demonstrated the frequency dependence of age-related differences in fatigability during stimulated contractions. It appears that this frequency dependence is due to the slowing of contractile properties in the elderly, which results in a leftward shift in the force–frequency curve (Narici *et al.*, 1991; Ng and Kent-Braun, 1999).

In addition to these studies of fatigue during stimulated contractions, researchers began to focus increasingly on the response of young and older muscle to fatigue during voluntary contractions (Table 5.3). Lindstrom *et al.* (1997) developed an approach to the study of knee extensor fatigue that involved dynamic voluntary contractions. In this study, groups of young (28 years) and older (73 years) men and women performed 100 maximal knee extensions at $90°·s^{-1}$. An important contribution of this study was the approach used to analyse the data, which fully characterised the muscle fatigue responses. In addition to peak torque (Nm), the data were analysed for the initial rate of fatigue, overall fatigue relative to baseline, and endurance level (absolute torque level maintained in the latter portion of the protocol). Their results (Figure 5.2) showed that, while the older group was weaker than the young group and thus produced lower absolute force throughout the protocol, they had a similar initial rate of fatigue, and at the end of the protocol had developed no more fatigue on a relative basis than did the young. The authors concluded that the older group was similar to the young in its capacity to resist fatigue. Recently, Petrella *et al.* (2005) focused on changes in the velocity of contraction during ten bilateral knee extensions, and found greater slowing in older compared with young adults. This work was followed by that of McNeil and Rice (2007), who observed greater ankle dorsiflexor muscle fatigue in a "very old" group of men (mean age of 77 years) compared to young and "old" (64 years) groups during rapid isotonic contractions at 20% MVC.

To date, few investigators have evaluated multiple contraction modes in the same cohort of old and young individuals. However, in the studies in which this has been done, the results are also mixed (Larsson and Karlsson, 1978; Lanza *et al.*, 2004; Baudry *et al.*, 2007). Lanza *et al.* (2004) reported that older men (72 years) fatigued relatively less than young men

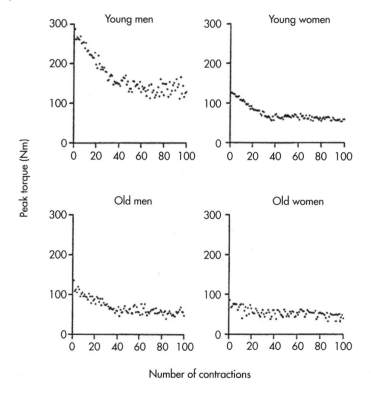

Figure 5.2 Fatigue in young and older men and women during dynamic knee extension exercise. Individual responses for young male, young female, older male, older female. As shown, the older adults were weaker than the young, but had a similar relative decline in dynamic torque during contractions at $90°·s^{-1}$

Source: With permission from Lindstrom *et al.*, 1997.

(26 years) during intermittent MVCs of the ankle dorsiflexor muscles, regardless of whether the contraction mode was isometric or isovelocity (at $90°·s^{-1}$). Notably, the groups were matched for habitual physical activity level, which minimised any confounding effects due to differing activity levels across groups. In a comparison of the responses to concentric and eccentric dorsiflexor contractions (both at $50°·s^{-1}$), Baudry *et al.* (2007) observed that older (77 years) men and women fatigued more than young (31 years) during intermittent MVCs using both contraction modes. In this study, the dynamic contractions were "pre-loaded" with isometric contractions, which may have had an effect on the results. Most recently, we have reported that, in the knee extensor muscles, healthy older men and women fatigued less than young during isometric contractions, but that this difference was abolished during isovelocity contractions at $120°·s^{-1}$ (unpublished data). It was suggested that the progressive, age-related deficit in the torque–velocity relationship in the knee extensor muscles (Lanza *et al.*, 2004) may

be responsible for the loss of fatigue resistance in older adults at $120°·s^{-1}$ compared to isometric contractions. These results support those of others (Petrella et al., 2005; McNeil and Rice, 2007) in suggesting that the slowing of voluntary contraction velocity may be an important factor in the fatigability of older adults. Collectively, these studies suggest that the muscle group studied may impact the effect of contraction mode on fatigue in the elderly.

Effect of muscle strength and contraction intensity in studies of endurance

Scientists studying muscle function in old age also have been interested in determining whether fatigue resistance in the elderly is affected by the intensity of the contractions performed. The rationale for this interest relates to the hypothesis – often applied to studies of sex-based differences in fatigue – that older, smaller muscles will generate lower intramuscular pressures during contractions compared to younger, larger muscles. By extension, it was hypothesised that older muscles should be better perfused than young during submaximal isometric endurance trials, thus explaining their relatively greater fatigue resistance. However, studies have shown that blood flow is occluded during isometric contractions above ~60% MVC (Barnes, 1980; Wigmore et al., 2004), regardless of absolute force level. Additional evidence against perfusion as a mechanism for age-related differences in muscle fatigue is that older adults generally have greater endurance than young adults regardless of the contraction intensity or muscle group, as indicated in Table 5.2. Further, studies of young and older adults who were matched for similar baseline strength also found increased endurance time (Hunter et al., 2005a) and fatigue resistance (Chung et al., 2007) in the older groups. These results are supported by studies showing less fatigue in healthy elderly, compared to young, men and women during intermittent, isometric dorsiflexor MVCs under ischemic conditions (Chung et al., 2007; Lanza et al., 2007). Therefore, it appears that any effect of muscle size on fatigue in the elderly is modest, and that the resistance to fatigue is independent of blood flow.

Does muscle fatigue vary by muscle group?

A large cross-sectional study of adults aged 20–74 years indicated that fatigue resistance varied by muscle group, but not by age (Bemben et al., 1996). However, in recent years the question of whether the effects of ageing are muscle-group specific has been the topic of renewed interest, as several studies have suggested that some muscles may be more "susceptible" to the ageing process than others (Houmard et al., 1998; Pastoris et al., 2000). For example, it has been suggested that the knee extensor muscles of older adults are more fatigable than the ankle dorsiflexor muscles, particularly during dynamic contraction protocols (Figure 5.3 and Table 5.3). In general, the

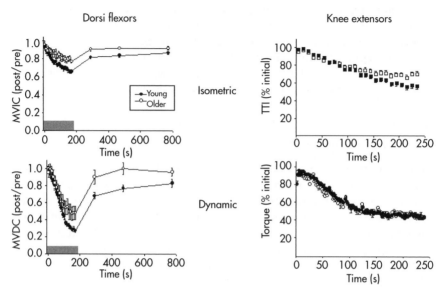

Figure 5.3 Fatigue may differ by muscle group and contraction mode. A study of healthy young and older adults indicated that, in the ankle dorsiflexor muscles (left panels), the older group (open symbols) fatigued less than the young (closed symbols) during both isometric and isoveloc-ity (90°·s⁻¹) contractions (Lanza *et al.*, 2004, with permission). In contrast, in the knee extensor muscles (right panels), the fatigue resistance observed in the elderly (open symbols) during iso-metric contractions was lost during dynamic contractions at 120°·s⁻¹ (Callahan *et al.*, 2009). These data suggest that age-related changes in the torque–velocity relationship that results in progressively less power in older compared to young muscle as velocity increases may be particularly important in fatigue resistance during dynamic knee extension in the elderly

ageing process may affect postural muscles somewhat less than muscles designed to generate rapid, powerful contractions, as these latter muscles may be more affected by the loss of type II fibre volume and therefore be less able to generate force quickly during dynamic contractions. This possibility leads to the hypothesis that the pattern of muscle use may be important in preserving function (see "Effects of activity level", below). That is, some degree of moderate-to-vigorous physical activity may be necessary to maintain adequate neuromuscular function as we age.

MECHANISMS OF MUSCLE FATIGUE IN ELDERLY ADULTS

As the body of literature regarding age-related differences in muscle fatigue has expanded, attention has turned to investigating the mechanisms of fatigue in ageing muscle. The next section presents a discussion of some of the major factors thought to be involved in muscle fatigue in the elderly.

Bioenergetic basis of fatigue resistance in the elderly

The energetics of ageing skeletal muscle has been a topic of increasing interest over the past decade (McCully et al., 1993; Houmard et al., 1998; Kent-Braun et al., 2002; Lanza et al., 2005; Lanza et al., 2007). In general, these investigations have included biopsy studies of enzyme activities (Coggan et al., 1992; McCully et al., 1993; Houmard et al., 1998; Pastoris et al., 2000) and magnetic resonance spectroscopy (MRS) studies of in vivo metabolism (Coggan et al., 1993; McCully et al., 1993; Taylor et al., 1997; Chilibeck et al., 1998; Kent-Braun et al., 2002; Lanza et al., 2006; Lanza et al., 2007). It has been reported that oxidative capacity declines or is maintained, depending on the muscle (Houmard et al., 1998), subject population and method of measurement (Russ and Kent-Braun, 2004). Less is known about glycolytic function. Early work by Larsson and Karlsson (1978) indicated a decrease in lactate dehydrogenase activity in the vastus lateralis muscle of older compared with younger men, although such a decrease was not observed by Coggan et al. (1992). Recent work using MRS has provided the first evidence of unimpaired glycolytic function in healthy elders in vivo (Lanza et al., 2007). In fact, the robust response of older muscle to various training interventions (Meredith et al., 1989; Coggan et al., 1993; Marsh et al., 1993; Berthon et al., 1995; Proctor et al., 1995; Jubrias et al., 2001) suggests that these metabolic changes occur due to changes in lifestyle, such as reduced physical activity, rather than ageing itself.

More recently, [31]phosphorus MRS has been applied to the study of fatigue in ageing skeletal muscle (Kent-Braun et al., 2002; Lanza et al., 2007). Analysis of muscle bioenergetics in healthy young and older persons matched by accelerometer for similar habitual physical activity level has revealed an age-related difference in the metabolic response to various contraction protocols. In the first of a series of studies (Kent-Braun et al., 2002), we found that older adults fatigued less than young adults during incremental contractions of the ankle dorsiflexor muscles. This fatigue resistance was associated with less acidosis and accumulation of fatigue-producing metabolites such as diprotonated inorganic phosphate. During this fatigue protocol, which produced moderate amounts of fatigue, failure of central activation and excitation–contraction coupling were eliminated as explanatory factors in the age-related fatigue resistance that was observed. Rather, the metabolic data suggested that the older adults relied relatively more on oxidative metabolism than young adults during this submaximal protocol.

To more directly test this new hypothesis regarding age-related changes in the sources of ATP during muscular work, this study was followed by one in which ATP production from oxidative phosphorylation, glycolysis and the creatine kinase reaction (CK) were quantified in young and older men in vivo (Lanza et al., 2005). The study groups were again matched for similar physical activity levels. During a 60 s MVC, the older men had similar ATP production from oxidative phosphorylation and CK, but lower glycolytic flux than the young men. Because the total ATP cost of the contraction was also lower in the older men, who were ~30% weaker than the young men, relatively more ATP was generated oxidatively in the older men. Notably, there was no difference between groups in the oxidative capacity of the muscle. Rather, the differences were in the utilisation of the pathways of ATP production during this maximal contraction.

Finally, to quantify ATP production during fatiguing contractions and to determine whether the age-related reliance on oxidative metabolism was a result of impaired glycolytic function, young and older men and women performed repeated dorsiflexor MVCs under free-flow and ischemic conditions (Lanza *et al.*, 2007). The rates of ATP production through oxidative phosphorylation, CK and glycolysis are shown in Figure 5.4 for the free-flow and ischemic protocols. Total ATP production is also given. Oxidative and CK production of ATP was similar in young and older groups. During free-flow contractions, the young group again generated more ATP through glycolysis than did the older group. However, there was no age-related difference in glycolytic flux during ischemic contractions, indicating that glycolytic function was unimpaired in the older muscle. As in earlier studies, oxidative capacity was similar across age groups (data not shown). Relative to total ATP produced, the older adults had greater ATP flux through oxidative phosphorylation compared to the young during free-flow conditions.

Surprisingly, when the "oxidative advantage" of the older adults was removed by the use of ischemia, the older group again fatigued less than the young. In both protocols, the young group had greater acidosis than the old, and fatigue was strongly associated with the accumulation of $H_2PO_4^-$ (Figure 5.5). Thus, as was the case during submaximal contractions (Kent-Braun *et al.*, 2002), there was a metabolic basis for the fatigue difference between young and old during these MVCs. In a separate study, Chung *et al.* (2007) also found less fatigue in old compared to young during both free-flow and ischemic contractions. This was the case for the full study groups, as well as for young and older sub-groups matched for similar strength. Measures of activation were made in this study, and the results indicated greater central and peripheral activation failure in the young compared to the older group during the ischemic protocol. Together, these studies suggest a metabolic basis for the age-related fatigue resistance observed in the ankle dorsiflexor muscles during isometric contractions. This difference appears to be independent of strength or blood flow, and may be related to differences in the energetic cost of contractions per unit of muscle, or metabolic economy, between young and old. Further, these results suggest that an age-related difference in the muscle metabolic response may provide a feedback mechanism by which intramuscular metabolites affect neural activation more in young adults during fatigue (Bigland-Ritchie *et al.*, 1986).

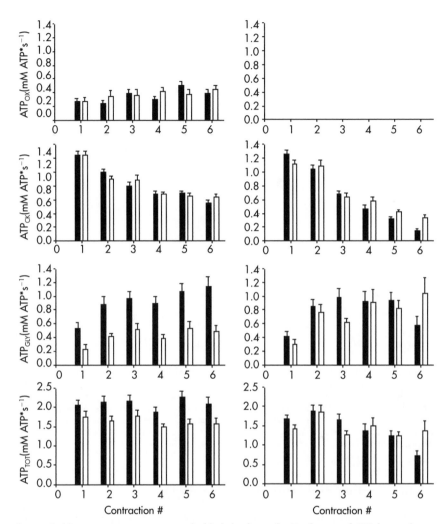

Figure 5.4 Bioenergetics in young and old skeletal muscle. Production of ATP by oxidative phosphorylation (top panels), the creatine kinase reaction (upper middle panels) and glycolysis (lower middle panels) during intermittent maximal isometric contractions of the ankle dorsiflexor muscles. Metabolites were measured by ^{31}P MRS in young (closed bars) and older (open bars) men and women. Under free-flow conditions (left-hand panels), production of ATP by oxidative phosphorylation and the creatine kinase reaction was similar in young and old, while glycolytic ATP production and total ATP flux (bottom panels) was lower in the older group. Thus, the skeletal muscle of healthy older adults relied relatively more on oxidative metabolism during free-flow. During ischemia (right-hand panels), glycolytic flux is similar in young and old, indicating that glycolytic function was unimpaired in the old

Source: Lanza *et al.*, 2007, with permission.

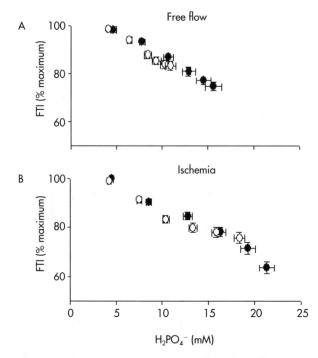

Figure 5.5 Metabolic basis of fatigue resistance in the elderly. The fall of maximal voluntary force was linearly related to changes in [$H_2PO_4^-$] during intermittent maximal contractions under free-flow (A) and ischemic (B) conditions in young (solid symbols) and older (open symbols) adults. Both fatigue and the accumulation of $H_2PO_4^-$ were greater in the young group than in the older group during these protocols. Data were obtained from the tibialis anterior muscles of healthy young and older adults, and groups were matched for similar physical activity levels

Source: Lanza *et al.*, 2007, with permission.

Neural factors and fatigue in the elderly

Central activation, while typically similar between young and older adults at baseline (Connelly *et al.*, 1999; Kent-Braun and Ng, 1999; Roos *et al.*, 1999; Kent-Braun *et al.*, 2002; Klass *et al.*, 2005; Chung *et al.*, 2007; Russ *et al.*, 2008), has been shown to play a role in fatigue during intermittent (Stackhouse *et al.*, 2001; Russ *et al.*, 2008) and sustained (Chan *et al.*, 2000) contractions at both low and high contraction intensities (Yoon *et al.*, 2008). However, the extent to which diminished neural drive, or central fatigue, mediates age-related differences in fatigue resistance is not clear. Central activation is typically measured by superimposing an electrical stimulation on an MVC (Kent-Braun and Le Blanc, 1996; Kent-Braun, 1999; Stackhouse *et al.*, 2001; Lanza *et al.*, 2004; Chung *et al.*, 2007). Additional force produced by the superimposed stimulus indicates impaired neural activation. Stackhouse *et al.* (2001) demonstrated greater central activation failure in older volunteers than in young volunteers during maximal, intermittent contractions of the knee extensors, although fatigue was not different between the groups. Other investigators have reported no age difference in the role of central activation failure during fatigue in the plantar flexors (Mademli and Arampatzis, 2008a) and dorsiflexors (Kent-Braun et al., 2002). Differences between studies may be due in part to variations in study populations and techniques (Herbert and Gandevia, 1999; Stackhouse et al., 2000). Likewise, muscle group may be a factor, as Jakobi and Rice (2002) have shown that the ability to fully activate a muscle can depend upon the muscle group studied. While the majority of studies to date suggest that central activation failure has a relatively minor impact on age-related differences in fatigue, limitations in the techniques generally used to measure central activation make definitive conclusions difficult at this time.

In addition to complete recruitment of all motor units in the contracting muscle, production of maximal force or power requires the generation of high motor unit discharge rates (MUDR). Roos et al. (1999) observed similar MUDR in the unfatigued knee extensor muscles of young and older men across a full range of voluntary contraction intensities. However, reduced maximal MUDR have been found in older adults by Kamen et al. and others (Kamen et al., 1995; Connelly et al., 1999; Patten et al., 2001; Rubinstein and Kamen, 2005). The degree to which changes in MUDR behaviour affect fatigue in the elderly is of growing interest. Recently, Rubinstein and Kamen (2005) demonstrated that while MUDR declined in both young and old during intermittent, maximal isometric contractions, the decrease in MUDR was smaller in the older adults than in the young, and this difference was reflected by an age-related difference in fatigue (Figure 5.6). Thus, there is evidence that changes in motor unit behaviour that occur in old age may aid the older neuromuscular system in resisting muscle fatigue during isometric contractions.

Under some conditions, however, lower MUDR in older adults may become a disadvantage. For example, reduced MUDR in older individuals, which may be beneficial for the maintenance of force production during isometric contractions, could limit the rapid production of power during dynamic contractions (Klass et al., 2008) to a greater extent than in young people, who generally have higher maximal MUDRs (Kamen et al., 1995; Connelly et al., 1999; Rubinstein and Kamen, 2005). In addition, the leftward shift in the force–velocity curve of older individuals (Narici et al., 1991; Ng and Kent-Braun, 1999; Allman and Rice, 2004), which would not impact fatigue resistance during isometric contractions, may also limit torque production during high-velocity dynamic contractions, as the ability to rapidly generate torque would become increasingly limited as target contraction velocities increase. Thus, age-related neural differences could combine with the slowed contractile properties of older muscle to limit power production beyond that which would be expected due to sarcopenia alone. Clearly, additional work in this area will be necessary to fully understand the mechanisms of fatigue in the elderly.

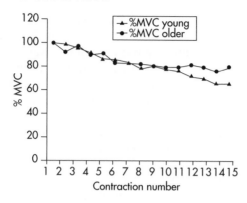

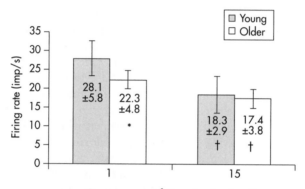

Figure 5.6 Fatigue and motor unit behaviour in young and older adults. At baseline, the tibialis anterior muscle of the older group was weaker and exhibited lower maximal motor unit discharge rates compared to the young. During intermittent maximal isometric contractions, the older group fatigued less than the young. At the same time, there was less of a decrease in maximal motor unit discharge rates in the old, suggesting that maintenance of motor unit behaviour may have been a factor in the fatigue resistance of the older group. The connection between motor unit behaviour, muscle energetics and fatigue remain to be determined.

Source: With permission from Rubinstein and Kamen, 2005.

IMPORTANCE OF STUDY POPULATION ON MUSCLE FATIGUE: RETAINING THE FOCUS ON SENESCENCE

Effects of health

The health of study participants is an important consideration for investigations of muscle fatigue in the elderly. Studies of fatigue in populations with chronic disease such as cardiovascular disease, diabetes and cancer illustrate the debilitating effects of disease processes and treatments on neuromuscular function. Typically, greater fatigue is observed in these clinical populations compared to healthy, age-matched control groups (Minotti *et al.*, 1992; Johansen *et al.*, 2005; Almeida *et al.*, 2008). Thus, there is potential for comparisons of young and older adults to be confounded by the increased prevalence of disease in the elderly. Care must be taken in studies of ageing to distinguish the effects of old age from those of disease, thereby allowing the mechanisms of fatigue to be attributed to the correct sources, and remediation of excessive muscle fatigue can be more effectively promoted.

Management of chronic disease may further complicate the investigation of age-related changes in muscle fatigue. It is clear that a number of medications widely used to prevent or treat some of these diseases may have negative effects on neuromuscular performance. This problem is compounded by the fact that most drug trials are conducted in young and middle-aged adults. As a result, the effects of these drugs on fatigue in the elderly are not known. While these medications are highly effective in treating a variety of conditions, the possibility exists that some drugs may alter the processes involved in resisting fatigue. Perhaps the most well-known of these candidates are the statin drugs that are used to treat hypercholesterolemia. While statins can be effective in reducing cholesterol levels, they have been shown to induce myopathy in some patients. Likewise, treatment of hypertension with beta-blockade is common and effective, but the drug's action limits the pressor response to exercise, which could result in early fatigue under some conditions (Kaiser *et al.*, 1986). Medications that affect peripheral blood flow or alter cognition or motivation may also impair a person's ability to maintain force production during a fatigue protocol. Until the effects on neuromuscular function of medications commonly used to treat disease in the elderly are fully understood, matching groups by medications may be prudent for the study of fatigue in the elderly.

As age increases beyond 70 years, loss of functional ability and onset of frailty become more common. Even in the absence of clinically identified disease, significant sarcopenia and neuromuscular changes can result in poor coordination and mobility in older adults. To date, little is known about muscle fatigue in elderly persons with impaired physical function. Schwender *et al.* (1997) observed that, while there was no difference in knee extensor endurance between young women (22 years) and older women with no history of falls (71 years), older women (73 years) who had experienced at least one fall in the preceding 18 months (mean = 1.44) had poor endurance and slowed recovery compared with the other two groups. The question of the interactions between muscle fatigue and physical function in the elderly requires more attention.

Effects of activity level

As with health status, the habitual physical activity patterns of study participants can impart a tremendous influence on the outcome of research on muscle fatigue. In reality, disentangling the effects of health and activity can be a challenge. Since older adults generally have lower levels of physical activity than young adults (Meijer *et al.*, 2001; Davis and Fox, 2007; Troiano *et al.*, 2008), this influence may become magnified as many of the biological changes associated with old age can also be observed with disuse (Bortz, 1982; Hainaut and Duchateau, 1989). Independent of age, both disuse and exercise training affect neuromuscular function in a variety of ways. Demonstrations of the reversibility of many "age-related" changes in neuromuscular function in response to training (Hakkinen *et al.*, 1998; Hakkinen *et al.*, 2001; Jubrias *et al.*, 2001; Kamen and Knight, 2004) suggest the extent to which these alterations are related to activity status, rather than senescence. Thus, training status and physical activity level should be taken into account in the design of muscle fatigue studies (Laforest *et al.*, 1990), thereby ensuring that the influence of physical activity can be minimised and the focus on senescence can be maintained in studies of fatigue in the elderly.

A recent study by Davis and Fox (2007) quantified the significant decrease in total and moderate-to-vigorous physical activity in older (76 years) compared to younger (27 years) men and women. The concept that physical activity level is important is supported by recent work from our laboratory (unpublished data) suggesting that the time spent in moderate-to-vigorous physical activity each day may be more important for maintaining neuromuscular function than is the total amount of daily activity in the elderly. Presumably, higher-intensity activity that recruits larger portions of the muscle – including type II fibres – is needed to maintain optimal neuromuscular function in older adults. Thus, the interpretation of study results can be enhanced by incorporating information about activity patterns. For example, the observation that some muscles may be more fatigable than others (Bemben *et al.*, 1996) in the elderly may be a result of the way the muscles are used each day, rather than an effect of ageing alone. Lower amounts of moderate-to-vigorous daily physical activity on the part of older adults may affect some muscles more than others. That is, muscles used every day for posture and locomotion, such as the ankle dorsiflexors, may be less affected by changes in activity habits than are muscles such as the knee extensors, which have a more mixed fibre type composition and are important for generating power during high-intensity activities. This hypothesis, if supported, has important implications for the design of interventions in the elderly, as physical activity and its consequences are modifiable variables.

Effect of age and sex

Studies of the effects of old age on muscle fatigue generally have included volunteers aged 60–80 years as representatives of "older" adults (see Tables 5.2 and 5.3). However, recent work by McNeil and Rice (2007) highlights the importance of clearly defining what is meant by "old age". In this study, no difference in fatigue was observed between young (26 years) and "old" (64 years) men during rapid, isotonic contractions of the ankle dorsiflexor muscles. In contrast, a group of "very old" (84 years) men had greater fatigue than the young group during this protocol. Further, the primary mechanism of this fatigue difference appeared to be the slowing of contraction speed in the very old group.

The question of differences in muscle fatigue between men and women has been the focus of a number of investigations in recent years (Hicks and McCartney, 1996; Hunter and Enoka, 2001; Kent-Braun *et al.*, 2002; Hunter *et al.*, 2004; Russ *et al.*, 2008), and has received some attention in studies of ageing adults. Young women often fatigue relatively less than young men, and it appears that both neural and metabolic factors contribute to this difference (Kent-Braun *et al.*, 2002; Russ and Kent-Braun, 2003; Russ *et al.*, 2005; Russ *et al.*, 2008). An interesting question that arises from these studies is whether men and women age similarly with respect to fatigue resistance. Some recent studies suggest that they do (Kent-Braun *et al.*, 2002; Lanza *et al.*, 2007), while others suggest that the advantage younger women have with regard to fatigue resistance may lessen or dissipate in old age (Hunter *et al.*, 2004). Hunter *et al.* have shown that, while young women had longer endurance times than young men, there was no difference in endurance times between older men and women (Hunter *et al.*, 2004). The lack of clarity about this issue indicates that, for the time being, age groups should be balanced by sex in studies of fatigue in the elderly. Overall, the question of interactions between sex and age in muscle fatigue, and their mechanisms, clearly warrants further exploration (see Chapter 6).

The effects of many characteristics on the changes in muscle fatigue with age are still unknown. These include intrinsic characteristics of the participants, such as race, ethnicity and genetics, as well factors that may change over the lifespan, such as daily physical activity patterns. The interactions between age, sex, muscle fatigue and health status are not well understood. While the acute effects of many diseases on changes in muscle fatigue have been studied (Sharma *et al.*, 1994; Sharma *et al.*, 1995; Kent-Braun and Miller, 2000; Johansen *et al.*, 2005), the long-term effects of chronic disease or its treatment on muscle fatigue in older adults is unclear at this time.

MUSCLE FATIGUE AND PHYSICAL FUNCTION IN THE ELDERLY

Given the wide range of functional capacities exhibited by the population of ageing adults, there is a growing need to understand how and why fundamental changes in neuromuscular function, including muscle fatigue resistance, exhibit such variation in the population as a whole. In particular, we know little about the interactions between the loss of muscle fatigue resistance and the development of physical limitations in older adults. The triggers for the onset of the decline in neuromuscular function remain to be identified. The studies described in this chapter have highlighted the promising result that the neuromuscular system of healthy elders adapts in ways that may be beneficial under some conditions. Figure 5.7 illustrates a potential scenario that describes how an older person may proceed from healthy adaptations to physical impairment due to a sequence of events set in motion by some factor such as injury, illness or very old age. While the importance of sarcopenia is well-recognised, recent epidemiological studies suggest that strength, more than muscle size alone, is predictive of future disability (Goodpaster *et al.*, 2006). Thus, more information is needed about the timing and mechanisms of changes in neuromuscular function in the elderly.

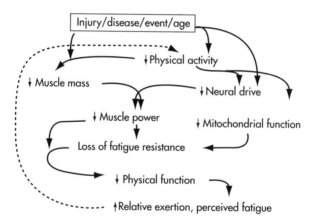

Figure 5.7 Hypothesised progression from fatigue resistance to physical impairment. While healthy elderly people often show better fatigue resistance than young people, there remains the question of which factors trigger the onset of decline in physical function in the elderly. A progression for the transition from fatigue resistance to limitations in physical function is illustrated. Potential triggers for this progression may include injury, illness or advanced age, any of which may result in a cascade of events leading to physical impairment, a sense of increased effort and the development of symptomatic fatigue. Once in place, this cascade can result in further declines in physical activity and loss of physical function.

CONCLUSION

It is clear that, under some conditions, the ability of human skeletal muscle to resist fatigue improves with old age. The basis of this difference appears to reside largely with intracellular energy metabolism, which in turn may be influenced by both neural and muscular changes in the elderly. However, there also exist conditions under which this enhanced fatigue resistance may be negated, such as during repeated, dynamic contractions. Under these conditions, the slowed contraction velocities of elderly adults, which likely have both neural and muscular sources, may limit their ability to keep pace with rapid contractions. Currently, the relative importance of central and peripheral factors in fatigue during dynamic contractions in the elderly is not known. The precise interactions between neural, contractile and energetic changes in older skeletal muscle remain to be determined. With this information, the design of interventions that delay or reverse the loss of physical capacity and the onset of mobility impairments in our ageing population will be possible.

FIVE KEY PAPERS THAT SHAPED THE TOPIC AREA

Study 1. Larsson, L. and Karlsson, J. (1978). Isometric and dynamic endurance as a function of age and skeletal muscle characteristics. *Acta Physiologica Scandinavica*, 104, 129–136.

This study is one of the earliest comprehensive assessments of the effects of age on muscular endurance. These investigators evaluated age-related differences in muscular endurance during dynamic and isometric knee extensions in men aged 22–65 years. Importantly, this investigation also identified morphologic alterations in skeletal muscle that may influence fatigue resistance. There were no significant differences in fatigue resistance by age. Type II muscle fibre area was associated with maximum dynamic strength, the relative force decline during dynamic contractions and isometric endurance. Further, the decline in absolute force was positively associated with lactate dehydrogenase activity. These results illustrated the importance of muscle morphology in fatigue, and established that older adults do not have a de facto decline in their ability to maintain force production during muscular activity.

Study 2. Lindstrom, B., Lexell, J., Gerdle, B. and Downham, D. (1997). Skeletal muscle fatigue and endurance in young and old men and women. *The Journals of Gerontology Series A: Biological Sciences and Medical Sciences*, 52, B59–66.

This study of knee extension fatigue in young (28 years) and old (73 years) men and women was notable for its use of isokinetic knee extensions and a thorough examination of the torque data obtained during 100 contractions at $90°·s^{-1}$. While baseline strength and the absolute force level maintained at the end of the protocol were lower in the older volunteers, there was no difference in the initial rate of fatigue or the fatigue index, which reflected the relative fall of torque during the protocol. Thus, weakness, but not excessive fatigue, was the primary deficit observed in the older subjects in this study.

Study 3. Kent-Braun, J.A., Ng, A.V., Doyle, J.W. and Towse, T.F. (2002). Human skeletal muscle responses vary with age and gender during fatigue due to incremental isometric exercise. *Journal of Applied Physiology*, 93, 1813–1823.

This study was the first to investigate the neural and metabolic contributions to ankle dorsiflexor muscle fatigue in young (33 years) and older (75 years) men and women *in vivo*. Measures of muscle energetics and neural activation were acquired during a fatigue protocol in which the intensity of isometric contractions was increased progressively. Older adults showed less fatigue, acidosis and change in muscle metabolites compared to the young adults, while neural activation was the same between groups. Thus, older subjects relied more on aerobic metabolism for the supply of ATP compared to young subjects. In all groups, fatigue was associated with the accumulation of diprotonated inorganic phosphate. This study established a metabolic basis for the greater fatigue resistance of older men and women.

Study 4. Allman, B.L. and Rice, C.L. (2004). An age-related shift in the force–frequency relationship affects quadriceps fatigability in old adults. *Journal of Applied Physiology*, 96, 1026–1032.

This study examined the impact of age-related changes in the force–frequency relationship on muscular fatigue in the quadriceps of young (27 years) and old (78 years) men. Stimulated fatigue trials consisting of either a constant frequency (14.3 Hz) or a frequency that elicited a contraction equal to 60% of that of a 100 Hz train were used to test the hypothesis that age-related changes in the force–frequency curve may affect the outcome of stimulated fatigue trials. In this study, old and young men had similar fatigue characteristics during constant frequency stimulation, but older subjects were more fatigue-resistant than young when stimulation was matched to their left-shifted force–frequency curves. This study shows the importance of chronic changes in contractile properties on the fatigue characteristics of ageing muscle.

Study 5. Rubinstein, S. and Kamen, G. (2005). Decreases in motor unit firing rate during sustained maximal-effort contractions in young and older adults. *Journal of Electromyography and Kinesiology*, 15, 536–543.

This investigation provided novel information about the changes in motor unit behaviour during fatiguing contractions of the tibialis anterior muscle in young (19 years) and older (73 years) women. MUDR were assessed during 15 maximal isometric contractions using a multi-channel, indwelling wire electrode. At baseline, older individuals had lower maximal MUDR compared to young individuals. Over the course of the fatigue protocol, MUDR declined to a greater extent in young than older women, such that firing rates during the last contraction were not different between groups. At the same time, older women had less fatigue than young. This study highlights the critical role of motor unit firing behaviour in age-related differences in muscle fatigue.

GLOSSARY OF TERMS

ATP	adenosine triphosphate
CK	creatine kinase
CNS	central nervous system
FTI	force–time integral
$H_2PO_4^-$	ion dihydrogenoorthophosphate
MRS	magnetic resonance spectroscopy
MUDR	motor unit discharge rates
MVC	maximal voluntary isometric contraction
TTI	time–tension integral
Type I fibres	slow twitch fibres
Type II fibres	fast twitch fibres

ACKNOWLEDGEMENT

This work was funded by NIH R01AG21094 and K02AG023582.

REFERENCES

Allman, B.L. and Rice, C.L. (2001). Incomplete recovery of voluntary isometric force after fatigue is not affected by old age. *Muscle and Nerve*, 24 (9), 1156–1167.

Allman, B.L. and Rice, C.L. (2004). An age-related shift in the force–frequency relationship affects quadriceps fatigability in old adults. *Journal of Applied Physiology*, 96 (3), 1026–1032.

Almeida, S., Riddell, M.C. and Cafarelli, E. (2008). Slower conduction velocity and motor unit discharge frequency are associated with muscle fatigue during isometric exercise in type 1 diabetes mellitus. *Muscle and Nerve*, 37 (2), 231–240.

Aniansson, A., Grimby, G., Hedberg, M., Rungren, A. and Sperling, L. (1978). Muscle function in old age. *Scandinavian Journal of Rehabilitative Medicine*, 6, 43–49.

Barnes, W.S. (1980). The relationship between maximum isometric strength and intramuscular circulatory occlusion. *Ergonomics*, 23 (4), 351–357.

Baudry, S., Klass, M., Pasquet, B. and Duchateau, J. (2007). Age-related fatigability of the ankle dorsiflexor muscles during concentric and eccentric contractions. *European Journal of Applied Physiology*, 100 (5), 515–525.

Bazzucchi, I., Marchetti, M., Rosponi, A., Fattorini, L., Castellano, V., Sbriccoli, P. and Felici, F. (2005). Differences in the force/endurance relationship between young and older men. *European Journal of Applied Physiology*, 93 (4), 390–397.

Bemben, M.G., Massey, B.H., Bemben, D.A., Misner, J.E. and Boileau, R.A. (1996). Isometric intermittent endurance of four muscle groups in men aged 20–74 yr. *Medicine and Science in Sports and Exercise*, 28 (1), 145–154.

Berthon, P., Freyssenet, D., Chatard, J.C., Castells, J., Mujika, I., Geyssant, A., Guezennec, C.Y. and Denis, C. (1995). Mitochondrial ATP production rate in 55 to 73-year-old men: effect of endurance training. *Acta Physiologica Scandinavica*, 154, 269–274.

Bigland-Ritchie, B.R., Dawson, N.J., Johansson, R.S. and Lippold, O.C. (1986). Reflex origin for the slowing of motoneurone firing rates in fatigue of human voluntary contractions. *Journal of Physiology (Lond)*, 379, 451–459.

Bilodeau, M., Erb, M.D., Nichols, J.M., Joiner, K.L. and Weeks, J.B. (2001a). Fatigue of elbow flexor muscles in younger and older adults. *Muscle and Nerve*, 24 (1), 98–106.

Bilodeau, M., Henderson, T.K., Nolta, B.E., Pursley, P.J. and Sandfort, G.L. (2001b). Effect of aging on fatigue characteristics of elbow flexor muscles during sustained submaximal contraction. *Journal of Applied Physiology*, 91 (6), 2654–2664.

Binzoni, T., Bianchi, S., Hanquinet, S., Kaelin, A., Sayegh, Y., Dumont, M. and Jequier, S. (2001). Human gastrocnemius medialis pennation angle as a function of age: from newborn to the elderly. *Journal of Physiology, Anthropology and Applied Human Science*, 20 (5), 293–298.

Bortz, W.M. (1982). Disuse and aging. *Journal of the American Medical Association*, 248 (10), 1203–1208.

Callahan, D.M., Foulis, S.A. and Kent-Braun, J.A. (2009) Age-related fatigue resistance in the knee extensor muscles is specific to contraction mode. *Muscle and Nerve*, 39 (5), 692–702.

Chan, K.M., Raja, A.J., Strohschein, F.J. and Lechelt, K. (2000). Age-related changes in muscle fatigue resistance in humans. *Canadian Journal of Neurological Science*, 27 (3), 220–228.

Chilibeck, P.D., Paterson, D.H., McCreary, C.R., Marsh, G.D., Cunningham, D.A. and Thompson, R.T. (1998). The effects of age on kinetics of oxygen uptake and phosphocreatine in humans during exercise. *Experimental Physiology*, 83, 107–117.

Chung, L.H., Callahan, D.M. and Kent-Braun, J.A. (2007). Age-related resistance to skeletal muscle fatigue is preserved during ischemia. *Journal of Applied Physiology*, 103 (5), 1628–1635.

Coggan, A.R., Abduljalil, A.M., Swanson, S.C., Earle, M.S., Farris, J.W., Mendenhall, L.A. and Robitaille, P.M. (1993). Muscle metabolism during exercise in young and older untrained and endurance-trained men. *Journal of Applied Physiology*, 75, 2125–2133.

Coggan, A.R., Spina, R.J., King, D.S., Rogers, M.A., Brown, M., Nemeth, P.M.

and Holloszy, J.O. (1992). Histochemical and enzymatic comparison of the gastrocnemius muscle of young and elderly men and women. *Journal of Gerontology*, 47, B71–76.

Coggan, A.R., Spina, R.J., Rogers, M.A., King, D.S., Brown, M., Nemeth, P.M. and Holloszy, J.O. (1990). Histochemical and enzymatic characteristics of skeletal muscle in master athletes. *Journal of Applied Physiology*, 68 (5), 1896–1901.

Connelly, D.M., Rice, C.L., Roos, M.R. and Vandervoort, A.A. (1999). Motor unit firing rates and contractile properties in tibialis anterior of young and old men. *Journal of Applied Physiology*, 87 (2), 843–852.

Cupido, C.M., Hicks, A.L. and Martin, J. (1992). Neuromuscular fatigue during repetitive stimulation in elderly and young adults. *European Journal of Applied Physiology and Occupational Physiology*, 65 (6), 567–572.

D'Antona, G., Pellegrino, M.A., Adami, R., Rossi, R., Carlizzi, C.N., Canepari, M., Saltin, B. and Bottinelli, R. (2003). The effect of ageing and immobilization on structure and function of human skeletal muscle fibres. *Journal of Physiology*, 552 (Pt 2), 499–511.

Davies, C.T., Thomas, D.O. and White, M.J. (1986). Mechanical properties of young and elderly human muscle. *Acta Medica Scandinavica*, 711, 219–226.

Davies, C.T., White, M.J. and Young, K. (1983). Electrically evoked and voluntary maximal isometric tension in relation to dynamic muscle performance in elderly male subjects, aged 69 years. *European Journal of Applied Physiology and Occupational Physiology*, 51 (1), 37–43.

Davis, M.G. and Fox, K.R. (2007). Physical activity patterns assessed by accelerometry in older people. *European Journal of Applied Physiology*, 100 (5), 581–589.

De Serres, S.J. and Enoka, R.M. (1998). Older adults can maximally activate the biceps brachii muscle by voluntary command. *Journal of Applied Physiology*, 84, 284–291.

Ditor, D.S. and Hicks, A.L. (2000). The effect of age and gender on the relative fatigability of the human adductor pollicis muscle. *Canadian Journal of Physiology and Pharmacology*, 78 (10), 781–790.

Essen-Gustavsson, B. and Borges, O. (1986). Histochemical and metabolic characteristics of human skeletal muscle in relation to age. *Acta Physiologica Scandinavica*, 126 (1), 107–114.

Goodpaster, B.H., Park, S.W., Harris, T.B., Kritchevsky, S.B., Nevitt, M., Schwartz, A.V., Simonsick, E.M., Tylavsky, F.A., Visser, M. and Newman, A.B. (2006). The loss of skeletal muscle strength, mass, and quality in older adults: the health, aging and body composition study. *The Journals of Gerontology Series A: Biological Sciences and Medical Sciences*, 61 (10), 1059–1064.

Hainaut, K. and Duchateau, J. (1989). Muscle fatigue, effects of training and disuse. *Muscle and Nerve*, 12, 660–669.

Hakkinen, K. (1995). Neuromuscular fatigue and recovery in women at different ages during heavy resistance loading. *Electromyography and Clinical Neurophysiology*, 35 (7), 403–413.

Hakkinen, K., Kallinen, M., Izquierdo, M., Jokelainen, K., Lassila, H., Malkia, E., Kraemer, W.J., Newton, R.U. and Alen, M. (1998). Changes in agonist-antagonist EMG, muscle CSA, and force during strength training in middle-aged and older people. *Journal of Applied Physiology*, 84, 1341–1349.

Hakkinen, K., Kraemer, W.J., Newton, R.U. and Alen, M. (2001). Changes in electromyographic activity, muscle fibre and force production characteristics during heavy resistance/power strength training in middle-aged and older men and women. *Acta Physiologica Scandinavica*, 171 (1), 51–62.

Herbert, R.D. and Gandevia, S.C. (1999). Twitch interpolation in human muscles: mechanisms and implications for measurement of voluntary activation. *Journal of Neurophysiology*, 82 (5), 2271–2283.

Hicks, A.L. and McCartney, N. (1996). Gender differences in isometric contractile properties and fatigability in elderly human muscle. *Canadian Journal of Applied Physiology*, 21 (6), 441–454.

Houmard, J.A., Weidner, M.L., Gavigan, K.E., Tyndall, G.L., Hickey, M.S. and Alshami, A. (1998). Fiber type and citrate synthase activity in the human gastrocnemius and vastus lateralis with aging. *Journal of Applied Physiology*, 85 (4), 1337–1341.

Hunter, S.K. and Enoka, R.M. (2001). Sex differences in the fatigability of arm muscles depends on absolute force during isometric contractions. *Journal of Applied Physiology*, 91 (6), 2686–2694.

Hunter, S.K., Critchlow, A. and Enoka, R.M. (2004). Influence of aging on sex differences in muscle fatigability. *Journal of Applied Physiology*, 97 (5), 1723–1732.

Hunter, S.K., Critchlow, A. and Enoka, R.M. (2005a). Muscle endurance is greater for old men compared with strength-matched young men. *Journal of Applied Physiology*, 99 (3), 890–897.

Hunter, S.K., Rochette, L., Critchlow, A. and Enoka, R.M. (2005b). Time to task failure differs with load type when old adults perform a submaximal fatiguing contraction. *Muscle and Nerve*, 31 (6), 730–740.

Jakobi, J.M. and Rice, C.L. (2002). Voluntary muscle activation varies with age and muscle group. *Journal of Applied Physiology*, 93 (2), 457–462.

Jakobsson, A., Borg, K. and Edstrom, L. (1990). Fiber-type composition, structure and cytoskeletal protein location of fibres in anterior tibial muscle. *Acta Neuropathologica*, 80, 459–468.

Jasperse, J.L., Seals, D.R. and Callister, R. (1994). Active forearm blood flow adjustments to handgrip exercise in young and older healthy men. *Journal of Physiology*, 474 (2), 353–360.

Johansen, K.L., Doyle, J., Sakkas, G.K. and Kent-Braun, J.A. (2005). Neural and metabolic mechanisms of excessive muscle fatigue in maintenance hemodialysis patients. *American Journal of Physiology Regulatory, Integrative and Comparative Physiology*, 289 (3), R805–813.

Jubrias, S.A., Esselman, P.C., Price, L.B., Cress, M.E. and Conley, K.E. (2001). Large energetic adaptations of elderly muscle to resistance and endurance training. *Journal of Applied Physiology*, 90 (5), 1663–1670.

Kaiser, P., Tesch, P.A., Frisk-Holmberg, M., Juhlin-Dannfelt, A. and Kaijser, L. (1986). Effect of beta 1-selective and non-selective beta-blockade on work capacity and muscle metabolism. *Clinical Physiology*, 6 (2), 197–207.

Kamen, G. and Knight, C.A. (2004). Training-related adaptations in motor unit discharge rate in young and older adults. *The Journals of Gerontology Series A: Biological Sciences and Medical Sciences*, 59 (12), 1334–1338.

Kamen, G., Sison, S.V., Duke Du, C.C. and Patten, C. (1995). Motor unit discharge behavior in older adults during maximal-effort contractions. *Journal of Applied Physiology*, 79 (6), 1908–1913.

Kent-Braun, J.A. (1999). Central and peripheral contributions to muscle fatigue in humans during sustained maximal effort. *European Journal of Applied Physiology and Occupational Physiology*, 80 (1), 57–63.

Kent-Braun, J.A. and Le Blanc, R. (1996). Quantitation of central activation failure during maximal voluntary contractions in humans. *Muscle and Nerve*, 19 (7), 861–869.

Kent-Braun, J.A. and Miller, R.G. (2000). Central fatigue during isometric exercise in amyotrophic lateral sclerosis. *Muscle and Nerve*, 23 (6), 909–914.

Kent-Braun, J.A. and Ng, A.V. (1999). Specific strength and voluntary muscle activation in young and elderly women and men. *Journal of Applied Physiology*, 87 (1), 22–29.

Kent-Braun, J.A., Ng, A.V., Doyle, J.W. and Towse, T.F. (2002). Human skeletal muscle responses vary with age and gender during fatigue due to incremental isometric exercise. *Journal of Applied Physiology*, 93 (5), 1813–1823.

Klass, M., Baudry, S. and Duchateau, J. (2005). Aging does not affect voluntary activation of the ankle dorsiflexors during isometric, concentric, and eccentric contractions. *Journal of Applied Physiology*, 99 (1), 31–38.

Klass, M., Baudry, S. and Duchateau, J. (2008). Age-related decline in rate of torque development is accompanied by lower maximal motor unit discharge frequency during fast contractions. *Journal of Applied Physiology*, 104 (3), 739–746.

Klein, C.S., Rice, C.L. and Marsh, G.D. (2001). Normalized force, activation, and coactivation in the arm muscles of young and old men. *Journal of Applied Physiology*, 91 (3), 1341–1349.

Kubo, K., Kanehisa, H., Azuma, K., Ishizu, M., Kuno, S.Y., Okada, M. and Fukunaga, T. (2003). Muscle architectural characteristics in women aged 20–79 years. *Medicine and Science in Sports and Exercise*, 35 (1), 39–44.

Kutsuzawa, T., Shioya, S., Kurita, D., Haida, M. and Yamabayashi, H. (2001). Effects of age on muscle energy metabolism and oxygenation in the forearm muscles. *Medicine and Science in Sports and Exercise*, 33 (6), 901–906.

Laforest, S., St Pierre, D.M., Cyr, J. and Gayton, D. (1990). Effects of age and regular exercise on muscle strength and endurance. *European Journal of Applied Physiology and Occupational Physiology*, 60 (2), 104–111.

Lanza, I.R., Befroy, D.E. and Kent-Braun, J.A. (2005). Age-related changes in ATP-producing pathways in human skeletal muscle in vivo. *Journal of Applied Physiology*, 99 (5), 1736–1744.

Lanza, I.R., Larsen, R.G. and Kent-Braun, J.A. (2007). Effects of old age on human skeletal muscle energetics during fatiguing contractions with and without blood flow. *Journal of Physiology*, 583 (Pt 3), 1093–1105.

Lanza, I.R., Russ, D.W. and Kent-Braun, J.A. (2004). Age-related enhancement of fatigue resistance is evident in men during both isometric and dynamic tasks. *Journal of Applied Physiology*, 97 (3), 967–975.

Lanza, I.R., Towse, T.F., Caldwell, G.E., Wigmore, D.M. and Kent-Braun, J.A. (2003). Effects of age on human muscle torque, velocity, and power in two muscle groups. *Journal of Applied Physiology*, 95 (6), 2361–2369.

Lanza, I.R., Wigmore, D.M., Befroy, D.E. and Kent-Braun, J.A. (2006). In vivo ATP production during free-flow and ischaemic muscle contractions in humans. *Journal of Physiology*, 577 (Pt 1), 353–367.

Larsson, L. and Karlsson, J. (1978). Isometric and dynamic endurance as a function of age and skeletal muscle characteristics. *Acta Physiologica Scandinavica*, 104 (2), 129–136.

Larsson, L., Grimby, G. and Karlsson, J. (1979). Muscle strength and speed of movement in relation to age and muscle morphology. *Journal of Applied Physiology*, 46 (3), 451–456.

Lawrenson, L., Poole, J.G., Kim, J., Brown, C., Patel, P. and Richardson, R.S. (2003). Vascular and metabolic response to isolated small muscle mass exercise: effect of age. *American Journal of Physiology, Heart and Circulation Physiology*, 285 (3), H1023–1031.

Lennmarken, C., Bergman, T., Larsson, J. and Larsson, L.E. (1985). Skeletal muscle function in man: force, relaxation rate, endurance and contraction time – dependence on sex and age. *Clinical Physiology*, 5 (3), 243–255.

Lexell, J., Taylor, C.C. and Sjostrom, M. (1988). What is the cause of the ageing atrophy? Total number, size and proportion of different fiber types studied in whole vastus lateralis muscle from 15- to 83-year-old men. *Journal of Neurological Science*, 84 (2–3), 275–294.

Lindstrom, B., Lexell, J., Gerdle, B. and Downham, D. (1997). Skeletal muscle fatigue and endurance in young and old men and women. *The Journals of Gerontology Series A: Biological Sciences and Medical Sciences*, 52, B59–66.

McCully, K.K., Fielding, R.A., Evans, W.J., Leigh Jr, J.S. and Posner, J.D. (1993). Relationships between in vivo and in vitro measurements of metabolism in young and old human calf muscles. *Journal of Applied Physiology*, 75 (2), 813–819.

McNeil, C.J. and Rice, C.L. (2007). Fatigability is increased with age during velocity-dependent contractions of the dorsiflexors. *The Journals of Gerontology Series A: Biological Sciences and Medical Sciences*, 62 (6), 624–629.

McNeil, C.J., Doherty, T.J., Stashuk, D.W. and Rice, C.L. (2005). Motor unit number estimates in the tibialis anterior muscle of young, old, and very old men. *Muscle and Nerve*, 31 (4), 461–467.

McNeil, C.J., Vandervoort, A.A. and Rice, C.L. (2007). Peripheral impairments cause a progressive age-related loss of strength and velocity-dependent power in the dorsiflexors. *Journal of Applied Physiology*, 102 (5), 1962–1968.

Mademli, L. and Arampatzis, A. (2008a). Effect of voluntary activation on age-related muscle fatigue resistance. *Journal of Biomechanics*, 41 (6), 1229–1235.

Mademli, L. and Arampatzis, A. (2008b). Mechanical and morphological properties of the triceps surae muscle–tendon unit in old and young adults and their interaction with a submaximal fatiguing contraction. *Journal of Electromyography and Kinesiology*, 18 (1), 89–98.

Magnusson, G., Kaijser, L., Isberg, B. and Saltin, B. (1994). Cardiovascular responses during one- and two-legged exercise in middle-aged men. *Acta Physiologica Scandinavica*, 150 (4), 353–362.

Marsh, G.D., Paterson, D.H., Thompson, R.T., Cheung, P.K., MacDermid, J. and Arnold, J.M.O. (1993). Metabolic adaptations to endurance training in older individuals. *Canadian Journal of Applied Physiology*, 18 (4), 366–378.

Meijer, E.P., Goris, A.H., Wouters, L. and Westerterp, K.R. (2001). Physical inactivity as a determinant of the physical activity level in the elderly. *International Journal of Obesity and Related Metabolic Disorders*, 25 (7), 935–939.

Meredith, C.N., Frontera, W.R., Fisher, E.C., Hughes, V.A., Herland, J.C., Edwards, J. and Evans, W.J. (1989). Peripheral effects of endurance training in young and older subjects. *Journal of Applied Physiology*, 66 (6), 2844–2849.

Minotti, J.R., Pillay, P., Chang, L., Wells, L. and Massie, B.M. (1992). Neurophysiological assessment of skeletal muscle fatigue in patients with congestive heart failure. *Circulation*, 86 (3), 903–908.

Morse, C.I., Thom, J.M., Birch, K.M. and Narici, M.V. (2005). Changes in triceps surae muscle architecture with sarcopenia. *Acta Physiologica Scandinavica*, 183 (3), 291–298.

Narici, M.V., Bordini, M. and Cerretelli, P. (1991). Effect of aging on human adductor pollicis muscle function. *Journal of Applied Physiology*, 71 (4), 1277–1281.

Narici, M.V., Maganaris, C.N., Reeves, N.D. and Capodaglio, P. (2003). Effect of aging on human muscle architecture. *Journal of Applied Physiology*, 95 (6), 2229–2234.

Ng, A.V. and Kent-Braun, J.A. (1999). Slowed muscle contractile properties are not associated with a decreased EMG/force relationship in older humans. *The Journals of Gerontology Series A: Biological Sciences and Medical Sciences*, 54 (10), B452–458.

Parker, B.A., Ridout, S.J. and Proctor, D.N. (2006). Age and flow-mediated dilation: a comparison of dilatory responsiveness in the brachial and popliteal arteries. *American Journal of Physiology, Heart Circulation Physiology*, 291 (6), H3043–3049.

Pastoris, O., Boschi, F., Verri, M., Baiardi, P., Felzani, G., Vecchiet, J., Dossena, M. and Catapano, M. (2000). The effects of aging on enzyme activities and metabolite concentrations in skeletal muscle from sedentary male and female subjects. *Experimental Gerontology*, 35 (1), 95–104.

Patten, C., Kamen, G. and Rowland, D.M. (2001). Adaptations in maximal motor unit discharge rate to strength training in young and older adults. *Muscle and Nerve*, 24 (4), 542–550.

Petrella, J.K., Kim, J.S., Tuggle, S.C., Hall, S.R. and Bamman, M.M. (2005). Age differences in knee extension power, contractile velocity, and fatigability. *Journal of Applied Physiology*, 98 (1), 211–220.

Proctor, D.N., Le, K.U. and Ridout, S.J. (2005). Age and regional specificity of peak limb vascular conductance in men. *Journal of Applied Physiology*, 98 (1), 193–202.

Proctor, D.N., Sinning, W.E., Walro, J.M., Sieck, G.C. and Lemon, P.W. (1995). Oxidative capacity of human muscle fiber types: effects of age and training status. *Journal of Applied Physiology*, 78 (6), 2033–2038.

Roos, M.R., Rice, C.L., Connelly, D.M. and Vandervoort, A.A. (1999). Quadriceps muscle strength, contractile properties, and motor unit firing rates in young and old men. *Muscle and Nerve*, 22 (8), 1094–1103.

Rubinstein, S. and Kamen, G. (2005). Decreases in motor unit firing rate during sustained maximal-effort contractions in young and older adults. *Journal of Electromyography and Kinesiology*, 15 (6), 536–543.

Russ, D.W. and Kent-Braun, J.A. (2003). Sex differences in human skeletal muscle fatigue are eliminated under ischemic conditions. *Journal of Applied Physiology*, 94 (6), 2414–2422.

Russ, D.W. and Kent-Braun, J.A. (2004). Is skeletal muscle oxidative capacity decreased in old age? *Sports Medicine*, 34 (4), 221–229.

Russ, D.W., Lanza, I.R., Rothman, D. and Kent-Braun, J.A. (2005). Sex differences in glycolysis during brief, intense isometric contractions. *Muscle and Nerve*, 32 (5), 647–655.

Russ, D.W., Towse, T.F., Wigmore, D.M., Lanza, I.R. and Kent-Braun, J.A. (2008). Contrasting influences of age and sex on muscle fatigue. *Medicine and Science in Sports and Exercise*, 40 (2), 234–241.

Schwendner, K.I., Mikesky, A.E., Holt Jr, W.S., Peacock, M. and Burr, D.B. (1997). Differences in muscle endurance and recovery between fallers and non-fallers, and between young and older women. *The Journals of Gerontology Series A: Biological Sciences and Medical Sciences*, 52 (3), M155–160.

Sharma, K.R., Kent-Braun, J.A., Majumdar, S., Huang, Y., Mynhier, M., Weiner, M.W. and Miller, R.G. (1995). Physiology of fatigue in amyotrophic lateral sclerosis. *Neurology*, 45 (4), 733–740.

Sharma, K.R., Kent-Braun, J.A., Mynhier, M.A., Weiner, M.W. and Miller, R.G. (1994). Excessive muscular fatigue in the postpoliomyelitis syndrome. *Neurology*, 44, 642–646.

Short, K.R., Bigelow, M.L., Kahl, J., Singh, R., Coenen-Schimke, J., Raghavakai-mal, S. and Nair, K.S. (2005). Decline in skeletal muscle mitochondrial function with aging in humans. *Proceedings of the National Academy of Science*, 102 (15), 5618–5623.

Stackhouse, S.K., Dean, J.C., Lee, S.C. and Binder-Macleod, S.A. (2000). Measurement of central activation failure of the quadriceps femoris in healthy adults. *Muscle and Nerve*, 23 (11), 1706–1712.

Stackhouse, S.K., Stevens, J.E., Lee, S.C., Pearce, K.M., Snyder-Mackler, L. and Binder-Macleod, S.A. (2001). Maximum voluntary activation in nonfatigued and fatigued muscle of young and elderly individuals. *Physical Therapy*, 81 (5), 1102–1109.

Taylor, D.J., Kemp, G.J., Thompson, C.H. and Radda, G.K. (1997). Ageing: effects on oxidative function of skeletal muscle in vivo. *Molecular and Cellular Biochemistry*, 174 (1–2), 321–324.

Thelen, D.G., Ashton-Miller, J.A., Schultz, A.B. and Alexander, N.B. (1996). Do neural factors underlie age differences in rapid ankle torque development? *Journal of the American Geriatrics Society*, 44 (7), 804–808.

Thom, J.M., Morse, C.I., Birch, K.M. and Narici, M.V. (2005). Triceps surae muscle power, volume, and quality in older versus younger healthy men. *The Journals of Gerontology Series A: Biological Sciences and Medical Sciences*, 60 (9), 1111–1117.

Troiano, R.P., Berrigan, D., Dodd, K.W., Masse, L.C., Tilert, T. and McDowell, M. (2008). Physical activity in the United States measured by accelerometer. *Medicine and Science in Sports and Exercise*, 40 (1), 181–188.

van Schaik, C.S., Hicks, A.L. and McCartney, N. (1994). An evaluation of the length–tension relationship in elderly human ankle dorsiflexors. *Journal of Gerontology*, 49 (3), B121–127.

Vandervoort, A.A. and McComas, A.J. (1986). Contractile changes in opposing muscles of the human ankle joint with aging. *Journal of Applied Physiology*, 61 (1), 361–367.

Welle, S., Bhatt, K., Shah, B., Needler, N., Delehanty, J.M. and Thornton, C.A. (2003). Reduced amount of mitochondrial DNA in aged human muscle. *Journal of Applied Physiology*, 94 (4), 1479–1484.

Wigmore, D.M., Damon, B.M., Pober, D.M. and Kent-Braun, J.A. (2004). MRI measures of perfusion-related changes in human skeletal muscle during progressive contractions. *Journal of Applied Physiology*, 97 (6), 2385–2394.

Yoon, T., De Lap, B.S., Griffith, E.E. and Hunter, S.K. (2008). Age-related muscle fatigue after a low-force fatiguing contraction is explained by central fatigue. *Muscle and Nerve*, 37 (4), 457–466.

CHAPTER 6

SEX DIFFERENCES IN MUSCLE FATIGUE

David W. Russ

OBJECTIVES

The objectives of this chapter are:

- to present an overview of research directed at addressing the role of sex in muscle fatigue;
- to describe the potential mechanisms of fatigue most commonly studied in addressing the role of sex in muscle fatigue;
- to highlight the task- and muscle-specific nature of fatigue and how these characteristics may contribute to discrepancies with regard to the role of sex in muscle fatigue.

INTRODUCTION

Skeletal muscle fatigue has been a topic of interest to physiologists since the turn of the twentieth century (Di Giulio *et al.*, 2006). Specific examination of the role of sex in fatigue however, has occurred only in comparatively recent years, with the first study designed to systematically address this factor published in the 1970s (Petrofsky *et al.*, 1975). This initial study evaluated fatigue of submaximal, isometric (40% of maximum volitional contraction (MVC)), handgrip exercise in a fairly large number of women (n = 83) of ages 19–65 years, and compared their responses to those recorded in a similar population of men during an earlier study (Petrofsky and Lind, 1975). These investigators found that although the women were weaker than the men, they were less fatigable, a finding they reported as "unexpected".

Despite the novel findings of this seminal paper, evaluation of the role of sex in muscle fatigue was sporadic and limited over the ensuing years. This might have been due, in part, to the persistence of baseless societal attitudes that women were unable to tolerate the stress of heavy physical exertion and were unfit for participation in sports (Thein and Thein, 1996). Beginning in the mid-1980s, study of differences in physical performance between the sexes began to increase. Coincident with the increased focus on female performance were the debut of the women's marathon in the Olympic Games and the National Collegiate Athletic Association taking over the administration of intercollegiate athletics for both men and women in the United States.

As the number of investigations comparing skeletal muscle fatigue in men and women has increased, so have the attempts to isolate the mechanisms that could potentially contribute to any sex-related differences in fatigability. Despite the proliferation of research in this area, the role of sex in fatigue remains unclear. Many studies have confirmed Petrofsky et al.'s findings of greater fatigue resistance in women, but others have not (see Table 6.1). Furthermore, a consistent mechanism for observed differences in fatigue between men and women has not been established. These discrepancies are likely the result of the different experimental protocols used to study fatigue, combined with the complex, multi-factorial nature of muscle fatigue itself.

A requirement for any study of muscle fatigue is that the phenomenon be defined so that it can be measured. A myriad of operational definitions of fatigue have been used (for recent reviews, cf. Barry and Enoka, 2007; Enoka and Duchateau, 2008), with components ranging from the perceived effort to spectral changes in electrical activity of the contracting muscles. Two different definitions have been employed in the majority of the studies of sex and muscle fatigue. The first is the transient decline in maximum force-generating capacity of the muscle, associated with recent muscle activity (Bigland-Ritchie and Woods, 1984; Kent-Braun et al., 2002; Pincivero et al., 2003; Russ et al., 2008). Typically, the MVC of the muscle or muscle group being studied is used as the measure of maximum force-generating capacity. Sometimes, electrical stimulation is superimposed during the MVC to assess central activation (see the section, "Central drive"). In these studies, the principal measure is the decline in force, relative to the pre-fatigue state, during and after the fatiguing activity (Bigland-Ritchie and Woods, 1984). The second frequently used definition is the "failure to achieve or maintain the expected force" (Edwards et al., 1977). In such instances, the main outcome is the time at which such failure occurs, often referred to as the time to task failure. In

Table 6.1 Summary listing of sex and fatigue studies

Study	Subject/sample size	Muscle/muscle groups	Fatiguing task	Fatigue	Other measurements
Petrofsky et al. (1975)	100 M (18–65 yrs) and 83 W (19–65 yrs), healthy subjects	Wrist and finger flexors and extensors	Sustained 40% handgrip MVC to failure	**M > W**	HR↑: M < W BP ↑: M = W Cardiac work: M = W
Maughan et al. (1986)	25 M (26 yrs) and 25 W (25 yrs), healthy subjects	Knee extensors Elbow flexors	Sustained 20%, 50% and 80% knee extensor isometric MVC to task failure Dynamic elbow flexion at 50%, 60%, 70%, 80% and 90%, 1-RM to task failure	Knee extensors: **M > W** at 20%; M = W at 50% and 80% Elbow flexors: **M > W** at 50–70%; M = W at 80% and 90%	
West et al. (1995)	7 M and 7 W, (22 ± 3 years), all healthy	Wrist and finger flexors and extensors	Sustained 30%, 50% and 75% handgrip MVC to failure	**M > W** at all three intensities	Average rectified EMG↑: M = W Baseline strength: unrelated to fatigue
Hicks and McCartney (1996)	44 M (60–80 years) and 65 W (60–80 years)	Elbow flexors Ankle dorsiflexors	3 min intermittent isometric MVC (5 s contract, 2 s rest)	Elbow flexors: **M > W** Ankle dorsiflexors: **M > W**	Increasing age: Sex difference in fatigue maintained ↓ Supramaximal twitch force: M > W
Fulco et al. (1999)	9 M (29 years) and 9 W (24 years), strength matched, activity matched by questionnaire	Adductor pollicis	Intermittent isometric at 50% MVC (5 s contract, 5 s rest), to task failure	**M > W**	Recovery from fatigue: M < W

Continued

Table 6.1 Continued

Study	Subject/sample size	Muscle/muscle groups	Fatiguing task	Fatigue	Other measurements
Ditor and Hicks (2000)	24 M [12 young (25 years) and 12 older (72 years)] and 24 W [12 young (23 years) and 12 older (70 years)]	Adductor pollicis	3 min intermittent isometric MVC (5 s contract, 2 s rest)	M=W (trend for M>W (P=0.06))	Baseline HRT: M<W ΔHRT: M=W ↓ Supramaximal twitch force: M=W Δ CMAP: ↓ M>↓W
Fulco et al. (2001)	12 M (29 years) and 21 W (22 years), strength matched, activity matched by questionnaire	Adductor pollicis	Intermittent isometric at 50% MVC (5 s contract, 5 s rest), to task failure Normoxic and hypoxic conditions	Normoxia: **M>W** Hypoxia: **M>W** (Hypoxia increased fatigue for M but not W)	Hypoxic Δ V̇O₂: M=W Baseline Strength: unrelated to fatigue
Hunter and Enoka (2001)	7 M (26 years) and 7 W (27 years)	Elbow flexors	Sustained isometric at 20% MVC to task failure	**M>W**	HRT↑: amplitude, M>W; rate M>W MAP↑: amplitude, M>W; rate M>W MVC↓: M=W Average rectified EMG↑: M>W Torque fluctuation: M>W RPE: M=W Differences in fatigue, HR and MAP were eliminated when baseline load was co-varied

Study	Subjects	Muscle group	Protocol	Result	Findings
Kent-Braun et al. (2002)	21 M (ten 25–45 years; ten 65–85 years) and 20 W (ten 25–45 years; ten 65–85 years). Activity matched by accelerometry	Ankle dorsiflexors	16 min intermittent, isometric contractions (4 s contract, 6 s relax), starting at 10% MVC, increasing by 10% every 2 min	M = W; O < Y; No age × sex interaction	Δ Central activation: M = W; Δ CMAP: M = W; End-exercise inorganic phosphate: M > W; End-exercise pH: M > W
Clark et al. (2003)	10 M (22 years) and 10 W (22 years), healthy, "recreationally active"	Back extensors	Sustained isometric at 50% MVC to task failure; Repetitive isotonics at 50% MVC load to task failure	Isometric: **M > W**; Isotonic: M = W	EMG: median frequency shift M > W; Muscle activation pattern: M = W; Differences in fatigue were eliminated when baseline strength was co-varied
Pincivero et al. (2003)	19 M (25 years) and 20 W (24 years), healthy, "recreationally active"	Knee extensors; Knee flexors	Reciprocal isokinetic knee extension/flexion at 3.14 rad•s^{-1}	Knee extensors: **M > W**; Knee flexors: **M > W**	Differences in fatigue eliminated when baseline strength was co-varied
Cheng et al. (2003)	12 M (70 years) and 12 W (70 years), all "active seniors"	Adductor pollicis	Intermittent isometric at 50% MVC (5 s contract, 5 s rest), to task failure	M = W	Δ Central activation: M = W; Δ Supramaximal twitch force: M = W; Baseline strength: unrelated to fatigue

Continued

Table 6.1 Continued

Study	Subject/sample size	Muscle/muscle groups	Fatiguing task	Fatigue	Other measurements
Russ and Kent-Braun (2003)	8 M (27 years) and 8 W (24 years), all healthy, activity matched by questionnaire	Ankle dorsiflexors	4 min intermittent isometric MVC (5 s contract, 5 s rest); free-flow and ischemic conditions	Free-flow: **M > W** Ischemia: M = W	Δ Central activation: M > W in free-flow; M = W in ischemia Δ CMAP: M = W in free-flow and ischemia Δ Stimulated force: W > M for twitch and 10 Hz, M = W for 50 Hz in free-flow; M = W for twitch, 10 Hz and 50 Hz in ischemia LFF: M = W in free-flow and ischemia
Hunter et al. (2004a)	10 M (22 years) and 10 W (22 years) Strength matched, activity assessed by questionnaire	Elbow flexors	Sustained, 20% isometric MVC to task failure;	M = W	HR↑: M = W MAP↑: M = W MVC↓: M = W Average rectified EMG↑: M > W EMG burst rate: M < W Torque fluctuation: M = W RPE: M = W

Continued

Study	Participants	Muscle	Task	Result	Findings
Hunter et al. (2004b)	10 M (21 years) and 10 W (21 years) Strength and activity matched	Elbow flexors	Intermittent, isometric at 50% MVC (6 s contract, 4 s rest) to task failure	**M > W**	HR↑: amplitude, M = W; rate M > W MAP↑: amplitude, M = W; rate M > W MVC↓: M = W Average rectified EMG↑: M > W Torque fluctuation: M > W RPE: M = W
Pincivero et al. (2004)	15 M (25 years) and 15 W (22 years), all physically active	Knee extensors	Repeated isotonic exercise at 50% 1-RM, to task failure	M = W	RPE: M = W
Clark et al. (2005)	11 M (26 years) and 11 W (24 years), healthy, activity matched by questionnaire	Knee extensors	Sustained, 25% isometric MVC to task failure; free-flow and ischemic conditions	Free-flow: **M > W** Ischemia: M = W	Average rectified EMG↑: M > W Torque fluctuation: W > M EMG at task failure correlated to time to fatigue in Free-flow and ischemia Baseline strength unrelated to fatigue

Table 6.1 Continued

Study	Subject/sample size	Muscle/muscle groups	Fatiguing task	Fatigue	Other measurements
Russ et al. (2005)	6 M (26 years) and 6 W (27 years), healthy, "recreationally active"	Ankle dorsiflexors	60 s sustained, isometric MVC	M = W	Δ Central activation: M = W Intramuscular glycolysis: M > W Intramuscular OxPhos: M = W Intramuscular PCr breakdown: M = W End-exercise pH: M > W Oxidative capacity: M = W
Lariviere et al. (2006)	16 M (39 years) and 15 W (31 years), healthy, activity determined by questionnaire	Back extensors	Intermittent, isometric 40% MVC (6.5 s contract; 1.5 s relax) to task failure	**M > W**	Alternation of synergist muscle activity: M < W Differences in fatigue eliminated when baseline strength was co-varied

Study	Participants	Muscle group	Task	Result	Details
Hunter et al. (2006a)	16 M (22 years) and 18 W (22 years); Subset of strength matched M and W (7 and 7, respectively)	Elbow flexors	Sustained, 20% isometric MVC to task failure	Overall: **M > W** Strength-matched: M = W	HR↑: amplitude, M = W; rate M > W MAP↑: M = W Active hyperemia: M = W; M > W at baseline Vascular conductance: M = W; M > W at baseline RPE: M = W, rate of increase: M > W
Hunter et al. (2006b)	9 M (25 years) and 8 W (26 years), healthy, activity determined by questionnaire	Elbow flexors	6 intermittent MVCs (22 s contract; 10 s relax)	**M > W**	Cortical activation (TMS): M = W Δ Central activation: M = W ↓ Supramaximal twitch relaxation: M > W
Gonzales and Scheuermann (2007)	11 M (24 years) and 11 W (23 years), healthy, strength-matched subsets of M and W (n = 5 each)	Wrist and finger flexors and extensors	Intermittent, isometric 50% handgrip MVC to task failure	Overall groups: M = W Strength-matched groups: M = W	
Jacobs et al. (2007)	15 M (24 years) and 15 W (23 years)	Hip abductors	Sustained isometric at 50% MVC to task failure	**M > W**	

Continued

Table 6.1 Continued

Study	Subject/sample size	Muscle/muscle groups	Fatiguing task	Fatigue	Other measurements
Martin and Rattey (2007)	8 M (23 years) and 8 W (21 years), healthy	Knee extensors	100 s, sustained, maximum isometric MVC	**M > W**	Δ Central activation: M > W Δ CMAP: M = W Δ Force/EMG relationship: M = W ↓ Supramaximal twitch force: M = W ΔHRT: M = W ΔRate of force development: M = W
Thompson et al. (2007)	18 M (23 years) and 20 W (23 years), all healthy	Wrist and finger flexors and extensors	Sustained, isometric handgrip at 20% MVC to failure Sustained, isometric handgrip at 50% MVC to failure	20% MVC: **M > W** 50% MVC: M = W	Relative forearm blood flow: 20% M = W; 50% M < W Absolute forearm blood flow: 20% M > W; 50% M = W Integrated EMG: 20% M = W; 50% M = W

| Yoon et al. (2007) | 9 M (22 years) and 9 W (22 years), healthy, physical activity determined by questionnaire | Elbow flexors | Sustained, isometric 20% MVC to failure
Sustained, isometric 80% MVC to failure | 20% MVC: **M > W**
80% MVC: M=W | Δ Central activation: M=W
↓ Supramaximal twitch force: M=W
Δ Force/EMG relationship: M=W
RPE↑: amplitude M=W; rate M>W
MAP↑: amplitude M>W; rate M>W; predicted fatigue for 20% MVC only
Torque fluctuation: 20% M>W; 80% M=W |
| Russ et al. (2008) | 15 M [8 young (24 years) and 7 older (74 years)] and 17 W [8 young (25 years) and 7 older (73 years)]; physical activity determined by questionnaire | Ankle dorsiflexors | 5 min intermittent isometric MVC (7 s contract; 3 s rest) | M=W
No age × sex interaction | Δ Central activation: M=W
Δ CMAP: M=W
Δ Stimulated force: M>W |

Notes
Results showing greater fatigue of men versus women in bold
M = men; W = women.

some instances, this definition of fatigue is also referred to as "muscular endurance", with greater endurance associated with a greater time to task failure. In each case, the typical experiment involves measuring various parameters thought to represent potential mechanisms of fatigue in men and women before and after a specific exercise task. Baseline differences and exercise-associated changes in the different parameters between men and women are then related to any observed sex-related differences in fatigue. Some studies combine both definitions by periodically assessing MVC during a submaximal exercise task (Cheng *et al.*, 2003).

The study of fatigue is made more difficult by the fact that the extent of fatigue and the mechanisms driving it vary with the nature of the fatiguing task or exercise. Thus, the type of fatiguing exercise chosen, the selected method of measuring fatigue and the physical fitness of the subjects tested are likely to influence the outcome of any study aimed at determining the influence of sex on muscle fatigue. This task-specific nature of fatigue, combined with continued use of differing definitions of fatigue, clearly contributes to the discrepancies in the literature. Recently, it has been suggested that studies using the second definition of fatigue use the term "task failure" rather than "fatigue" (Barry and Enoka, 2007). The same group advocates the use of task failure studies, rather than fatigue, as most functional tasks do not involve maximal contractions.

SEX AND MUSCLE FATIGUE

Over the past 30-plus years, the majority of studies have corroborated the greater fatigability of men versus women, as reported in Petrofsky's seminal paper (Petrofsky *et al.*, 1975), under at least some conditions (see Table 6.1). This finding is by no means universal however, with a substantial number of investigators reporting no difference in fatigue between men and women (Table 6.1). Likely contributing to these discrepancies are the different definitions of fatigue, the different fatigue tasks performed, the different muscles studied and the potential confounding factor of physical activity. One important factor to consider when studying sex-related differences in exercise is the potential effect of the menstrual phase in the female subjects. A number of excellent studies have demonstrated that the menstrual phase can affect exercise performance; however, this topic has not received as much attention with regard to the study of isolated muscle fatigue. One such study reported that no effect of the menstrual phase in healthy young women was found on fatigue produced by 50 dynamic knee extensions performed at $120°·s^{-1}$ (Friden *et al.*, 2003). Furthermore, greater fatigue resistance has been observed in post-menopausal women relative to age-matched men (Hicks and McCartney, 1996; Lindstrom *et al.*, 1997). Nevertheless, the relative lack of data regarding this subject should serve as a caution against considering the role of the menstrual cycle in sex-related differences in muscle fatigue closed.

Several factors have been implicated as causative agents in muscle fatigue, and they are addressed in detail elsewhere in this text. In the following sections, a select few of these mechanisms are addressed with regard to their potential roles in differences in fatigue between men and women. This list is not exhaustive, but does focus on those mechanisms that have been examined most often in studies of sex and muscle fatigue. Perhaps predictably, there are a number of conflicting studies in the existing literature. As noted earlier, the task specificity of muscle fatigue is likely a major contributor to the varying results reported over the years.

Strength and muscle perfusion

Differences in muscle mass and strength between men and women were among the earliest mechanisms suggested to account for greater resistance to fatigue in women (for a review, see Hicks *et al.*, 2001). Supporting this hypothesis is the fact that several studies show that the greater fatigue seen in men versus women is eliminated if subjects are matched with regard to strength (Hunter *et al.*, 2004b; Hunter *et al.*, 2006b) or if baseline strength is used as a covariate in data analysis (Hunter and Enoka, 2001; Clark *et al.*, 2003; Pincivero *et al.*, 2003; Lariviere *et al.*, 2006). The principal mechanism cited to support the role of strength in differences in fatigue between the sexes is muscle perfusion. Muscle contraction produces mechanical compression of the vascular supply, with the degree of compression increasing with increasing contraction intensity (Barnes, 1980; Wigmore *et al.*, 2006). It has been suggested that men, who typically have larger muscles than women, produce greater absolute forces during a given exercise task than women, resulting in a greater intramuscular pressure and hence, a higher degree of vascular occlusion. This greater occlusion has been suggested to accelerate fatigue in men. Because both men and women would be expected to generate sufficient forces to produce vascular occlusion during high-force tasks, this mechanism is also proposed to account for the observation that the sex-related difference in muscle fatigue is greater during exercise protocols involving low percentages of MVC (Hicks *et al.*, 2001).

However, other studies report that men fatigue more than women even if they are strength-matched (Fulco *et al.*, 1999; Hunter *et al.*, 2004c). Post hoc separation of subjects in a study showing greater fatigue in men versus women found that sub-samples of strength-matched men and women exhibited similar changes in fatigue and central activation. Moreover, sub-samples that maximised the strength difference between men and women showed similar differences to the strength-matched sub-samples (Figure 6.1). Together, these findings are consistent with several of the general observations related to fatigue and sex. It generally appears that sex-related differences in muscle fatigue generated by intermittent contractions or during sustained contractions at greater than 50% MVC are not influenced by differences in muscle strength. These findings are consistent with the perfusion hypothesis, as the lower absolute forces in the submaximal protocols and the rest periods between contractions during intermittent contractions likely allow sufficient reperfusion to maintain intramuscular oxygen levels during exercise (Wigmore *et al.*, 2006). It is worth noting, however, that a recent paper examining sustained contractions of the forearm muscles at both 20% and 50% of MVC found no difference in muscle blood flow between men and women at 20%, despite significantly greater time to task failure in women (Thompson *et al.*, 2007). In contrast, no difference in time to task failure occurred with the 50% MVC protocol, although the women actually exhibited greater muscle blood flow. Another laboratory has reported similar findings in the forearm muscles (Hunter *et al.*, 2006b), while a third found that changes in sex-related differences in

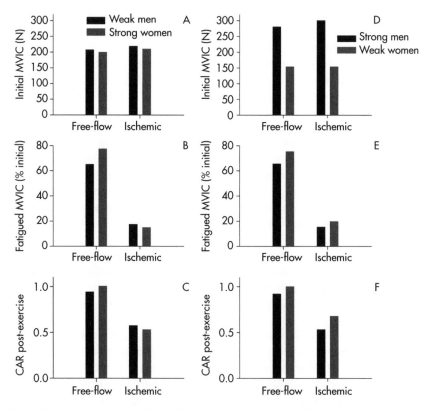

Figure 6.1 Strength-matched (A, B, C) and mismatched (D, E, F) men and women (n = 3/ group) from previously published data (Russ and Kent-Braun, 2003). Regardless of strength differences (A and D), sub-samples exhibited similar responses to the entire study sample, with greater fatigue (B and E) and central activation impairment (C and F) in men versus women

fatigue of the forearm muscles following immobilisation were not related to changes in force production (Clark *et al.*, 2008). Thus, while muscle strength does appear to play some role in sex-associated differences in muscle fatigue, particularly during sustained, submaximal contractions, the proposed mechanisms related to muscle perfusion remain questionable.

Muscle metabolism

Muscle perfusion can be viewed as one aspect of the metabolic supply in the supply–demand relationship of muscle contraction. Another aspect of the supply side of this function is metabolic substrate utilisation. Muscular ATP consumption is related to the overall myosin ATPase isoform composition of the contracting muscle (Sieck *et al.*, 1998). A few studies have suggested that muscles of men may contain a slightly greater amount of fast, type II myosin, as indicated by differences in the type II:type I fibre area ratio (Simoneau *et al.*, 1985; Jaworowski *et al.*, 2002; Holmback *et al.*, 2003). As the type II fibres have greater ATPase rates (Sieck *et al.*, 1998), this might be expected to elevate the rate of ATP consumption in men versus women, to some degree. Numerous studies show that ATP is well maintained during even high-intensity muscular activity (Kent-Braun *et al.*, 2002; Russ *et al.*, 2002; Lanza *et al.*, 2005), so the pathways of ATP re-synthesis are sufficient to compensate for any potential differences in metabolic demand between sexes. Perhaps of greater importance then, are several studies of *in vitro* enzyme activity, whole-body exercise and *in vivo* whole-muscle metabolism that report greater muscular glycolytic activity in men compared to women, and enhanced muscle oxidation of carbohydrate and lipid in women compared to men (Green *et al.*, 1984; Simoneau *et al.*, 1985; Coggan *et al.*, 1992; Hargreaves *et al.*, 1998; Esbjornsson-Liljedahl *et al.*, 1999; Jaworowski *et al.*, 2002; Kent-Braun *et al.*, 2002; Melanson *et al.*, 2002; Russ *et al.*, 2005). If myosin fibre type composition represents the metabolic demand of contraction and the metabolic enzyme activities and pathway utilisation represents metabolic supply, then it might be inferred that men could theoretically generate a greater mismatch between metabolic demand and oxidative supply than women. One potential measure of such a mismatch that has been proposed is the succinate dehydrogenase (SDH):ATPase ratio (Gregory *et al.*, 2005), although this analysis has not yet been applied to a comparison of men and women.

A greater utilisation of anaerobic glycolytic pathways of ATP synthesis by men has potential implications for sex-related differences in fatigue, as these non-oxidative pathways are associated with generation of metabolic by-products, including proton and lactate, that are frequently associated with increased fatigue (Nosek *et al.*, 1987; Kent-Braun, 1999; Dahlstedt *et al.*, 2001). Interestingly, greater anaerobic glycolysis in men versus women has been observed even when the men and women had comparable muscular oxidative capacities and utilised similar percentages of their capacities during a given task (Russ *et al.*, 2005). Thus, it does not appear that men activate their anaerobic pathways to compensate for deficits in the ability to replenish ATP via oxidative pathways. One possible mechanism contributing to the elevated glycolysis in men is a greater exercise-induced increase in epinephrine (Braun and Horton, 2001). Greater epinephrine levels and glucose fluxes in men versus women has been shown to occur, but only in prolonged, moderate cycling exercise, without assessment of fatigue (Horton *et al.*, 2006). Regardless of the specific mechanism, men do tend to exhibit greater acidosis and produce more lactate than women when performing comparable relative tasks (Esbjornsson-Liljedahl *et al.*, 1999; Russ *et al.*, 2005). Furthermore, a potential role for sex differences in metabolism is suggested by observations that the female resistance to fatigue is abolished under ischemic conditions (Russ and Kent-Braun, 2003; Clark *et al.*, 2005). These data suggest that women may have what we have previously termed an "oxidative advantage" (Russ and Kent-Braun, 2003) relative to men, whereby they are better able to utilise oxidative metabolism to support muscular activity. Another study found that the fatigue resistance of women relative to men was enhanced under hypoxic conditions (Fulco *et al.*, 2001). While this result would seem to contradict the ischemia studies, it is quite consistent with the concept of an oxidative advantage. In a reduced-oxygen environment, an oxidative advantage would enable one to maintain a higher level of activity, whereas it would be of no use in an oxygen-eliminated situation. This concept is supported by the finding that hypoxia induces greater fatigue and cellular dysfunction in rat muscles composed primarily of fibres with low versus high oxidative capacity (Howlett and Hogan, 2007).

From the standpoint of force generation at the level of the muscle itself, it appears unlikely that increases in proton (reduction in pH) or lactate would have much effect on the isometric tasks commonly used to evaluate human muscle fatigue. Lactate has not been shown to impair force production, and though reduced pH has been shown to impair isometric force production *in vitro*, the effect is temperature-dependent and is essentially absent at physiological temperatures (Pate *et al.*, 1995; Wiseman *et al.*, 1996). The effect of inorganic phosphate (P_i) accumulation has been linked to muscle fatigue, particularly the diprotonated form ($H_2PO_4^-$) (Nosek *et al.*, 1987; Kent-Braun, 1999; Kent-Braun *et al.*, 2002), which predominates at lower pH. The chief contributor to the production of P_i, assuming ATP is maintained, is the breakdown of phosphocreatine (PCr). Differences in the rate of PCr breakdown between men and women have not been demonstrated consistently however, although the greater acidosis typically observed in men would be expected to produce greater levels $H_2PO_4^-$ in men relative to women, given comparable levels of P_i. A temperature-dependence of the effects of P_i on force production similar to that described for pH has more recently been reported, however (Debold *et al.*, 2004), further complicating the issue. Recent data from the laboratory of Fitts has revived interest in the role of pH in fatigue by demonstrating negative effects of low pH on peak power production during dynamic contractions (Knuth *et al.*, 2006).

In addition, pH has been linked to impairments of calcium sensitivity of the muscle fibres (Godt and Nosek, 1989; Westerblad *et al.*, 2000). This type of dysfunction could contribute to fatigue by increasing the neural drive needed to maintain a given level of force, either by requiring greater motor unit recruitment or increased rate coding for already active motor units (Westerblad *et al.*, 2000). Some investigators have linked the phenomenon of low-frequency fatigue (LFF) to this type of impairment, but only two papers to date have reported on the role of sex in LFF (Russ and Kent-Braun, 2003; Martin *et al.*, 2004). Another avenue by which sex-related differences in muscle metabolism could affect neural activation during fatigue is through reflex inhibition. Both proton and lactate have been shown to activate metaboreceptors of Group III and IV afferent nerve fibres (Rotto and Kaufman, 1988; Rotto *et al.*, 1989). These fibres are thought to inhibit alpha motoneurons both at the spinal level and through activation of supraspinal structures (Rotto and Kaufman, 1988; Garland, 1991). Direct assessment Group III and IV-mediated inhibition of alpha motoneurons is not feasible in human studies of muscle fatigue. However, because activation of these afferents has also been shown to mediate the metaboreflex during exercise (Rowell and O'Leary, 1990; Ettinger *et al.*, 1996), some investigators have used changes in heart rate and mean arterial pressure to infer sex-related differences in neural feedback that could potentially influence central drive (Table 6.1).

As with other aspects of muscle fatigue, the role of potential metabolic differences between men and women is likely modified by the specific fatigue protocol employed in a given study. Because of the well-described Fenn Effect (Rall, 1982), dynamic, shortening contractions are more metabolically costly than isometric ones. Moreover, isometric force generation has been shown to be more metabolically demanding than maintaining that force. Thus, ATPase activity declines over the course of an isometric contraction, even if force is maintained (Russ *et al.*, 2002; West *et al.*, 2004). The metabolic demand of intermittent isometric contractions is therefore likely to be greater than that of sustained isometric contractions, assuming equal forces and/or work are performed. Thus, sex-related differences in anaerobic glycolysis might be expected to play more of a role in intermittent versus sustained protocols.

Central drive

While intramuscular factors certainly contribute to fatigue, as evidenced by the observation of fatigue under conditions in which the central nervous system (CNS) is not a factor (e.g. electrical stimulation), the influence of central drive on fatigue during volitional activity is undeniable. It is therefore not surprising that a number of experiments have examined the possible role of CNS activation of muscle in sex-related differences in muscle fatigue (Cheng *et al.*, 2003; Russ and Kent-Braun, 2003; Russ *et al.*, 2005; Hunter *et al.*, 2006a; Martin *et al.*, 2006; Yoon *et al.*, 2007; Russ *et al.*, 2008). These studies typically attempt to directly assess failure of central activation by one of two techniques: burst-superimposition or twitch interpolation. In the first, a short, high-frequency train of supramaximal pulses is delivered to the muscle(s) of interest via the peripheral nerve during an isometric MVC (Kent-Braun and Le Blanc, 1996). In the second, a supramaximal twitch is delivered to the muscle(s) via the peripheral nerve with the muscle at rest and during an isometric MVC (McKenzie *et al.*, 1992). The amplitude of the superimposed twitch is compared to the resting twitch. Although debate over the superiority of one method versus the other has persisted (Kent-Braun and Le Blanc, 1996; Behm *et al.*, 2001), the two methods actually share much in common. The key common element is the assumption that the electrical stimulation will increase force production in any muscle that is not fully recruited or being activated with a sufficiently high degree of rate coding. Both methods are typically used under isometric conditions, as attempts to apply them during non-isometric conditions have met with limited success, particularly at angular velocities of greater than $20°·s^{-1}$ (Newham *et al.*, 1991; Beelen *et al.*, 1995). Finally, both methods rely on the presence of an intact, healthy peripheral nervous system, without which the possibility of "false negatives", with regard to central activation deficits, exists (Scott *et al.*, 2007).

In a series of studies of fatigue in the ankle dorsiflexor muscles, one laboratory demonstrated that fatigue resistance was similar between men and women in submaximal, intermittent, isometric exercise of progressively increasing intensity (Kent-Braun *et al.*, 2002), in brief (60s) sustained MVCs (Russ *et al.*, 2005), and in intermittent MVCs with a 70% duty cycle (Russ *et al.*, 2008). Deficits in central activation observed with fatigue were similar for men and women in each of these studies. In contrast, this same group found that men fatigued more than women during intermittent MVCs with a 50% duty cycle, and that this difference was associated with greater failure of central activation in men (Russ and Kent-Braun, 2003). When the same subjects performed the fatigue protocol under ischemic conditions, the sex-related differences in fatigue and central activation were abolished. Together, these findings highlight the task-specific nature of fatigue, as a change as simple as altering the contraction duty cycle from 50% to 70% markedly altered the difference in fatigue between men and women. These data further suggest a role for central activation failure in sex-related differences in fatigue, when they occur. Because of the sensitivity of the central activation impairment to ischemia, it has been suggested that differences in the exercise-induced decline in pH could account for the differences in central activation (see the section, Muscle metabolism, above).

In recent years, the technique of transcranial magnetic stimulation

(TMS) has been used to examine the effects of central drive in fatigue. This technique uses pulsed electromagnetic fields to activate localised regions of the motor cortex, producing motor-evoked potentials that are recorded from the target muscles. Reductions in motor-evoked potentials during fatigue are taken as an indication that cortical activation is impaired (Martin *et al.*, 2006). Thus, it allows one to evaluate supraspinal contributions to fatigue. One recent paper has employed this technique to examine the influence of sex on fatigue, and demonstrated no differences in central activation impairments during intermittent, relatively long (22 s) elbow flexion MVCs, despite greater fatigue in men versus women (Hunter *et al.*, 2006a). It is worth noting that this method could not determine the degree to which any central activation failure occurred at the level of the spinal cord relative to supraspinal centres. Nevertheless, the results of that study reflect the fact that central activation differences between men and women are not universal and that sex-related differences in fatigue can occur without them. In addition, recent findings suggest that human extensor muscles appear to be much more vulnerable than flexor muscles to afferent inhibition of force production (Martin *et al.*, 2006). Thus, the different results reported in the elbow flexors and dorsiflexor muscles (which are embryologically extensors (Dudek and Fix, 2004)) may well be a result of muscle-specificity, as well as task-specificity, of fatigue mechanisms.

Central drive is related to sense of effort, and in many exercise models this is assessed through the use of scales of Ratings of Perceived Exertion (RPE) (Borg, 1962, 1982). Far fewer studies of isolated muscle fatigue incorporate the RPE than do more global exercise studies (running, cycling, etc.), and fewer still have specifically addressed the role of sex. Among the studies of sex and muscle fatigue that incorporate RPE, most have assessed task failure during submaximal contractile protocols. In these studies, it appears that perceived exertion in men and women is comparable whether or not differences in muscle fatigue between sexes are observed (see Table 6.1). However, one group has observed that men tend to exhibit a greater rate of increase in RPE, even when the final ratings are comparable in men and women (Hunter *et al.*, 2006a; Yoon *et al.*, 2007), and that the more rapid increase is associated with greater fatigue. It is important to note that peripheral, as well as central, factors are thought to contribute to RPE (Cafarelli, 1982; Robertson, 1982), and strong – though non-linear – relationships between RPE and muscle and blood lactate have been observed (Noble *et al.*, 1983). Thus, the more rapid rate of increase in perceived exertion could be related, at least in part, to the fibre type and metabolic differences noted in the preceding section.

Peripheral excitation

Changes in the compound muscle action potential (CMAP), or M wave, are typically used to assess changes in the excitability of the neuromuscular junction and sarcolemma associated with fatigue. Because the CMAP is electrically elicited, it is thought to bypass the CNS. Thus, any changes observed are typically attributed to a peripherally mediated reduction in motor unit activation. However, a recent paper describing the various factors that can influence the CMAP highlights the difficulties of assigning a direct mechanism to any observed changes in the CMAP (Keenan *et al.*, 2006). Whatever the cause of changes in the CMAP might be, sex-specific differences are not typically observed with fatigue (Russ and Kent-Braun, 2003; Martin and Rattey, 2007; Russ *et al.*, 2008). One study has reported greater reduction of CMAP amplitude in men versus women (Ditor and Hicks, 2000). Of note, these investigators did not report CMAP area. Maintenance of integrated CMAP area with reduced peak-to-peak amplitude is thought to indicate that conduction velocity may have slowed to different degrees across different motor units, without an overall loss of activation (Thomas *et al.*, 1989; Keenan *et al.*, 2006). Thus, it is possible that peripheral excitability was maintained despite the reduction in CMAP amplitude, which would be consistent with the absence of an observed difference in fatigue between men and women in the study (see Table 6.1).

SEX AND AGE IN MUSCLE FATIGUE

The specific role of age in muscle fatigue is discussed elsewhere in this text (Chapters 4 and 5). However, a few studies have examined age and sex simultaneously. In the dorsiflexor muscles it appears that the effect of age on fatigue is more robust than that of sex across a greater variety of tasks and conditions (Russ and Kent-Braun, 2003; Lanza *et al.*, 2004; Lanza *et al.*, 2005; Russ *et al.*, 2005; Russ *et al.*, 2008). Interestingly, several studies have demonstrated no age–sex interaction with regard to fatigue (Hicks and McCartney, 1996; Ditor and Hicks, 2000; Kent-Braun *et al.*, 2002; Russ *et al.*, 2008), but at least one group has reported that age exacerbates fatigue more in men than in women (Hunter *et al.*, 2004a). As the complexity of the separate effects of sex and age becomes increasingly apparent, it is no surprise that much more work will be needed to address any potential interactions between these two factors.

CONCLUSION

Since Petrofsky's initial findings of greater fatigability in men than women (Petrofsky *et al.*, 1975), many, though not all, subsequent studies have described similar sex-related differences. Attempts to isolate the mechanisms contributing to the relative fatigue resistance of women versus men have been complicated by the fact that any sex-related differences in fatigue are certainly influenced by the fitness and habitual physical activity of the subjects studied, the nature of the fatiguing task, the muscle studied and probably several other factors. For example, it has been demonstrated that sex-related differences in fatigue of the dorsiflexor muscle group, when they occur, are associated with differences in central activation (Russ and Kent-Braun, 2003) and when they do not, no differences in central activation are present (Kent-Braun *et al.*, 2002; Russ and Kent-Braun, 2003; Russ *et al.*, 2005; Russ *et al.*, 2008). This would suggest that better maintenance of central activation plays a key role in the fatigue resistance of women. However, greater fatigue of men versus women has been observed in the elbow flexors without any differences in central activation (Hunter *et al.*, 2006a; Yoon *et al.*, 2007). An attractive hypothesis for reconciling these differences is related to recent findings that extensor muscles are more susceptible to central activation impairment than flexors, although the question as to the mechanism contributing to the sex-related differences in fatigue of flexor muscles remains. However, the exercise protocols across the different studies varied with regard to contraction intensity, duty cycle and endpoint (task failure versus fixed time). It is worth noting that the dorsiflexor studies also performed the fatiguing task for set times, whereas the elbow flexor studies were performed until task failure.

While the bulk of data to date suggest that men are more fatigable than women, further research is clearly needed to address the issue of sex and muscle fatigue. A number of mechanisms appear to contribute to sex differences in muscle fatigue in specific circumstances, but none has yet been shown to play a role in every situation. Given the now well-established task-specificity of fatigue, this is perhaps not surprising. Furthermore, it now appears as though the role of sex in muscle fatigue may well be muscle-specific. Thus, future studies should probably address at least two different muscle groups within subjects to address this factor. As the study of fatigue continues to evolve, it is likely that more complex experiments aimed at uncovering the interactions among putative mechanisms will be needed to move the field forward.

FIVE KEY PAPERS THAT SHAPED THE TOPIC AREA

Study 1. Petrofsky, J.S., Burse, R.L. and Lind, A.R. (1975). Comparison of physiological responses of women and men to isometric exercise. *Journal of Applied Physiology*, 38, 863–868.

This study reported on a fairly large subject pool (83 women) and compared their data against those of 100 men tested in a previous study (Petrofsky and Lind, 1975). The investigators found that women exhibited a greater time to task failure (~23% greater) than men during a sustained handgrip exercise at 40% MVC. This difference persisted across a wide age range (19–65 years), and was coincident with greater handgrip strength in the men than the women, a difference that also persisted across the age range. This study was among the first to report greater muscular endurance in women versus men and the first to report it in such a large sample size. Furthermore, the findings of increased fatigue resistance with decreased strength in women led to the hypothesis which served as the basis for much subsequent research, that the greater muscular forces produced by men contributed to increased fatigue by limiting muscle perfusion during contraction.

Study 2. Green, H.J., Fraser, I.G. and Ranney, D.A. (1984). Male and female differences in enzyme activities of energy metabolism in vastus lateralis muscle. *Journal of Neurological Science*, 65, 323–331.

This paper reported on the sex-related differences of several important metabolic enzymes. Although a few previous studies had conducted similar experiments, this paper made a point of reporting ratios of certain enzymatic activities that are thought to be better indices of metabolic pathway utilisation than isolated enzymatic activities alone. Although this paper did not directly examine fatigue, the findings that women had reduced glycolytic potential and proportionately greater capacity for lipid oxidation, relative to men, provided support for a metabolic mechanism underlying sex-related differences in fatigue.

Study 3. Fulco, C.S., Rock, P.B., Muza, S.R., Lammi, E., Cymerman, A., Butterfield, G., Moore, L.G., Braun, B. and Lewis, S.F. (1999). Slower fatigue and faster recovery of the adductor pollicis muscle in women matched for strength with men. *Acta Physiologica Scandinavica*, 167, 233–239.

These investigators directly tested the assumption that greater fatigue in men versus women was due to the greater absolute forces typically generated by men. By matching men and women for isometric strength in the muscle of interest (adductor pollicis), the authors were able to control for this variable. Subjects performed intermittent contractions at 50% MVC, with full MVC tested every minute. Men exhibited greater rates of decline in MVC force, a reduced time to task failure (maintenance of 50% MVC) and a slower recovery of MVC. Although no other physiological data (EMG, central activation, etc.) were collected during the study, the use of strength matching was novel in the study of sex-related differences in fatigue, and the results indicated that differences in strength could not explain differences in fatigue, at least in some muscles or under certain conditions.

Study 4. Russ, D.W. and Kent-Braun, J.A. (2003). Sex differences in human skeletal muscle fatigue are eliminated under ischemic conditions. *Journal of Applied Physiology*, 94, 2414–2422.

Based on earlier work suggesting that women were better suited than men to support the metabolic demand of muscle contraction with oxidative metabolic pathways, these authors hypothesised that the fatigue resistance of women versus men would disappear under ischemic conditions. Healthy, young men and women performed 4 min of intermittent, isometric MVCs of the ankle dorsiflexors under free-flow and ischemic conditions on separate days. As hypothesised, the women exhibited less fatigue than the men under free-flow conditions, but comparable fatigue during ischemia. This result led the authors to suggest that women had an "oxidative advantage" with regard to muscle metabolism. Of the physiological measures taken, only central activation appeared to explain the change in the sex-related difference in fatigue. Impairment of central activation during free-flow contractions was greater in men than in women, but was similar during ischemia. This study was the first to examine the influence of ischemia on sex-related differences in fatigue and central activation, and the results were later confirmed in the knee extensor muscles (Clark *et al.*, 2005).

Study 5. Hunter, S.K., Schletty, J.M., Schlachter, K.M., Griffith, E.E., Polichnowski, A.J. and Ng, A.V. (2006b). Active hyperemia and vascular conductance differ between men and women for an isometric fatiguing contraction. *Journal of Applied Physiology*, 101, 140–150.

This study examined the contribution of muscle perfusion during contraction to differences in fatigue between men and women. Perfusion was determined indirectly through the use of venous occlusion plethysmography to measure post-contraction hyperaemia, which is inversely proportionate to the degree of muscle perfusion during contraction. Hyperaemia was measured in men and women during two different tasks: (1) a 20% isometric MVC sustained until task failure; and (2) a 20% isometric MVC sustained for 4 min. The protocols were also performed by a subset of subjects matched for muscle strength (elbow flexors) across sexes. Time to task failure was greater in women than men when men were stronger than the women, but not when the sexes were strength-matched. Muscle perfusion (hyperaemia) was greater in men than women, regardless of whether or not the sexes were matched for strength. These results suggested that while sex-related differences in muscle strength were associated with the differences in fatigue in sustained, submaximal contractions, differences in muscle perfusion were not likely to be the mechanism underlying the differences.

GLOSSARY OF TERMS

ATP	adenosine triphosphate
ATPase	adenosine triphosphatase
CMAP	compound muscle action potential
EMG	electromyogram
ES	electrical stimulation
LFF	low-frequency fatigue
MHC	myosin heavy chain
MVC	maximal voluntary isometric contraction
PCr	phosphocreatine
P_i	inorganic phosphate
RPE	rating of perceived exertion
SDH	succinate dehydrogenase
TMS	transcranial magnetic stimulation

ACKNOWLEDGEMENTS

Sincere gratitude and appreciation is due to Drs Sandra K. Hunter and Brian C. Clark for their criticism and helpful suggestions regarding earlier drafts of this chapter.

REFERENCES

Barnes, W.S. (1980). The relationship between maximum isokinetic strength and isokinetic endurance. *Research Quarterly for Exercise and Sport*, 51, 714–717.

Barry, B.K. and Enoka, R.M. (2007). The neurobiology of muscle fatigue: 15 years later. *Integrative and Comparative Biology*, 47, 465–473.

Beelen, A., Sargeant, A.J., Jones, D.A. and de Ruiter, C.J. (1995). Fatigue and recovery of voluntary and electrically elicited dynamic force in humans. *Journal of Physiology*, 484 (1), 227–235.

Behm, D., Power, K. and Drinkwater, E. (2001). Comparison of interpolation and central activation ratios as measures of muscle inactivation. *Muscle and Nerve*, 24, 925–934.

Bigland-Ritchie, B. and Woods, J.J. (1984). Changes in muscle contractile properties and neural control during human muscular fatigue. *Muscle and Nerve*, 7, 691–699.

Borg, G.A. (1962). *Physical Performance and Perceived Exertion*. Lund, Sweden: Gleerup.

Borg, G.A. (1982). Psychophysical bases of perceived exertion. *Medicine and Science in Sports and Exercise*, 14, 377–381.

Braun, B. and Horton, T. (2001). Endocrine regulation of exercise substrate utilization in women compared to men. *Exercise and Sport Sciences Reviews*, 29, 149–154.

Cafarelli, E. (1982). Peripheral contributions to the perception of effort. *Medicine and Science in Sports and Exercise*, 14, 382–389.

Cheng, A., Ditor, D.S. and Hicks, A.L. (2003). A comparison of adductor pollicis fatigue in older men and women. *Canadian Journal of Physiological Pharmacology*, 81, 873–879.

Clark, B.C., Collier, S.R., Manini, T.M. and Ploutz-Snyder, L.L. (2005). Sex differences in muscle fatigability and activation patterns of the human quadriceps femoris. *European Journal of Applied Physiology*, 94, 196–206.

Clark, B.C., Hoffman, R.L. and Russ, D.W. (2008). Immobilization-induced increases in fatigue resistance is not explained by changes in the muscle metaboreflex. *Muscle and Nerve*, 38, 1466–1473.

Clark, B.C., Manini, T.M., The, D.J., Doldo, N.A. and Ploutz-Snyder, L.L. (2003). Gender differences in skeletal muscle fatigability are related to contraction type and EMG spectral compression. *Journal of Applied Physiology*, 94, 2263–2272.

Coggan, A.R., Spina, R.J., King, D.S., Rogers, M.A., Brown, M., Nemeth, P.M. and Holloszy, J.O. (1992). Histochemical and enzymatic comparison of the gastrocnemius muscle of young and elderly men and women. *Journal of Gerontology*, 47, B71–76.

Dahlstedt, A.J., Katz, A. and Westerblad, H. (2001). Role of myoplasmic phosphate in contractile function of skeletal muscle: studies on creatine kinase-deficient mice. *Journal of Physiology*, 533, 379–388.

Debold, E.P., Dave, H. and Fitts, R.H. (2004). Fiber type and temperature dependence of inorganic phosphate: implications for fatigue. *American Journal of Physiology and Cellular Physiology*, 287, C673–681.

Di Giulio, C., Daniele, F. and Tipton, C.M. (2006). Angelo Mosso and muscular fatigue: 116 years after the first Congress of Physiologists: IUPS commemoration. *Advances in Physiological Education*, 30, 51–57.

Ditor, D.S. and Hicks, A.L. (2000). The effect of age and gender on the relative fatigability of the human adductor pollicis muscle. *Canadian Journal of Physiological Pharmacology*, 78, 781–790.

Dudek, R. and Fix, J. (2004). *Embryology*. Baltimore: Lippincott, Williams & Wilkins.

Edwards, R.H., Hill, D.K., Jones, D.A. and Merton, P.A. (1977). Fatigue of long duration in human skeletal muscle after exercise. *Journal of Physiology*, 272, 769–778.

Enoka, R.M. and Duchateau, J. (2008). Muscle fatigue: what, why and how it influences muscle function. *Journal of Physiology*, 586, 11–23.

Esbjornsson-Liljedahl, M., Sundberg, C.J., Norman, B. and Jansson, E. (1999). Metabolic response in type I and type II muscle fibers during a 30-s cycle sprint in men and women. *Journal of Applied Physiology*, 87, 1326–1332.

Ettinger, S.M., Silber, D.H., Collins, B.G., Gray, K.S., Sutliff, G., Whisler, S.K., McClain, J.M., Smith, M.B., Yang, Q.X. and Sinoway, L.I. (1996). Influences of gender on sympathetic nerve responses to static exercise. *Journal of Applied Physiology*, 80, 245–251.

Friden, C., Hirschberg, A.L. and Saartok, T. (2003). Muscle strength and endurance do not significantly vary across 3 phases of the menstrual cycle in moderately active premenopausal women. *Clinical Journal of Sports Medicine*, 13, 238–241.

Fulco, C.S., Rock, P.B., Muza, S.R., Lammi, E., Braun, B., Cymerman, A., Moore, L.G. and Lewis, S.F. (2001). Gender alters impact of hypobaric hypoxia on adductor pollicis muscle performance. *Journal of Applied Physiology*, 91, 100–108.

Fulco, C.S., Rock, P.B., Muza, S.R., Lammi, E., Cymerman, A., Butterfield, G., Moore, L.G., Braun, B. and Lewis, S.F. (1999). Slower fatigue and faster recovery of the adductor pollicis muscle in women matched for strength with men. *Acta Physiologica Scandinavica*, 167, 233–239.

Garland, S.J. (1991). Role of small diameter afferents in reflex inhibition during human muscle fatigue. *Journal of Physiology*, 435, 547–558.

Godt, R.E. and Nosek, T.M. (1989). Changes of intracellular milieu with fatigue or hypoxia depress contraction of skinned rabbit skeletal and cardiac muscle. *Journal of Physiology*, 412, 155–180.

Green, H.J., Fraser, I.G. and Ranney, D.A. (1984). Male and female differences in enzyme activities of energy metabolism in vastus lateralis muscle. *Journal of Neurological Sciences*, 65, 323–331.

Gregory, C.M., Williams, R.H., Vandenborne, K. and Dudley, G.A. (2005). Metabolic and phenotypic characteristics of human skeletal muscle fibers as predictors of glycogen utilization during electrical stimulation. *European Journal of Applied Physiology*, 95, 276–282.

Hargreaves, M., McKenna, M.J., Jenkins, D.G., Warmington, S.A., Li, J.L., Snow, R.J. and Febbraio, M.A. (1998). Muscle metabolites and performance during high-intensity, intermittent exercise. *Journal of Applied Physiology*, 84, 1687–1691.

Hicks, A.L., Kent-Braun, J. and Ditor, D.S. (2001). Sex differences in human skeletal muscle fatigue. *Exercise and Sport Sciences Reviews*, 29, 109–112.

Hicks, A.L. and McCartney, N. (1996). Gender differences in isometric contractile properties and fatigability in elderly human muscle. *Canadian Journal of Applied Physiology*, 21, 441–454.

Holmback, A.M., Porter, M.M., Downham, D., Andersen, J.L. and Lexell, J. (2003). Structure and function of the ankle dorsiflexor muscles in young and moderately active men and women. *Journal of Applied Physiology*, 95, 2416–2424.

Horton, T.J., Grunwald, G.K., Lavely, J. and Donahoo, W.T. (2006). Glucose kinetics differ between women and men, during and after exercise. *Journal of Applied Physiology*, 100, 1883–1894.

Howlett, R.A. and Hogan, M.C. (2007). Effect of hypoxia on fatigue development in rat muscle composed of different fibre types. *Experimental Physiology*, 92, 887–894.

Hunter, S.K. and Enoka, R.M. (2001). Sex differences in the fatigability of arm muscles depends on absolute force during isometric contractions. *Journal of Applied Physiology*, 91, 2686–2694.

Hunter, S.K., Butler, J.E., Todd, G., Gandevia, S.C. and Taylor J.L. (2006a). Supraspinal fatigue does not explain the sex difference in muscle fatigue of maximal contractions. *Journal of Applied Physiology*, 101, 1036–1044.

Hunter, S.K., Critchlow, A. and Enoka, R.M. (2004a). Influence of aging on sex differences in muscle fatigability. *Journal of Applied Physiology*, 97, 1723–1732.

Hunter, S.K., Critchlow, A., Shin, I.S. and Enoka, R.M. (2004b). Fatigability of the elbow flexor muscles for a sustained submaximal contraction is similar in men and women matched for strength. *Journal of Applied Physiology*, 96, 195–202.

Hunter, S.K., Critchlow, A., Shin, I.S. and Enoka, R.M. (2004c). Men are more fatigable than strength-matched women when performing intermittent submaximal contractions. *Journal of Applied Physiology*, 96, 2125–2132.

Hunter, S.K., Schletty, J.M., Schlachter, K.M., Griffith, E.E., Polichnowski, A.J. and Ng, A.V. (2006b). Active hyperemia and vascular conductance differ between men and women for an isometric fatiguing contraction. *Journal of Applied Physiology*, 101, 140–150.

Jacobs, C.A., Uhl, T.L., Mattacola, C.G., Shapiro, R. and Rayens, W.S. (2007). Hip abductor function and lower extremity landing kinematics: sex differences. *Journal of Athletic Training*, 42 (1), 76–83.

Jaworowski, A., Porter, M.M., Holmback, A.M., Downham, D. and Lexell, J. (2002). Enzyme activities in the tibialis anterior muscle of young moderately active men and women: relationship with body composition, muscle cross-sectional area and fibre type composition. *Acta Physiologica Scandinavica*, 176, 215–225.

Keenan, K.G., Farina, D., Merletti, R. and Enoka, R.M. (2006). Influence of motor unit properties on the size of the simulated evoked surface EMG potential. *Experimental Brain Research*, 169, 37–49.

Kent-Braun, J.A. (1999). Central and peripheral contributions to muscle fatigue in humans during sustained maximal effort. *European Journal of Applied Physiology and Occupational Physiology*, 80, 57–63.

Kent-Braun, J.A. and Le Blanc, R. (1996). Quantitation of central activation failure during maximal voluntary contractions in humans. *Muscle and Nerve*, 19, 861–869.

Kent-Braun, J.A., Ng, A.V., Doyle, J.W. and Towse, T.F. (2002). Human skeletal muscle responses vary with age and gender during fatigue due to incremental isometric exercise. *Journal of Applied Physiology*, 93, 1813–1823.

Knuth, S.T., Dave, H., Peters, J.R. and Fitts, R.H. (2006). Low cell pH depresses peak power in rat skeletal muscle fibres at both 30 degrees C and 15 degrees C: implications for muscle fatigue. *Journal of Physiology*, 575, 887–899.

Lanza, I.R., Befroy, D.E. and Kent-Braun, J.A. (2005). Age-related changes in ATP-producing pathways in human skeletal muscle in vivo. *Journal of Applied Physiology*, 99, 1736–1744.

Lanza, I.R., Russ, D.W. and Kent-Braun, J.A. (2004). Age-related enhancement of fatigue resistance is evident in men during both isometric and dynamic tasks. *Journal of Applied Physiology*, 97, 967–975.

Lariviere, C., Gravel, D., Gagnon, D., Gardiner, P., Bertrand Arsenault, A. and Gaudreault, N. (2006). Gender influence on fatigability of back muscles during intermittent isometric contractions: a study of neuromuscular activation patterns. *Clinical Biomechanics (Bristol, Avon)*, 21, 893–904.

Lindstrom, B., Lexell, J., Gerdle, B. and Downham, D. (1997). Skeletal muscle fatigue and endurance in young and old men and women. *Journal of Gerontology A: Biological Science and Medical Science*, 52, B59–66.

McKenzie, D.K., Bigland-Ritchie, B., Gorman, R.B. and Gandevia, S.C. (1992). Central and peripheral fatigue of human diaphragm and limb muscles assessed by twitch interpolation. *Journal of Physiology*, 454, 643–656.

Martin, P.G. and Rattey, J. (2007). Central fatigue explains sex differences in muscle fatigue and contralateral cross-over effects of maximal contractions. *Pflugers Archives*, 454, 957–969.

Martin, P.G., Smith, J.L., Butler, J.E., Gandevia, S.C. and Taylor, J.L. (2006). Fatigue-sensitive afferents inhibit extensor but not flexor motoneurons in humans. *Journal of Neuroscience*, 26, 4796–4802.

Martin, V., Millet, G.Y., Martin, A., Deley, G. and Lattier, G. (2004). Assessment of low-frequency fatigue with two methods of electrical stimulation. *Journal of Applied Physiology*, 97, 1923–1929.

Melanson, E.L., Sharp, T.A., Seagle, H.M., Horton, T.J., Donahoo, W.T., Grunwald, G.K., Hamilton, J.T. and Hill, J.O. (2002). Effect of exercise intensity on 24-h energy expenditure and nutrient oxidation. *Journal of Applied Physiology*, 92, 1045–1052.

Newham, D.J., McCarthy, T. and Turner, J. (1991). Voluntary activation of human quadriceps during and after isokinetic exercise. *Journal of Applied Physiology*, 71, 2122–2126.

Noble, B.J., Borg, G.A., Jacobs, I., Ceci, R. and Kaiser, P. (1983). A category-ratio perceived exertion scale: relationship to blood and muscle lactates and heart rate. *Medicine and Science in Sports and Exercise*, 15, 523–528.

Nosek, T.M., Fender, K.Y. and Godt, R.E. (1987). It is diprotonated inorganic phosphate that depresses force in skinned skeletal muscle fibers. *Science*, 236, 191–193.

Pate, E., Bhimani, M., Franks-Skiba, K. and Cooke, R. (1995). Reduced effect of pH on skinned rabbit psoas muscle mechanics at high temperatures: implications for fatigue. *Journal of Physiology*, 486 (Pt 3), 689–694.

Petrofsky, J.S. and Lind, A.R. (1975). Aging, isometric strength and endurance, and cardiovascular responses to static effort. *Journal of Applied Physiology*, 38, 91–95.

Petrofsky, J.S., Burse, R.L. and Lind, A.R. (1975). Comparison of physiological responses of women and men to isometric exercise. *Journal of Applied Physiology*, 38, 863–868.

Pincivero, D.M., Coelho, A.J. and Compy, R.M. (2004) Gender differences in perceived exertion during fatiguing knee extensions. *Medicine and Science in Sports and Exercise*, 36 (1), 109–117.

Pincivero, D.M., Gandaio, C.M. and Ito, Y. (2003). Gender-specific knee extensor torque, flexor torque, and muscle fatigue responses during maximal effort contractions. *European Journal of Applied Physiology*, 89, 134–141.

Rall, J.A. (1982). Sense and nonsense about the Fenn effect. *American Journal of Physiology*, 242, H1–6.

Robertson, R.J. (1982). Central signals of perceived exertion during dynamic exercise. *Medicine and Science in Sports and Exercise*, 14, 390–396.

Rotto, D.M. and Kaufman, M.P. (1988). Effect of metabolic products of muscular contraction on discharge of group III and IV afferents. *Journal of Applied Physiology*, 64, 2306–2313.

Rotto, D.M., Stebbins, C.L. and Kaufman, M.P. (1989). Reflex cardiovascular and ventilatory responses to increasing H+ activity in cat hindlimb muscle. *Journal of Applied Physiology*, 67, 256–263.

Rowell, L.B. and O'Leary, D.S. (1990). Reflex control of the circulation during exercise: chemoreflexes and mechanoreflexes. *Journal of Applied Physiology*, 69, 407–418.

Russ, D.W. and Kent-Braun, J.A. (2003). Sex differences in human skeletal muscle fatigue are eliminated under ischemic conditions. *Journal of Applied Physiology*, 94, 2414–2422.

Russ, D.W., Elliott, M.A., Vandenborne, K., Walter, G.A. and Binder-Macleod, S.A. (2002). Metabolic costs of isometric force generation and maintenance of human skeletal muscle. *American Journal of Physiology, Endocrinology and Metabolism*, 282, E448–457.

Russ, D.W., Lanza, I.R., Rothman, D. and Kent-Braun, J.A. (2005). Sex differences in glycolysis during brief, intense isometric contractions. *Muscle and Nerve*, 32, 647–655.

Russ, D.W., Towse, T.F., Wigmore, D.M., Lanza, I.R. and Kent-Braun, J.A. (2008). Contrasting influences of age and sex on muscle fatigue. *Medicine and Science in Sports and Exercise*, 40, 234–241.

Scott, W.B., Oursler, K.K., Katzel, L.I., Ryan, A.S. and Russ, D.W. (2007). Central activation, muscle performance, and physical function in men infected with human immunodeficiency virus. *Muscle and Nerve*, 36, 374–383.

Sieck, G.C., Han, Y.S., Prakash, Y.S. and Jones, K.A. (1998). Cross-bridge cycling kinetics, actomyosin ATPase activity and myosin heavy chain isoforms in skeletal and smooth respiratory muscles. *Comparative Biochemistry, Physiology B: Biochemical, Molecular Biology*, 119, 435–450.

Simoneau, J.A., Lortie, G., Boulay, M.R., Thibault, M.C., Theriault, G. and Bouchard, C. (1985). Skeletal muscle histochemical and biochemical characteristics in sedentary male and female subjects. *Canadian Journal of Physiology and Pharmacology*, 63, 30–35.

Thein, L.A. and Thein, J.M. (1996). The female athlete. *The Journal of Orthopaedic and Sports Physical Therapy*, 23, 134–148.

Thomas, C.K., Woods, J.J. and Bigland-Ritchie, B. (1989). Impulse propagation and muscle activation in long maximal voluntary contractions. *Journal of Applied Physiology*, 67, 1835–1842.

Thompson, B.C., Fadia, T., Pincivero, D.M. and Scheuermann, B.W. (2007). Forearm blood flow responses to fatiguing isometric contractions in women and men. *American Journal of Physiology, Heart and Circulatory Physiology*, 293, H805–812.

West, T.G., Curtin, N.A., Ferenczi, M.A., He, Z.H., Sun, Y.B., Irving, M. and Woledge, R.C. (2004). Actomyosin energy turnover declines while force remains constant during isometric muscle contraction. *Journal of Physiology*, 555, 27–43.

Westerblad, H., Bruton, J.D., Allen, D.G. and Lannergren, J. (2000). Functional significance of Ca^{2+} in long-lasting fatigue of skeletal muscle. *European Journal of Applied Physiology*, 83, 166–174.

Wigmore, D.M., Propert, K. and Kent-Braun, J.A. (2006). Blood flow does not limit skeletal muscle force production during incremental isometric contractions. *European Journal of Applied Physiology*, 96, 370–378.

Wiseman, R.W., Beck, T.W. and Chase, P.B. (1996). Effect of intracellular pH on force development depends on temperature in intact skeletal muscle from mouse. *American Journal of Physiology*, 271, C878–886.

Yoon, T., Schlinder Delap, B., Griffith, E.E. and Hunter, S.K. (2007). Mechanisms of fatigue differ after low- and high-force fatiguing contractions in men and women. *Muscle and Nerve*, 36, 515–524.

CHAPTER 7

FATIGUE AND TRAINING STATUS

Gregory C. Bogdanis

OBJECTIVES

The objectives of this chapter are to:

- identify and explain the differences in fatigability between individuals with different training background and daily levels of physical activity;
- understand the factors that determine differences in fatigability between individuals during high-intensity exercise;
- describe and explain the influence of changing the training status through systematic training on fatigue;
- analyse the effects of severe reduction of the daily activity level, as occurs in detraining or immobilisation, on muscle performance.

INTRODUCTION

In recent years, accumulating evidence has established the important role of training status in the ability to resist fatigue during high-intensity muscle contractions. Training status reflects the level of adaptation of the different physiological systems to chronic exercise or inactivity. Several studies have shown that different types of training result in distinct adaptations at the structural, metabolic, hormonal, neural and molecular level. Similarly, even a few weeks of inactivity in the form of detraining or immobilisation will decrease fatigue resistance due to both peripheral and central mechanisms.

The aim of this chapter is to demonstrate and explain the differences in fatigability between individuals with different training backgrounds. For this reason, the evidence from studies comparing untrained, endurance-trained and power/sprint-trained individuals will be critically evaluated. Also, the factors that determine the fatigue profile of an individual (e.g. muscle fibre type, aerobic fitness) will be analysed in an attempt to identify their relative contribution to the ability to resist fatigue.

An issue with both theoretical and practical interest concerns the changes that are observed in muscle performance and fatigue when the level of habitual activity is changed. This refers to either acute or chronic increases (training) or decreases in the activity level (injury-based immobilisation). Some athletes are required to stop their training or even immobilise a limb for several days or weeks due to injury. During this time, several morphological, metabolic and neurological changes take place, and muscle performance is reduced. Following a period of rehabilitation, the athlete gradually returns back to action and is "re-trained". To address the important issues of training–detraining and immobilisation, the second part of this chapter will examine the central and peripheral adaptations to these conditions in relation with their impact on muscle fatigue.

TRAINING BACKGROUND AND FATIGUE

Fatigue in untrained, endurance-trained and sprint/power-trained individuals

Single-sprint performance

The relationship between muscle fatigue and training status can be demonstrated by examining the decline of muscle performance during intense exercise in individuals with different training backgrounds. In most studies this was done by comparing sprint- or power-trained athletes with endurance athletes. It is well-known that sprint- and power-trained athletes are stronger and faster than endurance athletes and untrained individuals. For example, Paasuke et al. (1999), reported that power-trained athletes had 27% higher maximal voluntary contraction force (MVC) and 25% higher maximal rate of force development (RFD) compared to endurance athletes. Interestingly, there was no difference between the endurance-trained athletes and an untrained group of subjects in these parameters (Paasuke et al., 1999).

The ability to generate high power output during a short sprint is also superior in power athletes. Peak and mean power output measured on a 30 s cycle ergometer test were 38% and 27% higher, respectively, in sprint cyclists compared with endurance cyclists (peak power: $1,547 \pm 128\,\mathrm{W}$ versus $1,122 \pm 65\,\mathrm{W}$; mean power: $1,030 \pm 52\,\mathrm{W}$ versus $813 \pm 22\,\mathrm{W}$ (Calbet *et al.*, 2003)). Although part of these differences may be attributed to the larger body and muscle mass of power-trained athletes, peak and mean power were still 21% and 11% higher when expressed per kilogram of body mass (Calbet *et al.*, 2003). Comparisons of mean and peak power between the two groups of athletes reveals that their difference in mean power is almost 50% less than their difference in peak power (21% versus 11%). This is due to the ability of endurance athletes to better maintain their performance during the test, as shown by their lower fatigue index, calculated in that study as the rate of drop from peak to end power output (sprint cyclists: $0.46 \pm 0.12 \, W \cdot s^{-1} \cdot kg^{-1}$ versus endurance cyclists: $0.32 \pm 0.12 \, W \cdot s^{-1} \cdot kg^{-1}$ ($P < 0.05$)).

Repeated-sprint performance

Differences in fatigue between power-trained and endurance-trained athletes are more evident when repeated bouts of maximal exercise are performed with short recovery intervals. One of the early studies by Hamilton *et al.* (1991) showed that endurance runners had a smaller decrement in mean power output during ten 6 s all-out sprints with 30 s recovery periods, compared with games players ($14.2 \pm 11.1\%$ versus $29.3 \pm 8.1\%$ ($P < 0.05$). In the case of repeated short sprints, fatigue index is expressed as the drop of peak or mean power from the first to the last sprint (Hamilton *et al.*, 1991), or as the average decrement of power in all sprints relative to the first sprint (Fitzsimons *et al.*, 1993). According to this later calculation of fatigue, endurance runners had a 37% smaller power decrement over five 6 s maximal sprints departing every 30 s, compared with team-sports players (Bishop and Spencer, 2004). This was accompanied by smaller disturbances in blood homeostasis as reflected by lower post-exercise blood lactate concentration ($9.9 \pm 2.1 \, mmol \cdot l^{-1}$ versus $11.4 \pm 0.8 > mmol \cdot l^{-1}$) and higher blood pH ($7.27 \pm 0.04$ versus 7.20 ± 0.04 (Bishop and Spencer, 2004)). Furthermore, it has been shown that endurance athletes have higher oxygen uptake during a repeated-sprint test, indicating a greater contribution of aerobic metabolism to energy supply (Hamilton *et al.*, 1991). Taken together, these results suggest that the differences in fatigability between sprint/power and endurance athletes may be related to both quantitative and qualitative differences in a number of factors that influence the fatigue profile during intense exercise.

Factors that determine fatigue profile during high-intensity exercise

Muscle fibre composition and metabolic profile

It has long been known that muscle fibre composition is distinctly different between untrained individuals, sprint/power-trained and endurance-trained athletes (Costill *et al.*, 1976). Based on the myosin ATPase reaction, human skeletal muscle fibres can be divided into two main types: slow-contracting/ slow-fatiguing and fast-contracting/fast-fatiguing fibres. Recent analysis techniques have enabled a more detailed grouping of muscle fibres according to the expression of myosin heavy chain (MHC) isoforms. There is evidence to suggest that the functional heterogeneity of muscle fibres largely depends on MHC isoform content (Bottinelli, 2001; Malisoux *et al.*, 2007). This means that classification of fibre content of a muscle according to MHC can provide an informative picture about functional characteristics such as strength, power and fatigue resistance. According to the major MHC isoforms, three pure fibre types can be identified: slow type I and fast type IIA and IIX (Sargeant, 2007). Although these fibre types have similar force-per-unit cross-sectional area, they differ considerably in maximum shortening velocity (type I is about 4–5 times slower than IIX) and power-generating capacity (Sargeant, 2007). From a metabolic viewpoint, type IIX fibres have an enzymatic profile that favours anaerobic metabolism, i.e. high resting phosphocreatine (PCr) content (Casey *et al.*, 1996) and high concentration and activity of key glycolytic enzymes such as glycogen phosphorylase and phosphofructokinase (Pette, 1985). This profile makes the fibre more vulnerable to fatigue due to energy depletion or accumulation of metabolites (Fitts, 2008). On the other hand, type I fibres have a higher content and activity of oxidative enzymes that favour aerobic metabolism and fatigue resistance (Pette, 1985). Thus, it can be concluded that individuals with a greater proportion of type I fibres would be more fatigue-resistant compared with individuals with a greater proportion of type IIA and type IIX fibres. Table 7.1 shows the fibre type proportions in populations with different training backgrounds. As can be seen, endurance athletes, as opposed to sprinters and recreationally active individuals, have a higher percentage of type I slow and fatigue-resistant fibres. This is in accordance with the lower peak power output and force-generating capacity and the reduced muscle fatigue during intense exercise.

Table 7.1 Mean proportions of fibre types, according to MHC content, in populations with different training backgrounds. Hybrid fibres contain two MHC isoforms (I/IIA or IIA/IIX)

Population	Fibre type distribution (%)			
	I	*IIA*	*IIX*	*Hybrids*
Untrained (Widrick *et al.*, 2002)	42	30	2	26
Recreational runners (Harber and Trappe, 2008)	50	38	–	12
Endurance trained (Harber and Trappe, 2008)	65	32	–	3
Resistance trained (Widrick *et al.* 2002)	42	55	–	3
Body builders (D'Antona *et al.* 2006)	34	43	18	5
Plyometric trained (Malisoux *et al.* 2006)	29	42	2	27
Sprint trained (Korhonen *et al.* 2006)	40	50	10	–

The effects of muscle fibre type on fatigue have been examined by Hamada *et al.* (2003) in two groups of individuals who performed a series of 16 MVCs of the knee extensors. One group had a high percentage of type I fibres ($61.4 \pm 6.9\%$, group I) and the other group had a high percentage of type II fibres ($71.8 \pm 9.2\%$, group II). Group II had a greater decrease in MVC force than group I ($49.3 \pm 2.6\%$ versus $22.8 \pm 6.2\%$ ($P < 0.01$ (see Figure 7.1) and exhibited a greater impairment of twitch force parameters such as time to peak torque and half-relaxation time (Hamada *et al.*, 2003). It was concluded that muscle fibre composition plays a major role in the development of fatigue.

Another approach to demonstrate the important role of fibre type composition on muscle fatigue during high-intensity exercise is to measure changes in muscle metabolites in single-muscle fibre populations. Karatzaferi *et al.* (2001) has shown that there is a selective recruitment and selective fatigue of the fast fibres containing the IIX MHC isoform, as shown by the large (70%) decrease of ATP within 10 s of sprint exercise. On the other hand, type I fibres show no change in ATP at the same time (Figure 7.2). It has been proposed that this pattern of ATP depletion indicates that the contribution of the fastest and more powerful fibres (containing IIX isoform) is decreased after the first few seconds of sprint exercise (Sargeant, 2007). This decreased contribution is due to the greater metabolic disturbances in type II compared with type I fibres during maximal exercise. For example, during a 30 s sprint, the rate of PCr

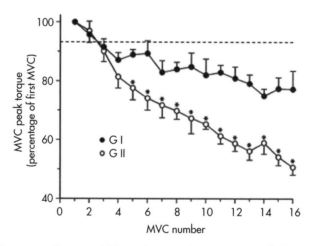

Figure 7.1 Percentage change in MVC peak torque during a 16-MVC fatigue protocol using the knee extensors. Each MVC lasted 5 s, with a 3 s rest in between. Values below the dashed line are significantly lower ($P < 0.05$) compared with the first MVC. Group GI: predominance of type I muscle fibres; group GII: predominance of type II muscle fibres in vastus lateralis. *$P < 0.05$ between groups GI and GII

Source: Reprinted from Hamada *et al.*, 2003. *Acta Physiol Scand*, 178, 167, Figure 1. Copyright 2003 by Blackwell Publishing. Used with permission.

degradation and glycogenolytic rates are 36% and 64% higher, respectively, in type II compared to type I fibres (Greehaff *et al.*, 1994; Casey *et al.*, 1996). This means greater lactate and H^+ accumulation, and thus increased fatigue. Furthermore, the metabolic characteristics of type II fibres, i.e. lower mitochondrial and aerobic enzyme content and lower capillary supply than type I fibres, result in slower PCr resynthesis and lactic acid removal (Tesch and Wright, 1983; Tesch *et al.*, 1985, 1989; Casey *et al.*, 1996), thus making this fibre type more vulnerable to fatigue in the subsequent exercise bout.

The idea of selective recruitment and fatigue of the fast fibres during maximal exercise may be extended to explain the greater fatigability of individuals with a high percentage of fast-twitch fibres. Colliander *et al.* (1988) have compared fatigue during three bouts of 30 MVC knee extensions between individuals with a high percentage of type I and II fibres. As expected, individuals with a high percentage of type II fibres produced higher peak torque, but fatigued significantly more during the exercise protocol. It is worth noting that when blood flow was occluded by means of a pneumatic cuff around the proximal thigh, the decrease in peak torque was greater (52% versus 10%) in the group with higher percentage of type I (slow) muscle fibres, probably indicating the reliance of these fibres on oxygen availability and aerobic metabolism (Colliander *et al.*, 1988).

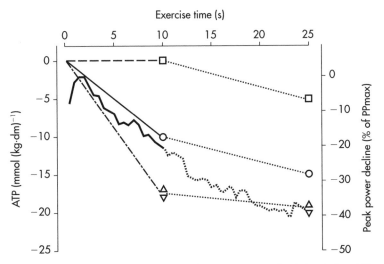

Figure 7.2 Mean decline in ATP concentration for type I (squares), IIA (circles) and IIAx and IIXa (upright and inverted triangles) during a 10 s and a 25 s maximal sprint on a cycle ergometer. Type IIAx and IIXa are hybrid fibres containing either less or more than 50% IIX MHC isoform. Power output data for the sprints are shown with the solid line for a typical subject

Source: Reprinted from Karatzaferi *et al.*, 2001. *Exp Physiol*, 86, 414, Figure 3. Copyright 2001 by Blackwell Publishing. Used with permission.

Influence of initial force/power output

One of the confounding factors when comparing fatigue calculated as a percentage decrement of force or power during a test between individuals of different training status, is the magnitude of the peak force or power that can be generated at the start of the test. Previous studies have reported that initial sprint performance is strongly correlated with power decrement during a repeated-sprint test (Hamilton *et al.*, 1991; Bishop *et al.*, 2003) and inversely related to maximal oxygen uptake (Bogdanis *et al.*, 1996). Thus, the greater fatigue resistance of endurance athletes may be more related to their low initial sprint performance rather than their high aerobic fitness.

The dependence of the calculated fatigue index on the initial power is demonstrated in the data presented by Calbet *et al.* (2003), comparing endurance and sprint cyclists (Figure 7.3). It should be noted that the differences in power output between the groups existed only for the first 10 s of the sprint, and thereafter the two groups generated the same power. Interestingly, the oxygen deficit of the sprint cyclists was much higher (+ 33%) for that first part of the sprint, indicating that these athletes rely more on anaerobic sources. In contrast, endurance-trained athletes had a higher oxygen uptake during the last 15 s of the test, suggesting a higher aerobic contribution.

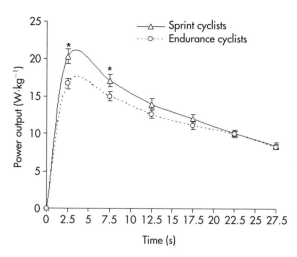

Figure 7.3 Time course of power output during a 30 s maximal cycle ergometer sprint performed by sprint- and endurance-trained cyclists. Power output is expressed per kilogram of body mass

Source: Redrawn with permission from Calbet *et al.*, 2003. *J Appl Physiol*, 94, 671, Figure. 1. Copyright 2003 by The American Physiological Society. Modified with permission.

Note
*$P < 0.05$ sprint versus endurance cyclists.

To resolve the issue about the influence of the initial power output on fatigue during high-intensity exercise, Tomlin and Wenger (2002), and later Bishop and Edge (2006), compared two groups of female team-sports athletes with different maximal oxygen uptake values (low $\dot{V}O_{2max}$: 34–36 ml·kg^{-1}·min^{-1} versus moderate $\dot{V}O_{2max}$: 47–50 ml·kg^{-1}·min^{-1}), but with the same peak and mean power output in a 6 s cycle ergometer sprint. These athletes were required to perform five 6 s sprints with 24 s recovery (Bishop and Edge, 2006) or ten 6 s sprints with 30 s recovery (Tomlin and Wenger, 2002). Even though the two groups were matched for initial sprint performance, the moderate $\dot{V}O_{2max}$ group had a smaller power decrement across the ten (low versus moderate: $18.0 \pm 7.6\%$ versus $8.8 \pm 3.7\%$ $(P = 0.02)$ or the five sprints (low versus moderate: $11.1 \pm 2.5\%$ versus $7.6 \pm 3.4\%$ $(P = 0.045)$. These results point to an important role of aerobic fitness on the ability to resist fatigue.

A recent study by Mendez-Villanueva et al. (2008) proposed an alternative point of view. Instead of examining the effect of the initial power output on subsequent fatigue, they proposed to explain differences in fatigue by looking at the anaerobic power reserve of each individual. This was quantified as the difference between the maximal anaerobic power measured during a 6 s sprint and the maximal aerobic power determined during a graded test to exhaustion. Individuals with a lower anaerobic power reserve, implying less reliance on anaerobic metabolism, showed a greater resistance to fatigue. This suggests that the relative contribution of the aerobic and anaerobic pathways to energy supply and not the absolute power generated may better explain power decrements during repeated sprints (Mendez-Villanueva et al., 2008).

Evaluating performance and fatigue during high-intensity exercise

The evaluation of performance during single or repeated bouts of high-intensity exercise is usually done using two indices: (a) total work done during the test; and (b) the percentage decrement of power output from the start to the end of the test. Both indices offer valuable (and complementary) information about the training status of an individual. However, confusion may arise when they are used to assess the effects of training and detraining. In most sprint-training studies, total work in single or multiple sprints is higher after sprint training. For example, Linossier et al. (1993) reported an increase in total work during a single 30 s sprint by 16%. As can be seen in Figure 7.4, the increase in total work was due to higher power output during the first half of the sprint, and thus, fatigue index, i.e. the drop of power from peak to end was higher after training. This "paradox", i.e. greater fatigue after training, has to be interpreted with caution because the overall performance suggests an increase rather than a decrease in power-generating capacity.

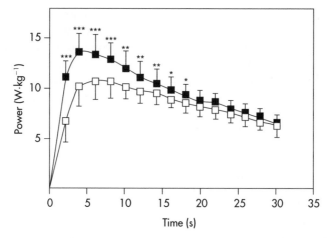

Figure 7.4 Power output during a 30 s maximal sprint on a cycle ergometer before and after seven weeks of sprint training (pre-training: open squares; post-training: closed squares)

Source: With kind permission from Springer Science+Business Media. Linossier *et al.*, 1993. Ergometric and metabolic adaptation to a 5-s sprint training programme. *Eur J Appl Physiol Occup Physiol, 67*, 1993, 410, Figure 3.

This argument also applies when comparing power-trained, endurance-trained and untrained individuals during a repeated-sprint test. As seen in Figure 7.5, the total work done and the percentage work decrement give a conflicting picture of repeated-sprint performance ability when sprint-trained and endurance-trained athletes are compared. The percentage work decrement is two-fold higher in the team-sport group compared with the endurance group (11.3 versus 5.6% ($P < 0.05$), implying greater fatigue and poorer repeated-sprint performance for the team players. In contrast, total work over the six sprints is significantly higher in the team players, and this clearly shows a better overall repeated-sprint performance of these players compared with the endurance athletes (Figure 7.5). These apparently contradictory results demonstrate the well-known characteristics of sprint/power athletes to produce high power over the initial part of a high-intensity effort, and then fatigue more. However, they also suggest that one has to select the appropriate index of repeated-sprint performance ability when comparing different groups of individuals.

Aerobic fitness: maximal oxygen uptake and aerobic contribution to energy supply

Several studies have shown that short bouts of all-out exercise, such as sprinting, are not as "anaerobic" as previously thought. Gaitanos *et al.* (1993) noted that while the decrease in peak and mean power during ten 6 s sprints with 30 s rest was 33% and 27%, respectively, anaerobic energy supply was reduced by about 70% due to a complete blocking of anaerobic glycolysis. They first suggested that power output during the last sprints

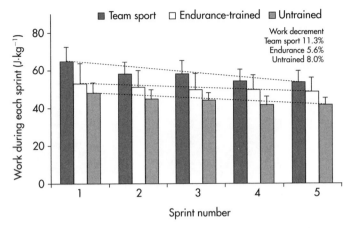

Figure 7.5 Total work per kilogram of body mass during each sprint of the five 6 s repeated-sprint test. Dotted lines denote the rate of work drop from the first to the last sprint. Work decrement is the average decrement of work in all sprints relative to the first sprint, according to Fitzsimons *et al.*, 1993

Source: With kind permission from Springer Science+Business Media. Edge *et al.*, 2006. The effects of training intensity on muscle buffer capacity in females. *Eur J Appl Physiol*, 96, 230, Figure 3, panel b.

Notes
*: Significantly different from untrained (P<0.05); §: significantly different from endurance-trained (P<0.05).

was sustained by increased contribution of aerobic metabolism. The magnitude of aerobic contribution to energy supply during sprint exercise was quantified in a later study using muscle metabolite and oxygen uptake measurements during repeated 30 s sprints (Bogdanis *et al.*, 1996). Indeed, when sprint exercise is repeated, aerobic contribution to energy supply is largely increased (Figure 7.6). This enhanced aerobic component during repeated-sprint exercise implies that individuals with higher $\dot{V}O_{2max}$ would have greater aerobic contribution to energy supply when performing this type of exercise. This increased aerobic contribution in endurance-trained individuals was found in several studies (Hamilton *et al.*, 1991; Tomlin and Wenger, 2002; Calbet *et al.*, 2003), with oxygen uptake reaching 80% and up to 100% $\dot{V}O_{2max}$ during the later stages of the test (Dupont *et al.*, 2005). Thus, the more endurance-trained an individual is, the higher the aerobic energy supply during single or repeated-sprint exercise.

In an attempt to quantify the effects of aerobic fitness on fatigue during intense exercise, some researchers have reported the percentage of variance in fatigue explained by the variance in $\dot{V}O_{2max}$ of the participants. This is done by squaring the correlation coefficient (r) between the two variables ($\dot{V}O_{2max}$ and per cent fatigue) and then multiplying by 100. The modest correlations between $\dot{V}O_{2max}$ and power decrement (typically between –0.5 and –0.65) give relatively low r^2 values, and thus only 25–42% of the variance in fatigue is explained by $\dot{V}O_{2max}$ (Tomlin and Wenger, 2002; Bishop

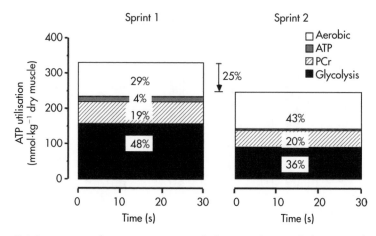

Figure 7.6 Per cent contribution to energy supply from aerobic metabolism, ATP degradation, PCr hydrolysis and anaerobic glycolysis during two 30 s cycle ergometer sprints separated by 4 min passive recovery

Source: Redrawn using data from Bogdanis *et al.*, 1996. *J Appl Physiol*, 80, 881, Figure 5. Copyright 1996 by The American Physiological Society. Modified with permission.

et al., 2004; Bishop and Edge, 2006). Moreover, when team-sport athletes and endurance runners were matched for $\dot{V}O_{2max}$, the power decrement during repeated 6 s sprints was still lower in the endurance athletes, suggesting that other factors in addition to $\dot{V}O_{2max}$ are important (Bishop and Spencer, 2004).

Aerobic fitness: oxygen uptake kinetics

The rate at which pulmonary oxygen uptake ($\dot{V}O_2$) increases following the onset of muscular exercise is potentially an important determinant of exercise tolerance. A faster $\dot{V}O_2$ adjustment at the onset of intense exercise leads to a greater contribution of oxidative phosphorylation and possibly a decreased reliance on anaerobic glycolysis, with positive effects on performance and reduction of muscle fatigue. The rate of increase in $\dot{V}O_2$ may be quantified by calculating the time constant ($\tau\dot{V}O_2$) of the $\dot{V}O_2$ response. Caputo *et al.* (2004) showed that $\tau\dot{V}O_2$ is dependent on training status, with untrained individuals having much longer time constants than endurance-trained athletes (e.g. 53 s versus 32 s). This may reflect differences in central oxygen delivery (heart rate kinetics), muscle blood flow and the ability to accelerate muscle oxidative metabolism, the so-called "metabolic inertia" that is related to the levels of cellular metabolic controllers and/or enzyme activation (Grassi, 2003). The dependence of $\dot{V}O_2$ kinetics on aerobic fitness and the metabolic profile of the muscle explains the fact that sprint-trained athletes have significantly slower $\dot{V}O_2$ response than their endurance-trained counterparts ($\tau\dot{V}O_2$: 32 ± 12 s versus 20 ± 6 s (Berger and

Jones, 2007)). The relationship between $\dot{V}O_2$ kinetics and performance in repeated-sprint exercise has been recently demonstrated by Dupont et al. (2005), who found that repeated-sprint performance was better correlated with $\tau\dot{V}O_2$ ($r = 0.80$ ($P < 0.01$)) than with $\dot{V}O_{2max}$ ($r = 0.48$ ($P < 0.05$)). These results show that individuals with faster $\dot{V}O_2$ kinetics have a faster adjustment of $\dot{V}O_2$ during repeated maximal exercise, leading to lower fatigue.

Aerobic fitness: effects on PCr resynthesis

One of the possible determinants of fatigue and performance recovery during repeated MVCs is PCr availability (Bogdanis et al., 1995, 1996). PCr is a major contributor to energy supply during the initial few seconds of maximal exercise, when it serves as a temporal buffer of free ADP (Sahlin et al., 1998). During repeated maximal exercise, the rate of PCr resynthesis determines its availability for the next exercise bout, and thus individuals with fast PCr kinetics are less prone to fatigue.

The dependence of PCr resynthesis on oxygen availability is well-known (Haseler et al., 1999, 2007). Phosphorus magnetic resonance spectroscopy (^{31}P MRS) has allowed the non-invasive measurement of PCr restoration, which has been widely used as an index of muscle oxidative capacity (Haseler et al., 2004). It is therefore not unexpected that individuals with higher muscle oxidative capacity have faster PCr resynthesis rate compared with untrained persons (Yoshida and Watari, 1993; Takahashi et al., 1995). The differences in PCr resynthesis and performance recovery between endurance-trained, sprint-trained and untrained individuals have been studied by Johansen and Quistroff (2003) using ^{31}P MRS. In that study, participants performed four MVCs of 30 s duration, interspersed by 60 s recovery intervals. Endurance-trained athletes showed almost twice as fast PCr recovery (half time, $t_{1/2}$: 12.5 ± 1.5 s) compared to sprint-trained ($t_{1/2}$: 22.5 ± 2.5 s) and untrained participants ($t_{1/2}$: 26.4 ± 2.8 s). This resulted in full restoration of PCr for the endurance athletes prior to each contraction, whereas the sprinters and untrained subjects started the subsequent contractions with a PCr level of about 80% PCr of resting.

The faster rate of PCr resynthesis in endurance athletes is probably unrelated to $\dot{V}O_{2max}$. Cooke et al. (1997) have demonstrated that individuals with high and low $\dot{V}O_{2max}$ (64.4 ± 1.4 ml·kg^{-1}·min^{-1} versus 46.6 ± 1.1 ml·kg^{-1}·min^{-1} ($P < 0.01$)) had similar PCr resynthesis rates. Also, individuals with an equal $\dot{V}O_{2max}$ may have remarkably different endurance capacity, due to peripheral muscle adaptations such as capillary density and blood lactate threshold (Coyle et al., 1988). Thus the faster rate of PCr resynthesis in endurance-trained individuals is most probably due to the specific adaptations of this type of training at the peripheral level, i.e. increased mitochondrial content, capillary density and oxidative enzyme content and activity (Andersen and Henriksson, 1977), that maximise oxygen delivery and aerobic ATP production during recovery. The role of

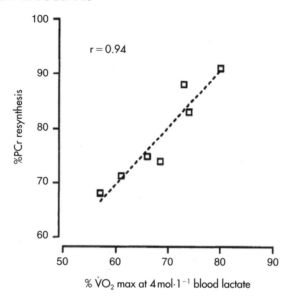

Figure 7.7 Relationship between percentage maximal oxygen uptake ($\dot{V}O_{2max}$) corresponding to a blood lactate concentration of $4\,mmol\cdot l^{-1}$ and percentage PCr resynthesis after $4\,min$ of passive recovery following a maximal $30\,s$ cycle ergometer sprint

Source: Reprinted with permission from Bogdanis et al., 1996. *J Appl Physiol*, 80, 880, Figure 4. Copyright 1996 by The American Physiological Society. Used with permission.

peripheral adaptations on PCr resynthesis has also been indirectly demonstrated by the high positive correlation ($r = 0.89$ ($P < 0.01$)) between percentage of PCr resynthesis and endurance fitness as determined from the percentage of $\dot{V}O_{2max}$ corresponding to a blood lactate concentration of $4\,mmol\cdot l^{-1}$ (Figure 7.7). As shown in several studies (Bogdanis *et al.*, 1996; Casey *et al.*, 1996; Johansen and Quistorff, 2003), increased PCr resynthesis is related to improved performance and reduced fatigue during repeated maximal exercise.

Muscle buffering capacity and ionic regulation

High-intensity static or dynamic exercise results in large changes in metabolites and ions within the working muscles (Bogdanis *et al.*, 1995). Disturbances in the concentration of muscle hydrogen (H^+), potassium (K^+) and calcium (Ca^{2+}) are linked with fatigue (McKenna *et al.*, 2008), and thus ionic regulation becomes critical for muscle membrane excitation, contraction and metabolism.

The rapid rates of glycolysis result in lactate and H^+ accumulation and a consequent reduction of muscle pH by about 0.5 (Bogdanis *et al.*, 1995). Although the negative effect of lactate and H^+ accumulation on muscle performance has been recently challenged (Allen and Westerblad, 2004; see

Chapter 3), there are a number of mechanisms by which muscle activation and metabolism may be impaired (Bangsbo and Juel, 2006). Therefore, the ability to regulate H^+ and lactate homeostasis during high-intensity exercise plays an important role in the fatigue process (Juel, 2008).

A number of mechanisms contribute to muscle pH regulation, including release of H^+ to the blood via different transport and buffering systems (Juel, 2008). The ability of the muscle to buffer the build-up of free H^+ is termed "muscle-buffering capacity" (βm) and is considered an important determinant of high-intensity exercise performance. βm is measured on a muscle biopsy sample using titration with acid, and is expressed as the number of micromoles of H^+ that are required to decrease pH by one unit from 7.1. The intracellular contents of protein (histidine residues), inorganic phosphate and the dipeptide carnosine, have been identified as important physicochemical buffers that can be affected by training status (Parkhouse and McKenzie, 1984). A high βm may favourably affect high-intensity performance by allowing anaerobic glycolysis to continue, resulting in a larger lactate (and thus energy) production without a concomitant increase in H^+ accumulation.

Bishop et al. (2004) were the first to report a correlation between work decrement (percentage fatigue) and βm during a test using five 6 s sprints ($r = -0.72$ ($P < 0.05$)) in untrained females. The correlation of βm with fatigue in that study was greater than that obtained from aerobic fitness indices, such as $\dot{V}O_{2max}$ ($r = -0.62$ ($P < 0.05$)) or lactate threshold ($r = -0.56$ ($P < 0.05$)), implying that the ability of the muscle to buffer H^+ is important for maintaining performance during brief repeated bouts. More recently, Edge et al. (2006b) measured βm and repeated-sprint ability in team-sport athletes, endurance athletes and untrained but physically active females. The authors hypothesised that team-sport players would have higher βm than the other two groups, because of the frequent high-intensity exercise bouts during training and competition. Indeed, the team-sport group had a significantly higher βm than either the endurance-trained or the untrained groups ($181 \pm 27 \mu mol\ H^+ \cdot g$ dry muscle$^{-1} \cdot pH^{-1}$ versus $148 \pm 11 \mu mol\ H^+ \cdot g$ dry muscle$^{-1} \cdot pH^{-1}$ versus $122 \pm 32 \mu mol\ H^+ \cdot g$ dry muscle$^{-1} \cdot pH^{-1}$ ($P < 0.05$)). Also, the team-sport group completed significantly more total work than the other two groups ($299 \pm 27 J \cdot kg^{-1}$ versus $263 \pm 31 J \cdot kg^{-1}$ versus $223 \pm 21 J \cdot kg^{-1}$ ($P < 0.05$). The authors confirmed their hypothesis, suggesting that either team-sport athletes develop high βm to cope with the high rates of lactic acid production and high concentrations of H^+, or alternatively, those with a greater βm are attracted to team sports that involve repeated high-intensity efforts (Edge et al., 2006b).

Neural factors

Training background has an impact on the functional organisation of the neuromuscular system, and power-trained athletes have been shown to be more affected by fatiguing exercise than endurance athletes (Paasuke *et al.*, 1999). The contribution of neural factors to muscle fatigue can be examined by measuring the electromyographic (EMG) activity of the agonist and antagonist muscles, as well as the level of voluntary activation of motor units by the twitch interpolation method (see Chapter 2). With the later method it is feasible to quantify the failure of an individual to mobilise all the motor units of a given muscle.

Mendez-Villanueva *et al.* (2008) measured changes in EMG activity of the vastus lateralis muscle during ten 6s sprints interspersed by 30s rest intervals. When compared with the first sprint, the decrease in the normalised EMG amplitude (root mean square: RMS) was 9.2% and 14.3% in the fifth and tenth sprint, respectively. There was a very high correlation ($r^2 = 0.972$) between total work and EMG amplitude in the tenth sprint, expressed as a percentage of first-sprint values, suggesting that fatigue is accompanied by reductions in neural drive and muscle activation, as confirmed in another similar experiment (decrease in voluntary activation from 95% to 91.5% ($P < 0.02$) (Racinais *et al.*, 2007). The opposite finding, i.e. an increase rather than a decrease in EMG RMS, was found in a similar study that measured force and EMG during an MVC after a ten 6s sprints protocol (Billaut *et al.*, 2006), suggesting peripheral rather than central mechanisms of fatigue. Although there is no consensus regarding the changes in EMG RMS during maximal repeated-sprint exercise, a common finding in many studies is that fatigue in maximal exercise is characterised by a shift in the EMG power spectrum of the muscles involved, suggesting selective fatigue of fast-twitch fibres (Kupa *et al.*, 1995; Billaut *et al.*, 2006). Consequently, it may be proposed that individuals with a higher percentage of fast fibres would have the largest shifts in EMG frequency.

Another neuromuscular characteristic that may vary with training status is voluntary activation. The level of voluntary activation of a muscle during an MVC can range between 80% and 100% (Behm *et al.*, 2002). If a muscle is activated only to 80% of its full capacity, fatigue is likely to develop at a slower rate than if it were fully activated. Nordlund *et al.* (2004) reported a wide range of voluntary activation (67.9–99.9%) for the plantar flexors of healthy habitually active males. A novel finding of that study was that the peripheral fatigue over nine bouts of ten brief MVCs (2s contraction, 1s rest, with 10s rest between bouts) was largely (by 58%; $r^2 = 0.58$) explained by the level of MVC torque and the percentage of voluntary activation during the first bout. This confirmed the suggestion that those who cannot fully activate their muscles fatigue less. An interesting suggestion made by Nordlund *et al.* (2004) was that individuals who do not fully activate their muscles do not recruit all of their fast-twitch motor units. This is based on the "size principle" of fibre activation, i.e. slow motor units are recruited before the fast ones as force is increased. Thus, failure to recruit all fast-twitch fibres results in less metabolic disturbances and less fatigue during a certain task.

The possible differences in voluntary activation between individuals of varied training status were examined by Lattier *et al.* (2003). An expected finding was that power-trained athletes had higher maximal rate of force development and MVC force of the knee extensors than the endurance athletes. However, power- and endurance-trained athletes had a similar maximal voluntary activation of the knee extensors (~80%), which was higher than that measured in a sedentary group (~65%). Thus, the results of that study suggest that there are no differences in muscle activation between well-trained athletes, irrespective of the type of training (endurance versus sprint training), but sedentary individuals may have a significant maximal activation deficit (Lattier *et al.*, 2003).

Fatigue during high-intensity dynamic contractions may be exacerbated by the contraction of the antagonist muscles. The level of coactivation of the antagonists would decrease the effective force or power generated by a joint, especially during faster movements where neuromuscular coordination is more important (Hautier *et al.*, 2000). Garrandes *et al.* (2007) reported that the coactivation level of the antagonist muscles during knee extension was increased by 31% after fatigue only in power-trained and not in endurance athletes. Also, another comparative study reported four times higher hamstrings coactivation during isokinetic knee extensions in sprinters compared with distance runners (57% versus 14%), probably indicating a sport-specific adaptation (Osternig *et al.*, 1986). Thus, it may be argued that a very important neural factor that modifies fatigue in individuals with different training backgrounds is the antagonist coactivation level. The higher antagonist coactivation in sprint/power-trained individuals may partly explain their greater fatigue during dynamic exercise, since part of the agonist force/power is lost to overcome antagonist muscle activity.

Key points (training background and fatigue)

- Endurance-trained athletes fatigue less and sprint/power-athletes fatigue more during high-intensity exercise.
- Individuals with a greater proportion of type I fibres would be more fatigue-resistant compared with individuals with a greater proportion of type IIA and type IIX fibres.
- Aerobic fitness plays an important role in the ability to resist fatigue during high-intensity exercise by enhancing aerobic contribution to energy supply and the rate of resynthesis of PCr during repeated bouts of high-intensity exercise.
- To deal with the higher rates of lactate and H^+ production, team-sport players have higher βm than endurance-trained athletes.
- Individuals who have a low percentage of maximal voluntary muscle activation (such as some sedentary individuals) do not recruit all of their fast-twitch motor units, and thus fatigue less during intense exercise.

- The coactivation of antagonist muscles during high-intensity exercise is much higher in sprint/power-trained athletes than in endurance athletes, and may partly explain their greater fatigue during dynamic exercise.

INCREASES IN ACTIVITY LEVEL AND FATIGUE: EFFECTS OF SYSTEMATIC TRAINING

Human skeletal muscle is a highly plastic tissue and adapts to the variable functional requirements by adjusting its size (D'Antona *et al.*, 2006), muscle fibre type expression and myosin phenotype (Malisoux *et al.*, 2007) and energy metabolism (Green and Pette, 1997). Changes in functional demands also result in neural adaptations that affect muscle performance, such as increased neural drive in untrained individuals.

The adaptations that may reduce fatigue during high-intensity exercise vary with the characteristics of the training programme, namely type (sprint, endurance, resistance training), intensity and duration. The most relevant adaptations regarding fatigue concern changes in fibre type characteristics, enzyme activity, regulation of ionic balance (Na^+-H^+, lactate) as well as changes in EMG activity.

Fibre type transformations after training

Sprint and strength training

As seen in Table 7.1, fibre type distribution according to MHC content is different between athletes of various sports and reflects the functional and metabolic requirements of each sport. For example, endurance athletes are characterised by a high percentage of type I fibres (65%), while sprinters have 60% of fibres containing fast (IIA and IIX) MHC isoforms (Table 7.1). One may argue that this may be the result of natural selection, i.e. individuals follow the sport that they are suitable for in terms of physical attributes. However, training studies show that it is possible to attain some degree of MHC transformation (Malisoux *et al.*, 2007). The transitions between MHC isoforms are done in a sequential, reversible order determined by the neural impulse patterns, the mechanical loading characteristics and by alterations in the metabolic homeostasis, as follows (Pette, 1998):

type I ↔ type IIA ↔ type IIX

The fibre types must be viewed as a spectrum that includes hybrid fibres co-expressing I and IIA or IIA and IIX MHC isoforms. Moreover, there is growing evidence suggesting that the functional properties of muscle fibres can change in several physiological and pathological conditions with no shift in myosin isoform. This does not negate the important role of muscle fibre composition on fatigue, but rather shows that a "fine tuning" of one or more characteristics of a given fibre may occur, according to functional demands (Malisoux et al., 2007).

Training for sprint and power sports involves short maximal exercise bouts of high power output. Thus, a switch towards the type IIX fibres that generate high power output would be expected. However, a consistent finding of sprint and strength training studies is that the MHC IIX isoforms are down regulated and there is usually a bidirectional change towards IIA at the expense of both I and IIX MHC isoforms (Andersen et al., 1994; Widrick et al., 2002). In the study of Andersen et al. (1994), a group of sprint athletes (n = 6) were trained intensively using strength and interval training for a period of three months. After training there was a decrease in fibres containing only MHC I isoform ($52.0 \pm 3.0\%$ versus $41.2 \pm 4.7\%$ ($P < 0.05$)) and an increase in the amount of fibres containing only MHC IIA isoform ($34.7 \pm 6.1\%$ versus $52.3 \pm 3.6\%$ ($P < 0.05$)). Fibres showing co-existence of MHC IIA and IIX isoforms decreased with training ($12.9 \pm 5.0\%$ versus $5.1 \pm 3.1\%$ ($P < 0.05$)). Only one out of 1,000 fibres analysed contained only MHC isoform IIX.

Studies examining the functional adaptations at the single-fibre level after sprint and strength training show that they depend mostly on increases in fibre cross-sectional area (CSA), since the force per CSA remains unchanged in most studies (Widrick et al., 2002; Malisoux et al., 2007). However, there are some suggestions for increased force per fibre CSA after body-building training (D'Antona et al., 2006). Maximal shortening velocity of single fibres also seems to be unchanged after resistance (Widrick et al., 2002) or sprint training (Harridge et al., 1998) in healthy young individuals. There is, however, some evidence suggesting that plyometric training may be effective in increasing maximal shortening velocity in single fibres (Malisoux et al., 2007).

Endurance training

The model of chronic low-frequency stimulation has been used over the past 40 years to demonstrate muscle plasticity. This protocol involves electrical stimulation of the nerve of an animal and shows a remarkable degree of transformation of fast, fatigable muscles towards slower, fatigue-resistant ones in terms of both fibre type and metabolism (Pette and Vrbova, 1999). This demonstrates that activity may have a large impact on the phenotype and fatigue profile of skeletal muscle. The closest equivalent to this type of activity in exercise training on humans is endurance exercise. However, the effects on muscle fibre type transitions after endurance training are less radical than those obtained by chronic electrical stimulation. Andersen and Henriksson (1977) reported no increase in the percentage of type I fibres, but an increase in type IIA at the expense of type IIX fibres after eight weeks of endurance training. Another study using previously untrained individuals who were trained for 13 weeks in order to complete a marathon, showed a decrease in slow (MHC I) and fast (MHC IIA) fibre CSA by about 20%, but an increase in the percentage of MHC I fibres (from $48 \pm 6\%$ to $56 \pm 6\%$ ($P < 0.05$)), while the percentage of MHC IIA fibres remained unchanged ($30 \pm 5\%$) (Trappe et al., 2006). A significant finding of that study was that single-fibre muscle power expressed per unit fibre volume as measured in vitro, was increased by >70% in both MHC I and IIA fibres. These increases of power demonstrate that high volume endurance training (30–60 km running per week) can modify the functional profile of the fibres that are most involved.

From the previous discussion, it seems that fibre type profile can be affected to some extent by sprint/power- and endurance-training in healthy individuals. The implications of changes in MHC isoform expression with training on muscle fatigue during high-intensity exercise can not been examined in isolation. The bidirectional shifts of fibre types towards IIA, with decreases in type I and IIX isoforms makes it difficult to predict if peak power will increase, decrease or remain unchanged, and whether fatigability will change in any direction. However, changes in fatigability observed after training cannot and should not be explained solely by fibre type transformations. Other factors such as muscle metabolic properties and neural activation patterns play an important role in fatigue resistance and should also be considered.

Muscle enzymes

It has long been shown that some metabolic characteristics such as the activities of aerobic and anaerobic enzymes may increase in the exercising muscle, without noticeable MHC-based fibre type transitions (Pette, 1998).

Sprint training with repeated short bouts (<10s)

The majority of investigations have noted increases in glycolytic enzyme activity after sprint training (Linossier et al., 1993, 1997; Dawson et al., 1998). Typically, these adaptations concern key enzymes of glycolysis, such as glycogen phosphorylase, phosphofructokinase (PFK) and lactate dehydrogenase (LDH). Linossier et al. (1993) trained young students for seven weeks with four sessions per week of short (5s) repeated cycle ergometer sprints with 55s resting intervals. The number of sprints was increased every week from 16 to 26 sprints per session. This programme resulted in increased energy production from anaerobic glycolysis, as indicated by the increased muscle lactate accumulation after compared to before training (Δ lactate 37.2 ± 17.9 versus 52.8 ± 13.5 ($P < 0.01$)) and the 20% higher PFK and LDH activity. A similar training study by Dawson *et al.* (1998) involving short running sprints of comparable duration to the previous study (30–60 m) found a 40% increase in phosphorylase, but no increase in PFK. A common finding of these two studies involving short sprints was that the activities of key oxidative enzymes involved in carbohydrate metabolism, e.g. citrate synthase (CS), or lipid oxidation, e.g. 3-hydroxyacyl-CoA dehydrogenase (HAD), were either unchanged (Linossier *et al.*, 1993, 1997) or decreased (Dawson *et al.*, 1998) with this type of training.

Sprint training with repeated long bouts (30 s)

However, sprint-training studies using longer sprint bouts, i.e. 30 s cycle ergometer sprints, showed increases in oxidative enzymes as well. In one study (MacDougall *et al.*, 1998), participants trained three times per week for seven weeks and performed 30 s sprints with 3–4 min rest in each session. The number of sprints increased progressively from four to ten per session. The training programme resulted in significant increase in the total work during the last three of the four 30 s sprints separated by 4 min rest. This was accompanied by a 49% increase in PFK activity ($P < 0.05$) and 36% and 65% increases in CS and succinate dehydrogenase (SDH) ($P < 0.05$). Also, an increase in $\dot{V}O_{2max}$ from 51.0 ± 1.8 to 54.5 ± 1.5 ($P < 0.05$) was found, suggesting that repeated long sprints (30 s) constitute a powerful aerobic stimulus. In a similar training study with repeated 30 s cycle ergometer sprints, there was a 7.1% increase in mean power over a 30 s sprint and an 8% increase in $\dot{V}O_{2max}$ (Barnett *et al.*, 2004). Interestingly, these authors reported a 42% increase in CS activity but no increase in PFK or anaerobic energy provision from PCr or glycolysis. They suggested that the improvement in 30 s sprint performance was probably mediated by increased energy provision from oxidative metabolism.

Sprint training versus traditional endurance training

More recent studies have focused on the repeated 30 s sprint protocol as an efficient way to increase oxidative potential of the muscle. A series of studies by Burgomaster *et al.* (2005, 2006, 2008) have shown that training with repeated 30 s sprints results in large increases in oxidative enzymes such as CS, cytochrome c oxidase (COX) and HAD. The most important outcome of these training studies is that the adaptations to the repeated-sprint training programme were similar to those resulting from traditional endurance training. For example, following the pioneering study by Burgomaster *et al.* (2005), the same group (Gibala *et al.*, 2006) compared the repeated-sprint training protocol (i.e. between four and six 30 s sprints with 4 min rest intervals, per session) with traditional endurance exercise (e.g. 90–120 min of continuous cycling at 65% $\dot{V}O_{2max}$) performed three times per week for two weeks. The two protocols resulted in similar increases in muscle oxidative capacity, as reflected by the activity of COX, and a similar improvement in an endurance time trial (by 10.1% and 7.5%). It should be noted that continuation of training for four more weeks did not result in any further changes in COX and CS, suggesting that the adaptations in oxidative enzymes occur early in the training process (Burgomaster et al., 2008).

In a subsequent study Burgomaster et al. (2006) reported that this type of repeated-sprint training resulted in an increased activity of the active form of pyruvate dehydrogenase (PDH), and this was accompanied by reduced glycogenolysis (100 ± 16 mmol·kg^{-1} dry weight versus 139 ± 11 mmol·kg^{-1} dry weight ($P = 0.03$) and lower lactate accumulation probably due to a greater mitochondrial pyruvate oxidation. The lower level of acidification due to decreased glycogenolysis will contribute to reduced fatigability following this type of training.

Molecular signalling for mitochondrial adaptations

A key regulator of oxidative enzyme expression in skeletal muscle is peroxisome proliferator-activated receptor-γ coactivator-1α (PGC-1α). PGC-1α coordinates mitochondrial biogenesis, by interacting with various nuclear genes encoding for mitochondrial proteins. Previous work has shown that muscle-specific overexpression of PGC-1α results in the conversion of the muscle from glycolytic to oxidative, with a dramatic upregulation of typical oxidative genes/proteins like COX. This indicates that simply by overexpressing PGC-1α, muscles obtain the characteristics and fatigue resistance found in the endurance-trained state. Calvo et al. (2008) demonstrated these characteristics in PGC-1α transgenic mice which exhibited greatly improved exercise performance and 20% higher peak oxygen uptake compared with wild-type control mice. The fact that PGC-1α protein content in the vastus lateralis muscle was equally increased with the repeated-sprint training protocol as with a prolonged cycling protocol (40–60 min at 65% $\dot{V}O_{2max}$) after six weeks of training (Burgomaster et al., 2008) demonstrates the potential of the repeated-sprint protocol for mitochondrial adaptations. As suggested by Coyle (2005), one of the advantages of the repeated-sprint protocol over the traditional endurance exercise lays on the high level of type II muscle fibre recruitment that is not achieved in the traditional low-intensity endurance exercise. Thus, the mitochondrial adaptations that occur in type II fibres after sprint training are absent when endurance training is performed. These adaptations of type II fibres would increase their fatigue resistance, but the exact effect of this mechanism of adaptation has yet to be determined.

Capillarisation and blood flow

Capillary supply and fatigue resistance

The model of chronic low frequency stimulation of animal muscles has shown that fatigue resistance increases in parallel with an enhanced oxidative capacity. However, fatigue resistance reaches its maximum while CS activity continues to rise (Simoneau et al., 1993). The dissociation between these changes points to additional factors, such as enhanced perfusion and proliferation of capillaries playing an important role in fatigue resistance (Pette and Vrbova, 1999).

The role of capillary supply for lactate elimination after intense exercise has long been suggested (Tesch and Wright, 1983). In addition to the role of the different lactate and H^+ transport mechanisms out of the exercising muscle, improved perfusion contributes to the increased release from muscle to the blood (Juel, 2008). This may be accomplished by improved microcirculation following training. The influence of intermittent intense leg extension training for seven weeks (1 min exercise, 3 min rest for one hour at ~150% of leg $\dot{V}O_{2max}$, 3–5 times weekly) on capillary growth was examined by Jensen et al. (2004). They reported an increase of capillary-to-fibre ratio from 1.74 ± 0.10 to 2.37 ± 0.12 capillaries per fibre, and a 17% increase in capillary density. A similar (20%) increase of capillaries in contact with type I and type II muscle fibres was found after training. This would increase oxygen extraction and facilitate aerobic metabolism during exercise, as well as PCr resynthesis during the recovery intervals. As shown in another similar study by Krustrup et al. (2004), the positive effects of increased muscle oxygen uptake, blood flow and vascular conductance are observed in the initial phase of high-intensity exercise, facilitating a faster adjustment of aerobic metabolism.

Peripheral adaptations involving increased mitochondrial function and capillary supply may also explain the faster time constant of PCr resynthesis (from 25.4 ± 1.7 s to 21.2 ± 1.7 s ($P < 0.05$)) after two weeks of endurance training, reported in one of the few studies examining PCr resynthesis after training (McCully et al., 1991).

Regulation of ionic balance

Improvement of K^+ regulation with sprint training

The marked increases in extracellular K^+ that are commonly observed during high-intensity exercise contribute to muscle fatigue by causing depolarisation of the sarcolemmal and t-tubular membranes (McKenna et al., 2008). An increase in Na–K^+ adenosine triphosphatase (ATPase) activity caused by systematic training has been shown to contribute to the control of K^+ homeostasis and reduce fatigue.

In a recent study Iaia *et al.* (2008) took endurance athletes who were regularly training using long-distance running, and switched their training into repeated sprinting (between eight and 12 30 s running sprints at 90–95% of maximal speed, with 3 min rest intervals), four times per week for four weeks. A matched control group continued training for that four-week period with endurance training, as before the study with 40–60 min continuous runs. This switch to sprint training resulted in an increase of the muscle Na^+–K^+ ATPase α1-subunit by 29%, with a concomitant decrease in plasma K^+ concentration during exercise. Also, performance in a 30 m sprint test and running time to exhaustion at ~130% $\dot{V}O_{2max}$ were increased by 7% and 27% ($P < 0.05$). No changes in $\dot{V}O_{2max}$ or 10 km running time were seen.

In another study, Mohr *et al.* (2007) compared the effects of two different intense training regimens on changes in muscle ATPase subunits and fatigue. Participants were divided into a sprint-training group (15 6 s sprints with 1 min rest intervals) and a speed endurance group (eight 30 s runs at 130% $\dot{V}O_{2max}$, with 1.5 min rest intervals). Training was performed 3–5 times per week and lasted for eight weeks. The fatigue index during a repeated-sprint performance test (five 30 m sprint running test with 25 s active recovery), calculated as the difference between the best sprint time and the time for the fifth sprint, was significantly reduced (by 54%) only in the speed endurance group, and remained unchanged in the sprint group. The reduction in fatigue was accompanied by a 68% increase in Na^+–K^+ ATPase isoform α2 and a 31% increase in the amount of the Na^+/H^+ exchanger isoform, only in the speed endurance group. It must be noted that during each session of speed endurance training, blood lactate (peak values: 14.5–16.5 mmol·l⁻¹) and plasma K^+ (peak value ~6.4 mmol·l⁻¹) were higher compared to the sprint-training responses (peak blood lactate: ~8.5 mmol·l⁻¹ and peak K^+: ~5.5 mmol·l⁻¹).

Collectively, the results of the above studies suggest that high-intensity exercise in the form of repeated 30 s bouts constitutes a better stimulus for K^+ regulation during exercise, compared with both continuous endurance training and short (~6 s) sprint training. This is probably related to the significant disturbance of muscle and blood ion homeostasis caused by this type of high-intensity exercise and may explain part of the greater increase in fatigue resistance after training (Mohr *et al.*, 2007).

Lactate and H⁺ regulation

The pH regulating systems in skeletal muscles are very responsive to high-intensity training (Juel, 2008). During high-intensity exercise and the subsequent recovery period, muscle pH is regulated by three systems: (a) lactate/H^+ co-transport by two important monocarboxylate transporter proteins: MCT1 and MCT4; (b) Na^+/H^+ exchange by a specific exchanger protein; and (c) Na^+/bicarbonate transporters (Juel, 2008). Of those systems, the MCT1 and MCT4 transporters are the most important during exercise, and thus their changes following training have been studied in animal and human muscle. Thomas *et al.* (2007) reported that high-intensity training of rats for five weeks resulted in changes of 30% and 85% in the MCT1 and Na^+/bicarbonate transporter, respectively, while MCT4 remained unchanged. In humans, changes in the Na^+/H^+ exchanger protein levels by 30% have been reported in the four-week high-intensity sprint-training study of Iaia *et al.* (2008), mentioned in the previous section. Moreover, significant increases in MCT1 and Na^+/H^+ exchanger protein densities have been found after high-intensity training, especially when training bouts cause a significant accumulation of H^+ in the muscle (Mohr *et al.*, 2007). Increased expression of lactate and H^+ transporters results in faster H^+ and lactate release (see Juel *et al.*, 2004, key paper 5). The improved lactate and H^+ transport out of the muscle results in higher muscle lactate production measured at exhaustion (Mohr *et al.*, 2007), suggesting that adaptations of these acid-balance regulatory systems improve performance by allowing greater contribution of glycolysis to energy supply.

Changes in muscle buffering capacity

The ability to buffer the build-up of free H^+ in the muscle during high-intensity exercise is an important determinant of fatigue resistance during sprint exercise (Bishop *et al.*, 2004) and is greater in sprint-trained athletes and team-sport players compared to endurance-trained athletes (Edge *et al.*, 2006b). Thus, it may be inferred that the intensity of training may be an important stimulus to increase in βm.

To test this hypothesis, Edge *et al.* (2006a) trained recreationally active female team-sport players for five weeks (three days per week), using two protocols with different intensity, but matched for total work. The high-intensity group performed between six and ten 2 min bouts of cycling with 1 min rest intervals at an intensity that was 120–140% of that corresponding to the blood lactate threshold. The moderate-intensity group performed continuous exercise at 80–95% of that corresponding to the lactate threshold for 20–30 min, so that the total work was the same as the high-intensity group. Blood lactate at the end of a typical training session was $16.1 \pm 4.0 \, mmol \cdot l^{-1}$ for the high-intensity group and only $5.1 \pm 3.0 \, mmol \cdot l^{-1}$ for the moderate-intensity exercise group.

Although $\dot{V}O_{2max}$ and the intensity corresponding to lactate threshold were equally improved in the two groups by about 10–14%, only the high-intensity group showed a significant increase in βm measured *in vitro* at 25% (from $123 \pm 5 \, \mu mol \, H^+ \cdot g$ dry muscle$^{-1} \cdot pH^{-1}$ to $153 \pm 7 \, \mu mol \, H^+ \cdot g$ dry muscle$^{-1} \cdot pH^{-1}$ ($P < 0.05$)), while no changes were observed in the moderate-intensity group. The improvement in repeated-sprint exercise (five 6 s cycling sprints with 24 s rest) was also greater in the high-intensity compared with the moderate-intensity group (13.0% versus 8.5% ($P < 0.05$))

(Edge *et al.*, 2005). These results highlight the contribution of βm to the increase in fatigue resistance observed after training, and emphasise the importance of training intensity for favourable adaptations to occur.

Key points (effects of systematic training)

- Fibre type transformation during exercise training is usually towards the intermediate type IIA isoform at the expense of both type I and type IIX MHC isoforms.
- Sprint training with repeated short bouts (<10s) results in increased activity of glycolytic enzymes such as glycogen phosphorylase, PFK and LDH, with little or no increase in oxidative enzymes.
- Training with longer sprints (30s) results in significant increases not only in glycolytic enzymes, but also in oxidative enzymes, such as CS, COX and PDH.
- A remarkable increase in endurance performance is attained when training with four to six 30s sprints with 4min rest intervals, even after six training sessions over two weeks.
- The advantages of the repeated-sprint protocol over the traditional endurance exercise, lays on the high level of type II muscle fibre recruitment that is not achieved in the traditional low-intensity endurance exercise.
- Muscle capillarisation of both slow and fast fibres is increased with high-intensity training, and this results in enhanced fatigue resistance due to improved oxygen extraction, PCr resynthesis and removal of H^+ and lactate.
- Repeated 30s bouts of high-intensity exercise are an effective method of increasing fatigue resistance, through expression of specific proteins (MCT1, $Na^+–H^+$ exchanger, $Na^+–K^+$ ATPase) that improve regulation of K^+, H^+ and lactate ions.
- High-intensity training that results in high rates of muscle and blood lactate accumulation is necessary to increase m and delay fatigue during repeated sprinting.

REDUCTIONS IN ACTIVITY LEVEL AND FATIGUE: EFFECTS OF DETRAINING AND IMMOBILISATION

Effects of detraining

Detraining is a period of insufficient or reduced training stimulus that causes various adaptations to both skeletal muscle and the nervous system.

By definition, during a detraining period, there is a partial or complete reversal of the training-induced adaptations, thus compromising performance. However, there is evidence to suggest that muscular and neural adaptations may be reversed at different rates, while muscle fibre phenotype may be altered towards an unexpected direction, i.e. an overshoot of the MHC IIX isoform relative to the pre-training levels (Andersen and Aagard, 2000).

Neuromuscular adaptations and fibre type shifts

The changes in force–velocity relationship, EMG activity and muscle CSA following the cessation of heavy resistance training were examined by Andersen et al. (2005). The participants were sedentary young males who trained their knee extensors three times per week for three months. The exercises performed in every session for 4–5 sets each were leg press, hack squat, knee extensions and leg curls, while loads ranged from 10–12 RM in the first sessions, to 6–8 RM in the later training phase. Testing was performed before the start of training, after three months of training and again three months after detraining. In response to training, anatomical quadriceps CSA and EMG were increased by 10%. Also, isokinetic muscle strength at $30°·s^{-1}$ and $240°·s^{-1}$ was increased by 18% ($P < 0.01$) and 10% ($P < 0.05$), but power, velocity and acceleration of unloaded knee extension was unchanged. The proportion of MHC IIX decreased from $5.6 \pm 0.8\%$ to $0.8 \pm 0.3\%$ ($P < 0.001$), whereas that of MHC IIA increased from $34.0 \pm 2.5\%$ to $39.4 \pm 2.0\%$ ($P < 0.001$). After three months of detraining, isokinetic CSA, EMG and muscle strength and power at $30°·s^{-1}$ and $240°·s^{-1}$ returned to pre-training levels. However, unloaded knee extension angular velocity and power were increased remarkably by 14% and 44% in relation to pre- and post-training. This was accompanied by an increase in MHC IIX isoform from $0.8 \pm 0.3\%$ to $7.7 \pm 1.1\%$, which was significantly higher compared with both pre- and post-training levels ($P < 0.001$). Furthermore, the rate of force development during an electrically evoked twitch was increase by 23%, further indicating the changes in contractile properties of the muscle. A similar fibre type shift towards the MHC IIX isoform had been reported previously and to a larger extent (from $2.0 \pm 0.8\%$ to $17.2 \pm 3.2\%$, $P < 0.01$), following a similar protocol of training and detraining (Andersen and Aagard, 2000). These authors also noted a tendency for a selective decrease in type II fibre size with detraining.

Changes in muscle enzymes and capillarisation

One of the main characteristics of muscular detraining is a marked decrease in oxidative capacity, as indicated by a marked reduction in mitochondrial enzyme activities. Linossier et al. (1997) reported that after ten weeks of sprint training with two series of fifteen 5 s sprints separated by

55 s rest intervals, four times per week, peak power increased by 28%, while $\dot{V}O_{2max}$ by only 3%. This was accompanied by higher activities of glycolytic enzymes such as glycogen phosphorylase (9%), PFK (17%) and LDH (31%), while there were no changes in the oxidative enzymes CS and HAD. After seven weeks of detraining, the gains in peak cycling power were maintained, but there was a decrease in $\dot{V}O_{2max}$ by 4%, accompanied by a decrease in the oxidative markers CS and HAD at or below the pre-training values, with no change in glycolytic enzymes. Similar results suggesting that detraining of that length has negligible effects on glycolytic enzymes but exerts a significant effect on oxidative enzymes have also been reported following a mixed continuous and high-intensity interval 15-week training programme, interrupted for seven weeks (Simoneau *et al.*, 1987). However, some studies have reported decreases in glycolytic enzymes in highly trained athletes who stop training for 4–8 weeks (Mujika and Padilla, 2001).

As shown earlier, changes in muscle capillarisation constitute an important part of adaptation to both aerobic and high-intensity training. The effects of training cessation on capillary density have not been established clearly, with contradictory results being reported. Coyle *et al.* (1984) measured changes in aerobic performance and capillarisation in endurance-trained individuals who stopped training for a period of 84 days. They reported that although $\dot{V}O_{2max}$ and CS and SDH activity were decreased early in the detraining period (–7%, –17.1% and –18.5% in 12 days, respectively), skeletal muscle capillarisation was unchanged during the 84 day detraining, whether expressed as capillaries around a fibre, capillaries per fibre or capillaries per mm^2. However, another study by Klausen *et al.* (1981), using a period of endurance training followed by detraining in previously sedentary individuals, showed an increase in all indices of fibre capillarisation by 20–30% for all fibre types. During eight weeks of detraining, the number of capillaries around each fibre returned to pre-exercise values, but the number of capillaries per mm^2 remained at the post-training level due to a decrease in muscle fibre area. The authors argued that this may point to a favourable long-term effect on the average diffusion distance.

Thus, detraining seems to affect mostly aerobic metabolism, with decreases in oxidative enzymes and $\dot{V}O_{2max}$, while anaerobic performance and power seem to be maintained, or even increased, as seen in the case of unloaded knee extension (Andersen *et al.*, 2005), due to the shift of MHC composition towards the fast IIX isoform. However, one should keep in mind that muscle fibre area, especially of the fast fibres, is reduced during detraining, and this would reduce the power generated, especially when the movement is performed against a load (e.g. body weight) or an athletic implement (e.g. shot or javelin).

Detraining and muscle fatigue

Little is known about the effects of changes in muscle fibre composition and enzymatic profile on muscle fatigability following detraining. However, it may be speculated that the maintenance of the gains in peak power (Linossier *et al.*, 1997), together with the decreases in muscle oxidative capacity and the increases is the faster, more fatigable MHC IIX isoform, would increase fatigue during high-intensity exercise following a period of detraining. However, a short (approximately two-week) "tapering" period of decreased training volume (by 40–60%), without changes in training intensity and frequency, is commonly used by athletes to maximise performance gains (Bosquet *et al.*, 2007). This short period of reduced training volume would take advantage of the positive adaptations of detraining, while at the same time would avoid the negative long-term effects of reduced activity.

Effects of immobilisation/disuse

Changes in muscle mass and function

Short term immobilisation of a muscle or a limb may be necessary after an acute injury. In cases where bed rest is required due to illness, gravitational unloading, especially of the leg muscles, is an "unwanted" consequence. One of the first negative adaptations to immobilisation is a decrease in muscle size, i.e. muscle atrophy, accompanied with decreases in functional capacity. Many experimental models, such as space flight, hind limb suspension in animals, immobilisation of joints and denervation have been used to study the effects of disuse on skeletal muscle (Degens and Alway, 2006).

The loss of muscle strength during a period of 4–6 weeks of unloading has been largely attributed to the loss of contractile proteins (Degens and Alway, 2006). However, neural factors are also involved, and this is demonstrated by unloading interventions lasting less than two weeks. Deschenes *et al.* (2002) hypothesised that the loss of strength resulting from a two-week unilateral lower limb unloading was due to impaired neural activation of the affected muscle. The lower limb of healthy young college students was immobilised in a lightweight orthopaedic knee brace at an angle of 70°, so as to eliminate weight-bearing activity. After two weeks of immobilisation, peak isokinetic torque of the knee extensors across a range of velocities was reduced by an average of 17.2%, with greater losses in slow than in fast contraction velocities. The reduction in torque was coupled by reduced EMG activity, but the ratio of total torque/EMG was unchanged (neuromuscular efficiency ratio). Muscle fibre composition remained unchanged in the two-week unloading period.

Changes in fatigue resistance

In the immobilisation study of Deschenes *et al.* (2002), fatigue resistance was assessed as the difference in total work produced during the first ten repetitions compared with the last ten repetitions, during a 30 repetition set of isokinetic knee extensions at $3.14°·s^{-1}$. By calculating this percentage decrease of work, fatigue resistance was enhanced ($29.8 \pm 2.5\%$ versus $20.6 \pm 6.5\%$ ($P < 0.05$)). However, the total work generated over the 30 contractions was 15% less after immobilisation ($2,735.3 \pm 207.6$ J versus $2,339.0 \pm 163.3$ J ($P < 0.05$)). As mentioned earlier in the chapter, the decreased fatigue index after immobilisation should be interpreted with caution. As also reported by the authors (Deschenes *et al.*, 2002), this improvement was mainly due to the lower total work in the first ten repetitions after immobilisation, while total work during the last ten repetitions was unchanged.

The situation may be different when immobilisation is longer than four weeks, where there is an increased fatigability, possibly due to reductions in oxidative capacity due to decreases in CS and PDH. Indeed, Ward *et al.* (1986), showed that after five weeks of immobilisation, the proportion of PDH in the active form was only 52%, compared with 98% after five months of training. This resulted in greater lactate accumulation during exercise after the immobilisation period.

A decrease in capillary supply and blood flow during rest and exercise is common in the disused muscle. However, it should be mentioned that the capillary loss and reduction in maximal blood flow are largely proportional to the loss of muscle mass, maintaining blood flow per unit of muscle mass (Degens and Alway, 2006).

Key points (effects of detraining and immobilisation)

- Cessation of training for a long period (three months) following an equal period of systematic resistance training results in a complete loss of the gains in muscle CSA and EMG activity and muscle strength of concentric contractions.
- An increase in fast (MHC IIX) myosin expression above the pre-training levels is observed after detraining, and this is accompanied by increases in unloaded knee extension velocity and power. Due to muscle fibre atrophy, this shift in MHC isoform does not result in improved functional capacity, but may rather predispose towards increased fatigability.
- Detraining is characterised by large decreases in oxidative enzymes and $\dot{V}O_{2max}$, while the activity of glycolytic enzymes and gains in power output are better maintained during 4–8 weeks of reduced activity.
- The loss of muscle mass and strength is greater during immobilisation than detraining. However, a decrease in EMG is observed even after only two weeks of immobilisation.

- Fatigue resistance is seemingly increased following two weeks of immobilisation, but this is due to a decreased initial power output during the fatiguing test.
- Longer periods of immobilisation (greater than four weeks) result in increased fatigability due to reductions in muscle oxidative capacity and capillary supply.

CASE STUDY EXAMPLE: PROBLEM AND RESOLUTION – REDUCTION OF FATIGUE IN SOCCER

Background

Perhaps the most difficult, but also very fascinating, part of a theory development is when it is tested in real-life conditions. Fatigue during soccer play has an impact on both physical and technical performance, especially in the second half and towards the end of the match (Mohr *et al.*, 2003). This is, in many cases, the time when the opponents may take advantage of the reduced performance and win a match. Therefore, the ability of players to resist fatigue and maintain their high-intensity running ability towards the end of a match is an important goal of physical conditioning.

Advances in technology of match analysis using image recognition systems have allowed quantification of individual performances during a match. The distance covered with high-intensity running (speed: $14.4–19.8\,km\cdot h^{-1}$), very high-intensity running (speed: $19.8–25.2\,km\cdot h^{-1}$) and sprinting (speed: $>25.2\,km\cdot h^{-1}$), seem to be the most sensitive indicators of fatigue during the match, and differentiates top class from moderate-level players (Mohr *et al.*, 2003). On the other hand, total distance run during a match is not always indicative of performance, and teams covering the same distance in a match may have totally different speed profiles.

Problem

The questions then arise:

1 What is the appropriate training regimen for improving fatigue resistance in demanding and long-lasting repeated-sprint sports such as soccer?

2 How can we use the findings of the recent studies to assist coaches and fitness trainers to delay fatigue in repeated-sprint sports?

Resolution: repeated-sprint training

Repeated short bouts

One recently emerging training strategy concerning physical conditioning for football is the use of repeated sprints. A very recent study (Bravo *et al.*, 2008) compared the effects of the commonly used high-intensity aerobic interval training (four 4 min runs at 90–95% of maximal heart rate, with 3 min active recovery) with repeated-sprint training (three sets of six 40 m all-out "shuttle" sprints with 20 s passive recovery between sprints and 4 min between sets). A "shuttle" sprint involves changes in direction of sprinting by 180° every 10 m (first three training weeks) or every 20 m (remaining four weeks). Surprisingly, the repeated-sprint group, compared with the aerobic interval training group, showed a greater improvement, not only in repeated-sprint performance, but also in the soccer-specific "yo-yo" intermittent recovery test (28.1% versus 12.5% ($P < 0.01$)). This test consists of 20 m shuttle runs with increasing speed, with 10 s of active recovery between runs until exhaustion. This usually covers 1.8–2.5 km. A similar improvement in $\dot{V}O_{2max}$ (6%) was found for the two groups. These findings suggest that repeated-sprint training may be very effective to improve aerobic football-specific fitness.

Repeated longer (30 s) bouts

Repeated bouts of 30 s high-intensity running/sprinting with recovery intervals of 1.5–3 min can be used to achieve adaptations that improve ionic balance regulation (K^+, H^+, and lactate) and thus delay fatigue during high-intensity exercise. The results of two recent studies (Mohr *et al.*, 2007 and Iaia *et al.*, 2008) suggest that increases in ionic transport proteins and Na^+–K^+ ATPase enable faster recovery during repeated sprints. However, single-sprint performance is not increased with this method, and thus it has to be used in conjunction with short sprint training.

Caution

Although the above methods will result in improved performance and reduced fatigue within 4–8 weeks, trainers should be aware that the possibility of overreaching or overtraining and the risk of injuries are increased when repeated-sprint/high-intensity training load is increased.

CONCLUSION

Individuals with different training status may exhibit large differences in fatigability during high-intensity exercise. Factors that are related to increased fatigue resistance include a high proportion of type I fibres, high $\dot{V}O_{2max}$, increased concentration of oxidative enzymes and capillary density, as well as high coordination between agonist and antagonist muscles. Although these features were thought to be improved mainly by endurance exercise, recent studies show that repeated-sprint exercise is a very effective and time-efficient stimulus for both aerobic and anaerobic adaptations that improve fatigue resistance and possibly promote health. However, due to the high exercise intensity, the possible risks and negative effects of this method have to be evaluated before it is applied to unhealthy populations. Finally, cessation of training results in significant decreases in oxidative enzymes and $\dot{V}O_{2max}$ within 4–8 weeks, while glycolytic capacity and anaerobic performance are better maintained.

FIVE KEY PAPERS THAT SHAPED THE TOPIC AREA

Study 1. Pette, D. and Vrbová, G. (1985). Changes in muscle phenotypic expression by altered activity. *Muscle and Nerve*, 8, 676–689.

This highly cited review (more than 500 citations) summarised the results of chronic electrical stimulation experiments that demonstrated the plasticity of muscle. Experiments were performed on rabbit muscle *in situ*, by stimulating the nerve with electrodes either intermittently (for eight hours per day) or continuously (for 24 hours per day) for a long period of time. This type of chronically increased contractile activity by low-frequency stimulation induces a progressive transformation of fast- into slow-twitch muscle fibbers. Changes in the enzyme activity pattern occur already after four days and reach their maximum after 2–3 weeks. The contractile properties of the fast muscles are also modified with slowing of times to peak tension and relaxation and increased fatigue resistance. Adaptive changes also include membrane-associated proteins and transformation of the sarcoplasmic reticulum membranes. Oxygen and fuel supply and removal of metabolites and ions that may cause fatigue are also facilitated by increased capillary density, perfusion and myoglobin content. The fast to slow transformation is completed by an exchange of fast-type with slow-type myosin isoforms, which occurs relatively late in the chronically stimulated muscle. The qualitative similarity of changes evoked by long-term endurance training suggested that fast-to-slow transitions represent a regular response to increased contractile activity. Finally, the possibility that an increased activity, no matter at what frequency it is delivered, would cause a slowing of the fast muscle fibres was also proposed in that review. This suggestion was later confirmed in several studies.

Study 2. Gaitanos, G.C., Williams, C., Boobis, L.H. and Brooks, S. (1993). Human muscle metabolism during intermittent maximal exercise. *Journal of Applied Physiology, 75,* 712–719.

The metabolic pathways supporting energy supply during a repeated-sprint test play a major role in fatigue. Gaitanos *et al.* (1993) examined energy supply from anaerobic metabolism during ten 6s cycle ergometer sprints with 30s rest intervals. Anaerobic ATP production was calculated from changes in ATP, PCr, lactate and pyruvate measured in muscle biopsy samples from vastus lateralis on the first and tenth sprint. The main outcome of that study was that the contribution of anaerobic glycolysis to ATP productions was negligible in the tenth sprint, leaving PCr as the sole anaerobic energy source. The disproportionate fall in anaerobic energy supply (by 70%) and work done in the last sprint (by 27%) led to the suggestion that oxidative metabolism was probably providing the rest of the energy needed to support power generation, making the exercise much less "anaerobic" than previously thought, with significant implications for muscle fatigue.

Study 3. Bogdanis, G.C., Nevill, M.E., Boobis, L.H. and Lakomy, H.K. (1996). Contribution of phosphocreatine and aerobic metabolism to energy supply during repeated-sprint exercise. *Journal of Applied Physiology, 80,* 876–884.

The possible increase in oxidative metabolism to compensate for the decrease in anaerobic pathways during repeated-sprint exercise was examined by Bogdanis *et al.* (1996). Participants performed two 30s cycle ergometer sprints separated by 4min rests, and muscle biopsies from vastus lateralis were taken before and after each sprint, as well as 10s into the second sprint. As also reported by Gaitanos *et al.* (1993), a mismatch between anaerobic energy release and power output was observed in the second sprint. It was largely compensated for by an increased contribution of aerobic metabolism that amounted to 43% of the total energy (Figure 7.6). The importance of aerobic fitness for fatigue and performance recovery was indicated by the significant correlations between $\dot{V}O_{2max}$ and the percentage aerobic contribution to both the first and second sprint ($r = 0.79$ ($P < 0.05$) and $r = 0.87$ ($P < 0.01$), respectively). Also, the percentage PCr resynthesis relative to the resting value was highly correlated with an index of endurance fitness ($r = 0.94$ (see Figure 7.7)).

Study 4. Burgomaster, K.A., Hughes, S.C., Heigenhauser, G.J., Bradwell, S.N. and Gibala, M.J. (2005). Six sessions of sprint interval training increases muscle oxidative potential and cycle endurance capacity in humans. *Journal of Applied Physiology, 98,* 1985–1990.

Based on the increased aerobic contribution when a 30s sprint is repeated, it was proposed that this type of training can have a significant effect on aerobic metabolism and performance. Indeed, Burgomaster *et al.*

(2005) showed that only six training sessions performed over two weeks, with 1–2 days rest (between four and seven 30 s sprints per session, with 4 min rest) resulted in a significant increase in CS activity (by 38%) and a remarkable 100% increase in endurance capacity as defined by time to exhaustion at 80% $\dot{V}O_{2max}$ (from 26 ± 5 min to 51 ± 11 min ($P < 0.05$). The authors have proposed the repeated 30 s sprint method as a time-efficient training strategy to improve aerobic and anaerobic fitness and reduce fatigue. The extremely low time commitment (2.5 min per session for five 30 s sprints, or less than 20 min including the 4 min rest intervals) makes this method attractive, and further research is warranted to examine its possible applications in health and disease. This is because the exercise stimulus that promotes mitochondrial biogenesis also appears to stimulate other healthy metabolic adaptations in skeletal muscle, such as improved insulin action, improved lipoprotein lipase activity and greater clearance of plasma triglycerides (Coyle, 2005).

Study 5. Juel, C., Klarskov, C., Nielsen, J.J., Krustrup, P., Mohr, M. and Bangsbo, J. (2004). Effect of high-intensity intermittent training on lactate and H$^+$ release from human skeletal muscle. *American Journal of Physiology, Endocrinology and Metabolism*, 286, E245–251.

This study used the one-legged knee extensor exercise model to examine changes in muscle-pH regulating systems following intense training. After a seven-week training period with fifteen 1 min bouts of single knee extensions at 150% $\dot{V}O_{2max}$ per day, time to exhaustion was improved by 29%. The rate of lactate release at exhaustion was almost double (19.4 ± 3.6 mmol·min^{-1} versus 10.6 ± 2.0 mmol·min^{-1} ($P < 0.05$)) and the rate of H$^+$ release was ~50% higher (36.9 ± 3.1 mmol·min^{-1} versus 24.2 ± 1.5 mmol·min^{-1} ($P < 0.05$)) for the trained than for the untrained leg. The membrane contents of the MCT1 lactate/H$^+$ co-transporter and Na$^+$/H$^+$ exchanger proteins were increased by 15% and 16%, while blood flow was also increase by 16% in the trained compared to the untrained leg. This study showed that when muscle is stressed with training stimuli that cause high intramuscular lactate and H$^+$ concentration, it adapts by increasing the rate of lactate and H$^+$ transport out of the muscle. These adaptations are done by both changes in specific membrane proteins and structural changes, such as increased capillary density (Jensen *et al.*, 2004), that enhance blood flow and thus transport of lactate and H$^+$ away from the working muscle.

GLOSSARY OF TERMS

ADP	adenosine diphosphate
ATP	adenosine triphosphate
ATPase	adenosine triphosphatase
COX	cytochrome c oxidase
CS	citrate synthase
CSA	cross-sectional area
EMG	electromyogram
ES	electrical stimulation
HAD	hydroxy-acyl-CoA dehydrogenase
LDH	lactate dehydrogenase
MHC	myosin heavy chain
MVC	maximal voluntary isometric contraction
PCr	phosphocreatine
PDH	pyruvate dehydrogenase
PFK	phosphofructokinase
PGC-1α	peroxisome proliferator-activated receptor-γ coactivator-1α
RFD	rate of force development
SDH	succinate dehydrogenase
$\dot{V}O_{2max}$	maximal oxygen uptake
$\tau\dot{V}O_2$	time constant of $\dot{V}O_2$ response

REFERENCES

Allen, D. and Westerblad, H. (2004). Lactic acid: the latest performance-enhancing drug. *Science*, 305, 1112–1113.

Andersen, J.L. and Aagaard, P. (2000). Myosin heavy chain IIX overshoot in human skeletal muscle. *Muscle and Nerve*, 23, 1095–1104.

Andersen, J.L., Klitgaard, H. and Saltin, B. (1994). Myosin heavy chain isoforms in single fibres from m. vastus lateralis of sprinters: influence of training. *Acta Physiologica Scandinavica*, 151, 135–142.

Andersen, L.L., Andersen, J.L., Magnusson, S.P., Suetta, C., Madsen, J.L., Christensen, L.R. and Aagaard, P. (2005). Changes in the human muscle force–velocity relationship in response to resistance training and subsequent detraining. *Journal of Applied Physiology*, 99, 87–94.

Andersen, P. and Henriksson, J. (1977). Capillary supply of the quadriceps femoris muscle of man: adaptive response to exercise. *Journal of Physiology*, 270, 677–690.

Bangsbo, J. and Juel, C. (2006). Counterpoint: lactic acid accumulation is a disadvantage during muscle activity. *Journal of Applied Physiology*, 100, 1412–1413.

Barnett, C., Carey, M., Proietto, J., Cerin, E., Febbraio, M.A. and Jenkins, D. (2004). Muscle metabolism during sprint exercise in man: influence of sprint training. *Journal of Science and Medicine in Sport*, 7, 314–322.

Behm, D.G., Whittle, J., Button, D. and Power, K. (2002). Intermuscle differences in activation. *Muscle and Nerve*, 25, 236–243.

Billaut, F., Basset, F.A., Giacomoni, M., Lemaître, F., Tricot, V. and Falgairette, G. (2006) Effect of high-intensity intermittent cycling sprints on neuromuscular activity. *International Journal of Sports Medicine*, 27, 25–30.

Bishop, D. and Edge, J. (2006). Determinants of repeated-sprint ability in females matched for single-sprint performance. *European Journal of Applied Physiology*, 97, 373–379.

Bishop, D. and Spencer, M. (2004). Determinants of repeated-sprint ability in well-trained team-sport athletes and endurance-trained athletes. *Journal of Sports Medicine and Physical Fitness*, 44, 1–7.

Bishop, D., Edge, J. and Goodman, C. (2004). Muscle buffer capacity and aerobic fitness are associated with repeated-sprint ability in women. *European Journal of Applied Physiology*, 92, 540–547.

Bishop, D., Lawrence, S. and Spencer, M. (2003). Predictors of repeated-sprint ability in elite female hockey players. *Journal of Science and Medicine in Sport*, 6, 199–209.

Bogdanis, G.C., Nevill, M.E., Boobis, L.H. and Lakomy, H.K. (1996). Contribution of phosphocreatine and aerobic metabolism to energy supply during repeated sprint exercise. *Journal of Applied Physiology*, 80, 876–884.

Bogdanis, G.C., Nevill, M.E., Boobis, L.H., Lakomy, H.K. and Nevill, A.M. (1995). Recovery of power output and muscle metabolites following 30 s of maximal sprint cycling in man. *Journal of Physiology*, 482, 467–480.

Bosquet, L., Montpetit, J., Arvisais, D. and Mujika, I. (2007) Effects of tapering on performance: a meta-analysis. *Medicine and Science in Sports and Exercise*, 39 (8), 1358–1365.

Bottinelli, R. (2001). Functional heterogeneity of mammalian single muscle fibres: do myosin isoforms tell the whole story? *Pflugers Archives*, 443, 6–17.

Bravo, D.F., Impellizzeri, F.M., Rampinini, E., Castagna, C., Bishop, D. and Wisloff, U. (2008). Sprint vs. interval training in football. *International Journal of Sports Medicine*, 29, 668–674.

Burgomaster, K.A., Heigenhauser, G.J. and Gibala, M.J. (2006). Effect of short-term sprint interval training on human skeletal muscle carbohydrate metabolism during exercise and time-trial performance. *Journal of Applied Physiology*, 100, 2041–2047.

Burgomaster, K.A., Howarth, K.R., Phillips, S.M., Rakobowchuk, M., Macdonald, M.J., McGee, S.L. and Gibala, M.J. (2008). Similar metabolic adaptations during exercise after low volume sprint interval and traditional endurance training in humans. *Journal of Physiology*, 586, 151–160.

Burgomaster, K.A., Hughes, S.C., Heigenhauser, G.J., Bradwell, S.N. and Gibala, M.J. (2005). Six sessions of sprint interval training increases muscle oxidative potential and cycle endurance capacity in humans. *Journal of Applied Physiology*, 98, 1985–1990.

Calbet, J.A., De Paz, J.A., Garatachea, N., Cabeza de Vaca, S. and Chavarren, J. (2003). Anaerobic energy provision does not limit Wingate exercise performance in endurance-trained cyclists. *Journal of Applied Physiology*, 94, 668–676.

Calvo, J.A., Daniels, T.G., Wang, X., Paul, A., Lin, J., Spiegelman, B.M., Stevenson, S.C. and Rangwala, S.M. (2008) Muscle-specific expression of PPARgamma coactivator-1alpha improves exercise performance and increases peak oxygen uptake. *Journal of Applied Physiology*, 104 (5), 1304–1312.

Caputo, F. and Denadai, B.S. (2004). Effects of aerobic endurance training status and specificity on oxygen uptake kinetics during maximal exercise. *European Journal of Applied Physiology*, 93, 87–95.

Casey, A., Constantin-Teodosiu, D., Howell, S., Hultman, E. and Greenhaff, P.L. (1996). Metabolic response of type I and II muscle fibres during repeated bouts of maximal exercise in humans. *American Journal of Physiology*, 271, E38–43.

Colliander, E.B., Dudley, G.A. and Tesch, P.A. (1988). Skeletal muscle fibre type composition and performance during repeated bouts of maximal, concentric contractions. *European Journal of Applied Physiology and Occupational Physiology*, 58, 81–86.

Cooke, S.R., Petersen, S.R. and Quinney, H.A. (1997). The influence of maximal aerobic power on recovery of skeletal muscle following anaerobic exercise. *European Journal of Applied Physiology and Occupational Physiology*, 75, 512–519.

Costill, D.L., Daniels, J., Evans, W., Fink, W., Krahenbuhl, G. and Saltin, B. (1976). Skeletal muscle enzymes and fibre composition in male and female track athletes. *Journal of Applied Physiology*, 40, 149–154.

Coyle, E.F. (2005). Very intense exercise-training is extremely potent and time efficient: a reminder. *Journal of Applied Physiology*, 98, 1983–1984.

Coyle, E.F., Coggan, A.R., Hopper, M.K. and Walters, T.J. (1988). Determinants of endurance in well-trained cyclists. *Journal of Applied Physiology*, 64, 2622–2630.

Coyle, E.F., Martin III, W.H., Sinacore, D.R., Joyner, M.J., Hagberg, J.M. and Holloszy, J.O. (1984). Time course of loss of adaptations after stopping prolonged intense endurance training. *Journal of Applied Physiology*, 57, 1857–1864.

D'Antona, G., Lanfranconi, F., Pellegrino, M.A., Brocca, L., Adami, R., Rossi, R., Moro, G., Miotti, D., Canepari, M. and Bottinelli, R. (2006). Skeletal muscle hypertrophy and structure and function of skeletal muscle fibres in male body builders. *Journal of Physiology*, 570, 611–627.

Dawson, B., Fitzsimons, M., Green, S., Goodman, C., Carey, M. and Cole, K. (1998). Changes in performance, muscle metabolites, enzymes and fibre types after short sprint training. *European Journal of Applied Physiology and Occupational Physiology*, 78, 163–169.

Degens, H. and Alway, S.E. (2006). Control of muscle size during disuse, disease, and aging. *International Journal of Sports Medicine*, 27, 94–99.

Deschenes, M.R., Giles, J.A., McCoy, R.W., Volek, J.S., Gomez, A.L. and Kraemer, W.J. (2002). Neural factors account for strength decrements observed after short-term muscle unloading. *American Journal of Physiology, Regulatory, Integrative and Comparative Physiology*, 282, R578–583.

Dupont, G., Millet, G.P., Guinhouya, C. and Berthoin, S. (2005). Relationship between oxygen uptake kinetics and performance in repeated running sprints. *European Journal of Applied Physiology*, 95, 27–34.

Edge, J., Bishop, D. and Goodman, C. (2006a). The effects of training intensity on muscle buffer capacity in females. *European Journal of Applied Physiology*, 96, 97–105.

Edge, J., Bishop, D., Goodman, C. and Dawson, B. (2005). Effects of high- and moderate-intensity training on metabolism and repeated sprints. *Medicine and Science in Sports and Exercise*, 37, 1975–1982.

Edge, J., Bishop, D., Hill-Haas, S., Dawson, B. and Goodman, C. (2006b).

Comparison of muscle buffer capacity and repeated-sprint ability of untrained, endurance-trained and team-sport athletes. *European Journal of Applied Physiology*, 96, 225–234.

Fitts, R.H. (2008). The cross-bridge cycle and skeletal muscle fatigue. *Journal of Applied Physiology*, 104, 551–558.

Fitzsimons, M., Dawson, B., Ward, D. and Wilkinson, A. (1993). Cycling and running tests of repeated sprint ability. *Australian Journal of Science and Medicine in Sport*, 25, 82–87.

Gaitanos, G.C., Williams, C., Boobis, L.H. and Brooks, S. (1993). Human muscle metabolism during intermittent maximal exercise. *Journal of Applied Physiology*, 75, 712–719.

Garrandes, F., Colson, S.S., Pensini, M., Seynnes, O. and Legros, P. (2007). Neuromuscular fatigue profile in endurance-trained and power-trained athletes. *Medicine and Science in Sports and Exercise*, 39, 149–158.

Gibala, M.J., Little, J.P., van Essen, M., Wilkin, G.P., Burgomaster, K.A., Safdar, A., Raha, S. and Tarnopolsky, M.A. (2006) Short-term sprint interval versus traditional endurance training: similar initial adaptations in human skeletal muscle and exercise performance. *Journal of Physiology*, 575 (3), 901–911.

Grassi, B. (2003). Oxygen uptake kinetics: old and recent lessons from experiments on isolated muscle in situ. *European Journal of Applied Physiology*, 90, 242–249.

Green, H.J. and Pette, D. (1997). Early metabolic adaptations of rabbit fast-twitch muscle to chronic low-frequency stimulation. *European Journal of Applied Physiology and Occupational Physiology*, 75, 418–424.

Greenhaff, P.L., Nevill, M.E., Soderlund, K., Bodin, K., Boobis, L.H., Williams, C. and Hultman, E. (1994) The metabolic responses of human type I and II muscle fibres during maximal treadmill sprinting. *Journal of Physiology*, 478 (Pt 1), 149–155.

Hamada, T., Sale, D.G., MacDougall, J.D. and Tarnopolsky, M.A. (2003). Interaction of fibre type, potentiation and fatigue in human knee extensor muscles. *Acta Physiologica Scandinavica*, 178, 165–173.

Hamilton, A.L., Nevill, M.E., Brooks, S. and Williams, C. (1991). Physiological responses to maximal intermittent exercise: differences between endurance-trained runners and games players. *Journal of Sports Sciences*, 9, 371–382.

Harber, M. and Trappe, S. (2008). Single muscle fiber contractile properties of young competitive distance runners. *Journal of Applied Physiology*, 105, 629–636.

Harridge, S.D., Bottinelli, R., Canepari, M., Pellegrino, M., Reggiani, C., Esbjörnsson, M., Balsom, P.D. and Saltin, B. (1998). Sprint training, in vitro and in vivo muscle function, and myosin heavy chain expression. *Journal of Applied Physiology*, 84, 442–449.

Haseler, L.J., Hogan, M.C. and Richardson, R.S. (1999). Skeletal muscle phosphocreatine recovery in exercise-trained humans is dependent on O_2 availability. *Journal of Applied Physiology*, 86, 2013–2018.

Haseler, L.J., Lin, A., Hoff, J. and Richardson, R.S. (2007). Oxygen availability and PCr recovery rate in untrained human calf muscle: evidence of metabolic limitation in normoxia. *American Journal of Physiology: Regulatory, Integrative and Comparative Physiology*, 293, R2046–2051.

Haseler, L.J., Lin, A. and Richardson, R.S. (2004). Skeletal muscle oxidative metabolism in sedentary humans: ^{31}P-MRS assessment of O_2 supply and demand limitations. *Journal of Applied Physiology*, 97, 1077–1081.

Hautier, C.A., Arsac, L.M., Deghdegh, K., Souquet, J., Belli, A. and Lacour, J.R. (2000). Influence of fatigue on EMG/force ratio and cocontraction in cycling. *Medicine and Science in Sports and Exercise*, 32, 839–843.

Iaia, F.M., Thomassen, M., Kolding, H., Gunnarsson, T., Wendell, J., Rostgaard, T., Nordsborg, N., Krustrup, P., Nybo, L., Hellsten, Y. and Bangsbo, J. (2008). Reduced volume but increased training intensity elevates muscle Na$^+$-K$^+$ pump α1-subunit and NHE1 expression as well as short-term work capacity in humans. *American Journal of Physiology: Regulatory, Integrative and Comparative Physiology*, 294, R966–974.

Jensen, L., Bangsbo, J. and Hellsten, Y. (2004). Effect of high intensity training on capillarization and presence of angiogenic factors in human skeletal muscle. *Journal of Physiology*, 557, 571–582.

Johansen, L. and Quistorff, B. (2003). ^{31}P-MRS characterization of sprint and endurance trained athletes. *International Journal of Sports Medicine*, 24, 183–189.

Juel, C. (2008). Regulation of pH in human skeletal muscle: adaptations to physical activity. *Acta Physiologica*, 193, 17–24.

Juel, C., Klarskov, C., Nielsen, J.J., Krustrup, P., Mohr, M. and Bangsbo, J. (2004). Effect of high-intensity intermittent training on lactate and H$^+$ release from human skeletal muscle. *American Journal of Physiology, Endocrinology and Metabolism*, 286, E245–251.

Karatzaferi, C., de Haan, A., van Mechelen, W. and Sargeant, A.J. (2001). Metabolic changes in single human fibres during brief maximal exercise. *Experimental Physiology*, 86, 411–415.

Klausen, K., Andersen, L.B. and Pelle, I. (1981). Adaptive changes in work capacity, skeletal muscle capillarization and enzyme levels during training and detraining. *Acta Physiologica Scandinavica*, 113, 9–16.

Korhonen, M.T., Cristea, A., Alén, M., Mäkkinen, K., Supilä, S., Mero, A., Viitasalo, J.T., Larsson, L. and Suominen, K. (2006). Aging, muscle fiber type, and contractile function in sprint-trained athletes. *Journal of Applied Physiology*, 101 (3), 906–917.

Krustrup, P., Hellsten, Y. and Bangsbo, J. (2004). Intense interval training enhances human skeletal muscle oxygen uptake in the initial phase of dynamic exercise at high but not at low intensities. *Journal of Physiology*, 559, 335–345.

Kupa, E.J., Roy, S.H., Kandarian, S.C. and De Luca, C.J. (1995). Effects of muscle fibre type and size on EMG median frequency and conduction velocity. *Journal of Applied Physiology*, 79, 23–32.

Lattier, G., Millet, G.Y., Maffiuletti, N.A., Babault, N. and Lepers, R. (2003). Neuromuscular differences between endurance-trained, power-trained, and sedentary subjects. *Journal of Strength and Conditioning Research*, 17, 514–521.

Linossier, M.T., Denis, C., Dormois, D., Geyssant, A. and Lacour, J.R. (1993). Ergometric and metabolic adaptation to a 5-s sprint training programme. *European Journal of Applied Physiology and Occupational Physiology*, 67, 408–414.

Linossier, M.T., Dormois, D., Perier, C., Frey, J., Geyssant, A. and Denis, C. (1997). Enzyme adaptations of human skeletal muscle during bicycle short-sprint training and detraining. *Acta Physiologica Scandinavica*, 161, 439–445.

McCully, K.K., Kakihira, H., Vandenborne, K. and Kent-Braun, J. (1991). Noninvasive measurements of activity-induced changes in muscle metabolism. *Journal of Biomechanics*, 24 (Supplement 1), 153–161.

MacDougall, J.D., Hicks, A.L., MacDonald, J.R., McKelvie, R.S., Green, H.J. and Smith, K.M. (1998). Muscle performance and enzymatic adaptations to sprint interval training. *Journal of Applied Physiology*, 84, 2138–2142.

McKenna, M.J., Bangsbo, J. and Renaud, J.M. (2008). Muscle K$^+$, Na$^+$, and Cl disturbances and Na$^+$–K$^+$ pump inactivation: implications for fatigue. *Journal of Applied Physiology*, 104, 288–295.

Malisoux, L., Francaux, M., Nielens, H., Renard, P., Lebacq, J. and Theisen, D. (2006). Calcium sensitivity of human single muscle fibers following plyometric training. *Medicine and Science in Sports and Exercise*, 38, 1901–1908.

Malisoux, L., Francaux, M. and Theisen, D. (2007). What do single-fibre studies tell us about exercise training? *Medicine and Science in Sports and Exercise*, 39, 1051–1060.

Mendez-Villanueva, A., Hamer, P. and Bishop, D. (2008). Fatigue in repeated-sprint exercise is related to muscle power factors and reduced neuromuscular activity. *European Journal of Applied Physiology*, 103 (4), 411–419.

Mohr, M., Krustrup, P. and Bangsbo, J. (2003). Match performance of high-standard soccer players with special reference to development of fatigue. *Journal of Sports Science*, 21, 519–528.

Mohr, M., Krustrup, P., Nielsen, J.J., Nybo, L., Rasmussen, M.K., Juel, C. and Bangsbo, J. (2007). Effect of two different intense training regimens on skeletal muscle ion transport proteins and fatigue development. *American Journal of Physiology: Regulatory, Integrative and Comparative Physiology*, 292, R1594–1602.

Mujika, I. and Padilla, S. (2001) Muscular characteristics of detraining in humans. *Medicine and Science in Sports and Exercise*, 33 (8), 1297–1303.

Nordlund, M.M., Thorstensson, A. and Cresswell, A.G. (2004). Central and peripheral contributions to fatigue in relation to level of activation during repeated maximal voluntary isometric plantar flexions. *Journal of Applied Physiology*, 96, 218–225.

Osternig, L.R., Hamill, J., Lander, J.E. and Robertson, R. (1986). Co-activation of sprinter and distance runner muscles in isokinetic exercise. *Medicine and Science in Sports and Exercise*, 18, 431–435.

Pääsuke, M., Ereline, J. and Gapeyeva, H. (1999). Twitch contractile properties of plantar flexor muscles in power and endurance trained athletes. *European Journal of Applied Physiology and Occupational Physiology*, 80, 448–451.

Parkhouse, W.S. and McKenzie, D.C. (1984). Possible contribution of skeletal muscle buffers to enhanced anaerobic performance: a brief review. *Medicine and Science in Sports and Exercise*, 16, 328–338.

Pette, D. (1985). Metabolic heterogeneity of muscle fibres. *Journal of Experimental Biology*, 115, 179–189.

Pette D. (1998) Training effects on the contractile apparatus. *Acta Physiologica Scandinavica*, 162 (3), 367–376.

Pette, D. and Vrbová, G. (1999). What does chronic electrical stimulation teach us about muscle plasticity? *Muscle and Nerve*, 22, 666–677.

Racinais, S., Bishop, D., Denis, R., Lattier, G., Mendez-Villaneuva, A. and Perrey, S. (2007). Muscle deoxygenation and neural drive to the muscle during repeated sprint cycling. *Medicine and Science in Sports and Exercise*, 39, 268–274.

Sahlin, K., Tonkonogi, M. and Söderlund, K. (1998). Energy supply and muscle fatigue in humans. *Acta Physiologica Scandinavica*, 162, 261–266.

Sargeant, A.J. (2007). Structural and functional determinants of human muscle power. *Experimental Physiology*, 92, 323–331.

Simoneau, J.A., Kaufmann, M. and Pette, D. (1993). Asynchronous increases in oxidative capacity and resistance to fatigue of electrostimulated muscles of rat and rabbit. *Journal of Physiology*, 460, 573–580.

Simoneau, J.A., Lortie, G., Boulay, M.R., Marcotte, M., Thibault, M.C. and Bouchard C. (1987). Effects of two high-intensity intermittent training programs interspaced by detraining on human skeletal muscle and performance. *European Journal of Applied Physiology and Occupational Physiology*, 56, 516–521.

Takahashi, H., Inaki, M., Fujimoto, K., Katsuta, S., Anno, I., Niitsu, M. and Itai, Y. (1995). Control of the rate of phosphocreatine resynthesis after exercise in trained and untrained human quadriceps muscles. *European Journal of Applied Physiology and Occupational Physiology*, 71, 396–404.

Tesch, P.A. and Wright, J.E. (1983). Recovery from short term intense exercise: its relation to capillary supply and blood lactate concentration. *European Journal of Applied Physiology and Occupational Physiology*, 52, 98–103.

Tesch, P.A., Thorsson, A. and Fujitsuka, N. (1989). Creatine phosphate in fibre types of skeletal muscle before and after exhaustive exercise. *Journal of Applied Physiology*, 66, 1756–1759.

Tesch, P.A., Wright, J.E., Vogel, J.A., Daniels, W.L., Sharp, D.S. and Sjödin, B. (1985). The influence of muscle metabolic characteristics on physical performance. *European Journal of Applied Physiology and Occupational Physiology*, 54, 237–243.

Thomas, C., Bishop, D., Moore-Morris, T. and Mercier, J. (2007). Effects of high-intensity training on MCT1, MCT4, and NBC expressions in rat skeletal muscles: influence of chronic metabolic alkalosis. *American Journal of Physiology, Endocrinology and Metabolism*, 293, E916–922.

Tomlin, D.L. and Wenger, H.A. (2002). The relationships between aerobic fitness, power maintenance and oxygen consumption during intense intermittent exercise. *Journal of Science and Medicine in Sport*, 5, 194–203.

Trappe, S., Harber, M., Creer, A., Gallagher, P., Slivka, D., Minchev, K. and Whitsett, D. (2006) Single muscle fiber adaptations with marathon training. *Journal of Applied Physiology*, 101 (3), 721–727.

Widrick, J.J., Stelzer, J.E., Shoepe, T.C. and Garner, D.P. (2002). Functional properties of human muscle fibres after short-term resistance exercise training. *American Journal of Physiology: Regulatory, Integrative and Comparative Physiology*, 283, R408–416.

Yoshida, T. and Watari, H. (1993). Metabolic consequences of repeated exercise in long distance runners. *European Journal of Applied Physiology and Occupational Physiology*, 67, 261–265.

CHAPTER 8

ERGOGENIC AIDS AND FATIGUE DURING MULTIPLE-SPRINT EXERCISE

David Bishop

OBJECTIVES

The objectives of this chapter are to:

- define skeletal muscle fatigue with special reference to that which occurs during multiple-sprint exercise;
- define multiple-sprint exercise and to differentiate between repeated-sprint tests and intermittent-sprint tests;
- explain how to quantify fatigue during multiple-sprint exercise and the limitations associated with the various methods;
- describe the possible determinants of fatigue during multiple-sprint exercise;
- explain how the use of ergogenic aids may help to understand the mechanisms of fatigue during multiple-sprint exercise.

INTRODUCTION

Human skeletal muscle fatigue can be defined as a transient, exercise-induced reduction in the maximal force capacity of the muscle (Vollestad *et al.*, 1984; Barry and Enoka, 2007) (see Chapter 1). This is especially apparent during multiple-sprint exercise with limited recovery between sprints, where there is a decrease in performance following the first sprint

(Spencer *et al.*, 2005) (Figure 8.1). Despite numerous studies however, there is still no clear explanation for the mechanisms underlying this transient decrease in performance. A better understanding of the underlying mechanisms of fatigue would provide greater insight into how to delay the onset of fatigue and thus improve multiple-sprint performance. Various ergogenic aids have been used to minimise fatigue during multiple-sprint activities and provide a valuable means to experimentally assess possible determinants of fatigue during multiple-sprint tasks. Such information may assist in the design of training programmes which could enhance sports performance (especially in team sports), some occupational tasks, and also exercise capability in clinical situations.

One factor complicating a description of the effects of ergogenic aids on fatigue during multiple-sprint activities is the recent evidence that the mechanisms underlying force decline are likely to be highly task-specific (Enoka and Stuart, 1992; Gandevia, 2001) (see Chapter 1). This means that muscle fatigue can be induced by a combination of processes, contributing in different ways to the decline in force, according to the details of the task (e.g. intensity, duration, mode of contraction, muscle, etc). Therefore, where possible, this chapter will focus on research that has investigated fatigue and/or the use of ergogenic aids specifically during multiple-sprint activities.

Another important factor to consider is that human skeletal muscle fatigue has been reported to be influenced by the biological sex of the individual (see Chapter 6). It needs to be acknowledged however, that the vast majority of the research summarised in this chapter is based on the

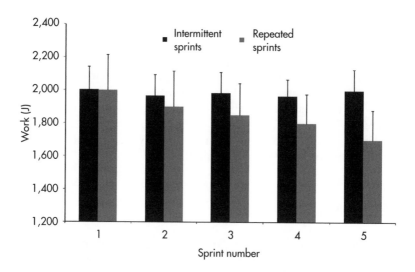

Figure 8.1 The effect of rest duration on maximal sprint performance on a cycle ergometer. Intermittent sprints were performed every 2 min, whereas repeated sprints were performed every 30 s

responses of young adult males. While it has been suggested that there is likely to be little difference in the fatigue response of males and females during multiple-sprint exercise (Billaut and Bishop, 2009), caution should be exercised when extrapolating the conclusions made in this chapter to females.

MULTIPLE-SPRINT EXERCISE

Terminology

While it is relatively easy to define continuous exercise, it is more difficult to arrive at a universally accepted definition for multiple-sprint exercise. Most authors agree that the main feature of multiple-sprint exercise is the alternation of sprint bouts with periods of recovery (consisting of complete rest or moderate-to-low-intensity activity). Where opinions may differ however, is in the definition of what constitutes sprint exercise. For example, it is common for some authors to associate "sprint" with exercise lasting 30s or more (Sharp *et al.*, 1986; Bogdanis *et al.*, 1996). For the purposes of this chapter, the definition of a "sprint" activity will be limited to brief exercise bouts, in general ≤10s, where peak intensity (power/velocity) can be maintained throughout the entire period with only a slight decrement. Longer duration maximal exercise, where there is a considerable decrease in performance, will be referred to as "all-out" exercise (Figure 8.2).

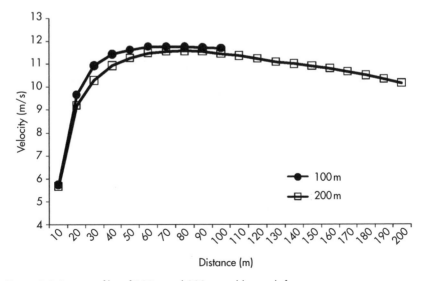

Figure 8.2 Sprint profiles of 100 m and 200 m world records for men

When sprints are repeated, it is also useful to define two different types of exercise – intermittent-sprint exercise, and repeated-sprint exercise. Intermittent-sprint exercise can be characterised by short sprints (≤10 s), performed at an intensity that can be defined as "maximal sprint velocity", interspersed with recovery periods long enough to allow near complete recovery of sprint performance. In comparison, repeated-sprint exercise can be characterised by maximal, short sprints (≤10 s) with a brief recovery (usually ≤60 s). Although subtle, the main difference between these two types of exercise is that during intermittent-sprint exercise there is little or no performance decrement throughout the entire test (Balsom *et al.*, 1992b; Bishop and Claudius, 2005), whereas during repeated-sprint exercise there is a marked performance decrement (Bishop *et al.*, 2004a) (Figure 8.1).

Quantifying fatigue during multiple-sprint exercise

As explained above, during multiple-sprint exercise fatigue is manifested as a progressive decline in power output, the magnitude of which is largely determined by the duration of the intervening recovery periods (Figure 8.1) (Holmyard *et al.*, 1987; Balsom *et al.*, 1992b). To quantify the amount of fatigue experienced during multiple-sprint exercise, researchers have tended to use one of two terms. The fatigue index (FI) has generally been calculated as the drop-off in performance from the best to worst performance during a set of repeated sprints (equation 1).

$$FI = 100 \times \frac{(S_1 - S_5)}{S_1} \qquad (1)$$

S refers to sprint performance on a cycle ergometer and can be calculated for either work or power scores. Note, if these calculations are performed for running (where there is an *increase* in time as subjects fatigue), then S_1 and S_5 should be replaced with the slowest and fastest times respectively. Using the data in Figure 8.1, obtained from sprints performed on a cycle ergometer, the fatigue index would be calculated as:

$$FI = 100 \times \frac{(2,000\,J - 1,700\,J)}{2,000\,J} = 15\%$$

In comparison, the percentage decrement score (S_{dec}) attempts to quantify fatigue by comparing actual performance to an imagined "ideal performance" (where the best effort would be replicated in each sprint) (equation 2) (Bishop *et al.*, 2001; Spencer *et al.*, 2006a; Spencer *et al.*, 2006b). A possible advantage of the per cent decrement method is that it takes into consideration all sprints, whereas the fatigue index will be influenced more by a particularly good or bad first or last sprint.

$$S_{dec}(\%) = \left\{ 1 - \frac{(S_1 + S_2 + S_3 + S_4 + S_5)}{5 \times S_1} \right\} \times 100 \qquad (2)$$

A slight modification of the formula is required for sprint running performance (as times will *increase* as subject's fatigue) (equation 3).

$$S_{dec}(\%) = \left\{ \left\{ 1 - \frac{(S_1 + S_2 + S_3 + S_4 + S_5)}{5 \times S_1} - 1 \right\} \right\} \times 100 \qquad (3)$$

Using the same data in Figure 8.1, obtained from repeated sprints performed on a cycle ergometer, the work decrement (W_{dec}) would be calculated as:

$$W_{dec}(\%) = \left\{ 1 - \frac{(2,000 + 1,900 + 1,850 + 1,800 + 1,700)}{5 \times 2000} \right\} \times 100 = 7.5\%$$

Reliability of fatigue scores

When assessing the effects of an intervention (e.g. administration of a proposed ergogenic aid), it is crucial that any observed changes in performance (or fatigue) can be attributable to the intervention. As such, it is important that performance scores are reliable and do not fluctuate unacceptably from one trial to the next. Both single and total (or mean) sprint performance have been reported to have good reliability, that is coefficients of variation (CVs) less than 4.0% (Capriotti *et al.*, 1999; Wragg *et al.*, 2000; Spencer *et al.*, 2006b; Mendez-Villanueva *et al.*, 2007a; Oliver, 2009). In contrast, calculated fatigue indices have been found to be much less reliable, with CVs ranging from 11% to 50% (Spencer *et al.*, 2006b; McGawley and Bishop, 2006; Oliver, 2009). This is because the subtraction of one variable from another results in relative increases in the within-subject variation with respect to the magnitude of the measure (Oliver, 2009). Consequently, several authors have concluded that any calculated fatigue scores should be viewed with caution (McGawley and Bishop, 2006), and that where possible changes in multiple-sprint performance should be discussed with respect to changes in mean or total sprint time (Oliver, 2009).

Interpretation of fatigue scores

An additional factor to consider when assessing the effects of ergogenic aids on fatigue scores during multiple-sprint exercise is that the various fatigue indices have been positively correlated with initial sprint performance ($0.57 < r < 0.89$ ($P < 0.05$)) (Bishop *et al.*, 2003; Yanagiya *et al.*, 2003; Bishop and Spencer, 2004). This can probably be attributed to the observation that subjects with greater initial sprint performance will have greater changes in muscle metabolites, which in turn have been related to greater performance decrements (Gaitanos *et al.*, 1993; Mendez-Villanueva *et al.*, 2008). This has important implications for the current chapter as ergogenic aids which enhance performance during initial sprint efforts may become ergolytic during the latter stages of exercise, possibly due to a greater increase in the by-products of anaerobic metabolism or other factors associated with the improved initial effort. In this instance, changes in mean or total work may not provide an accurate reflection of changes in multiple-sprint performance and it will be important to also comment on changes in one of the fatigue indices discussed above.

Summary

Given the large variability inherent in calculated fatigue scores, in this chapter the effects of ergogenic aids on fatigue during multiple-sprint exercise will be predominately discussed with respect to changes in mean or total sprint time. However, where the ergogenic aid improves the performance of the first sprint, changes in any fatigue scores will also be noted and discussed.

POTENTIAL MECHANISMS LIMITING PERFORMANCE DURING MULTIPLE-SPRINT EXERCISE

Several mechanisms have been proposed to contribute concurrently to fatigue during multiple-sprint exercise, and the classic approach used to identify the cause of muscle fatigue has been to distinguish between "central" and "peripheral" mechanisms. Typically, peripheral skeletal muscle fatigue involves processes occurring at or distal to the neuromuscular junction (Merton, 1954; Gibson and Edwards, 1985; Bigland-Ritchie et al., 1986; Fuglevand et al., 1993). Although not mutually exclusive, these peripheral factors can be broadly categorised as insufficient supply of energy (due to depletion of energy substrates or failure of the metabolic processes to resynthesise ATP at the required rate) and factors related to the accumulation of metabolic by-products (Figure 8.3). On the other hand, central fatigue is due to failure at a site within the central nervous system (Kent-Braun, 1999; Taylor et al., 2000; Gandevia, 2001). Studies applying an electrical stimulus to peripheral nerves have demonstrated that both "central" and "peripheral" mechanisms are involved during fatiguing contractions, and a number of good scientific reviews on this topic are available (Enoka and Stuart, 1992; Gandevia, 2001). For a more detailed explanation of central and peripheral fatigue, readers are referred to two comprehensive reviews (Fitts, 1994; Gandevia, 2001) (see Chapter 3).

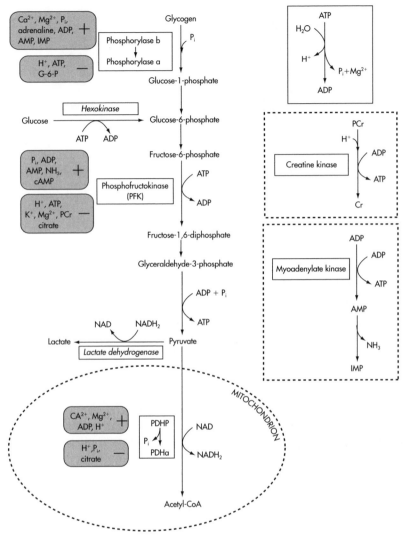

Figure 8.3 Schematic representation of the major metabolic pathways and a number of potential regulators

Peripheral fatigue

Energy supply

Adenosine triphosphate (ATP) is the immediate source of chemical energy for muscle contraction and subsequent movement. As intramuscular stores of ATP are small (20–25 mmol·kg^{-1} dry muscle (dm)), the continual regeneration of ATP is critical for the maintenance of muscle force

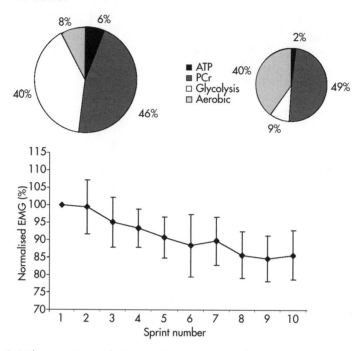

Figure 8.4 Changes in metabolism and neuromuscular activity during repeated-sprint exercise

output during exercise lasting more than a few seconds. At high power outputs (such as those observed during intermittent- or repeated-sprint exercise), this is largely achieved through non-mitochondrial (anaerobic) ATP production; i.e. the breakdown of phosphocreatine (PCr) or the degradation of muscle glycogen to lactate (Figure 8.4). It is important to recall however, that depending on the duration of the task, the number of sprints and the recovery between sprints, there may also be a considerable aerobic contribution (Figure 8.4). Muscular fatigue that develops during multiple-sprint exercise is associated with signs of energy deficiency, i.e. increased concentrations of inosine monophosphate (IMP), inosine, hypoxanthine, and uric acid (Balsom *et al.*, 1992a; Stathis *et al.*, 1999).

ADENOSINE TRIPHOSPHATE

Energy for muscular work is obtained from the hydrolysis of ATP, catalysed by the ATPase enzymes (equation 4), where Mg^{2+} is magnesium, ADP is adenosine triphosphate and P_i is inorganic phosphate.

$$ATP + H_2O + \leftrightarrow ADP + Mg^{2+} + P_i + energy \qquad (4)$$

During a single, short-duration sprint (≤ 6 s), the mean rate of ATP utilisation is ~15 mmol·kg^{-1}·s^{-1} dm (Boobis *et al.*, 1983; Gaitanos *et al.*, 1993), which, without resynthesis, would exhaust the muscle ATP stores within approximately 2 s. However, due to rapid rates of ATP resynthesis, it has been demonstrated that the ATP concentration in mixed muscle fibre samples is reasonably well protected (decreasing by ~30–40%) during both single- and repeated-sprint exercise (Boobis *et al.*, 1982; Cheetham *et al.*, 1986; Boobis, 1987; Gaitanos *et al.*, 1993; Dawson *et al.*, 1997; Edge *et al.*, 2005). While this would suggest that ATP depletion is unlikely to constitute a limiting factor on intermittent- or repeated-sprint performance, greater ATP depletion in fast-twitch (type II) fibres may limit the ability of these fibres to contribute to performance (Casey *et al.*, 1996; Karatzaferi *et al.*, 2001). Furthermore, critical reductions in ATP availability within certain key areas of the muscle cell (e.g. near ATP-dependent enzymes and within the calcium (Ca^{2+}) release channels of the sarcoplasmic reticulum) could potentially impair muscle contraction. Finally, it has been demonstrated that lowering the ATP concentration in skinned skeletal muscle fibres can impair excitation–contraction coupling and force production (Dutka and Lamb, 2004). Thus, ergogenic aids (e.g. ribose) that have been proposed to increase the rate of ATP resynthesis, and better protect ATP stores, have the potential to improve intermittent- and repeated-sprint performance.

PHOSPHOCREATINE

As the intramuscular stores of ATP become depleted, the ATP necessary for continued muscular work is supplied by the breakdown of PCr (equation 5) and the degradation of muscle glycogen to lactate. Phosphocreatine is particularly important during sprints, where a high rate of ATP resynthesis is required.

$$\text{PCr} + \text{ADP} + \text{H}^+ \xleftarrow{\quad\text{Creatine kinase}\quad} \text{ATP} + \text{Cr} \tag{5}$$

Intramuscular PCr stores total approximately 80 mmol·kg dm^{-1}. As maximal turnover rates of PCr degradation can approach 9 mmol·kg dm^{-1}·s^{-1} (Hultman and Sjoholm, 1983), maximal sprinting therefore results in a severe reduction in intramuscular PCr concentration. For example, PCr depletion after 6 s of sprinting has been reported to be around 35–55% of resting values (Boobis *et al.*, 1982; Gaitanos *et al.*, 1993; Dawson *et al.*, 1997; Parra *et al.*, 2000). Furthermore, some of the decrease in performance during repeated-sprint exercise has been attributed to the decrease in the absolute contribution of PCr to the total ATP production from the first sprint to the tenth (Figure 8.4). These decreases indicate that the ability to perform subsequent sprints is likely to be dependent, at least in part, on the degree to which PCr is resynthesised during the recovery periods between multiple sprints. This is supported by studies reporting close relationships ($0.84 < r < 0.86$, $P < 0.05$) between PCr resynthesis and the recovery of performance in different sprinting conditions (Bogdanis et al., 1995; Bogdanis et al., 1996). It has also been shown that active recovery between repeated sprints decreases both PCr resynthesis and performance recovery (Spencer et al., 2006a). Collectively, these results show those ergogenic aids (e.g. creatine supplementation) that can increase the stores of PCr and/or the rate of PCr resynthesis have the potential to improve intermittent and repeated-sprint performance.

MUSCLE GLYCOGEN

The large drop in intramuscular PCr content, along with the consequent rise in P_i, ADP and adenosine monophosphate (AMP), stimulates the rapid activation of anaerobic glycolysis during sprint exercise (Chasiotis et al., 1982; Crowther et al., 2002). Anaerobic glycogenolysis (or glycolysis) involves the breakdown of muscle glycogen (or glucose) to ATP and lactate (equation 6).

$$\text{Glycogen} + 3\ \text{ADP} + 3\ P_i \rightarrow 3\ \text{ATP} + 2\ \text{lactate}^- + 2\ H^+ \tag{6}$$

From muscle lactate concentrations, it has been estimated that ATP production from anaerobic glycolysis may reach maximal rates of around $6\text{–}9\ \text{mmol ATP·kg dm}^{-1}\text{·s}^{-1}$ after approximately 5 s (Hultman and Sjoholm, 1983; Jones et al., 1985; Parolin et al., 1999) and that anaerobic glycolysis supplies approximately 40% of the total energy during a single 6 s sprint (Figure 8.4) (Boobis et al., 1982; Gaitanos et al., 1993). As muscle glycogen content has been reported to range from $250\ \text{mmol·kg dm}^{-1}$ to $650\ \text{mmol·kg dm}^{-1}$, this suggests that glycogen stores are unlikely to represent a limiting factor during most types of multiple-sprint exercise, especially as glycogen use decreases with subsequent sprints (Figure 8.5). However, it should be noted that significant muscle glycogen depletion has been reported during multiple-sprint exercise over the duration of a soccer match (Gaitanos et al., 1993; Rico-Sanz et al., 1999), and that about half of individual muscle fibres are depleted or almost depleted of glycogen following a soccer match (Krustrup et al., 2006). Furthermore, complete depletion of muscle glycogen stores may not be necessary to impair performance as muscle glycogen levels of $\sim200\ \text{mmol·kg dm}^{-1}$ appear necessary to maintain maximal glycolytic rate (Bangsbo et al., 1992a). Therefore, ergogenic aids (e.g. carbohydrate ingestion) that increase or maintain muscle glycogen stores could potentially be important to reduce fatigue during prolonged, multiple-sprint exercise.

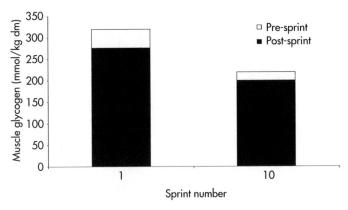

Figure 8.5 Muscle glycogen levels (mmol glucosyl units·kg dm⁻¹) pre- and post-sprint 1 and 10 of a ten 6 s repeated-sprint test (with 30 s of recovery between sprints)

AEROBIC METABOLISM

While the resynthesis of ATP during sprint exercise is largely achieved through non-mitochondrial (anaerobic) ATP production, there is also a small contribution from aerobic metabolism; this is primarily achieved through the oxidation of glucose (equation 7).

$$C_6H_{12}O_6 \text{ (glucose)} + 6O_2 + 38 \text{ ADP} + 38 \text{ P}_i \rightarrow 6 \text{ CO}_2 + 6 \text{ H}_2\text{O} + 38 \text{ ATP}$$

$$(7)$$

There are several methodological problems associated with accurately estimating the aerobic contribution to sprint exercise (Glaister, 2005). Nonetheless, during the first 6 s of a 30 s all-out exercise, the mean rate of aerobic ATP turnover has been estimated to be 1.3 mmol ATP·kg dm⁻¹·s⁻¹ (approximately 10% of the total energy supply (Figure 8.4)) (Parolin *et al.*, 1999). As sprints are repeated, the level of aerobic ATP provision progressively increases such that aerobic metabolism may contribute as much as 40% of the total energy supply during the final sprints of a repeated-sprint test (McGawley and Bishop, 2008). Furthermore, subjects may reach their $\dot{V}O_{2max}$ during the latter sprints. This suggests that aerobic contribution during repeated sprints may be limited by $\dot{V}O_{2max}$ and that increasing $\dot{V}O_{2max}$ may allow for a greater aerobic contribution during the latter sprints, potentially improving performance. This hypothesis may explain why subjects with greater aerobic fitness have an enhanced repeated-sprint ability (RSA), and is supported by correlations between aerobic fitness and RSA (Bishop *et al.*, 2003). While these results suggest that ergogenic aids that increase $\dot{V}O_{2max}$ may improve multiple-sprint performance (e.g. erythropoietin (EPO) (Balsom *et al.*, 1994)), this chapter will be limited to the discussion of legal ergogenic aids.

Metabolite accumulation

Activation of the metabolic pathways that produce ATP also results in increased muscle and plasma levels of numerous metabolic by-products that potentially contribute to fatigue during multiple-sprint exercise (see Figure 8.3). These include Mg^{2+}, ADP, P_i, ammonia (NH_3), hydrogen ions (H^+), potassium (K^+) and reactive oxygen species (ROS).

Mg^{2+}, ADP, P_i

As a result of the rapid breakdown of ATP and PCr during sprinting, there is increased intramuscular accumulation of Mg^{2+}, ADP, and P_i. Increased levels of both Mg^{2+} and P_i can reduce Ca^{2+} release from the sarcoplasmic reticulum (SR) (Dutka and Lamb, 2004), while intramuscular accumulation of ADP can reduce force and slow muscle relaxation by adversely affecting the contractile myofilaments and decreasing Ca^{2+} re-uptake into the SR (Macdonald and Stephenson, 2004). As Ca^{2+} release and uptake from the sarcoplasmic reticulum controls actin–myosin cross-bridge interactions, and thereby regulates force production, any metabolites that influence Ca^{2+} exchange are likely to impair multiple-sprint performance. In addition, increases in both ADP and P_i can allosterically inhibit many metabolic enzymes (Figure 8.3) and thus reduce the resynthesis of ATP from various metabolic pathways (Sahlin et al., 1998). These findings have been used to propose that ergogenic aids which can reduce the accumulation of Mg^{2+}, ADP, P_i have the potential to improve intermittent- and repeated-sprint performance.

AMMONIA (NH_3)

As a by-product of the breakdown of either ATP or amino acids there is an increased release of NH_3 from contracting skeletal muscle into the blood and a corresponding rise in plasma NH_3 levels during sprint exercise. Ammonia in the plasma can then cross the blood–brain barrier and may influence brain neurotransmitters and contribute to central fatigue (Nybo and Secher, 2004). While more research is required to better understand the role of NH_3 accumulation in the aetiology of fatigue, these findings suggests that ergogenic aids that can attenuate the rise in plasma NH_3 levels have the potential to improve multiple-sprint performance. One such ergogenic aid may be carbohydrate ingestion which has been demonstrated to attenuate plasma NH_3 accumulation (Snow et al., 2000) and cerebral NH_3 uptake (Nybo and Secher, 2004) during prolonged, submaximal exercise.

HYDROGEN ION (H⁺) ACCUMULATION

The intense muscle contractions required during sprinting result in large ionic changes and an increased non-mitochondrial ATP turnover, contributing to the accumulation of H^+ (decreased pH). While recent findings have questioned the role of H^+ accumulation in the fatigue process (Pedersen et al., 2004), the accumulation of H^+ may impair multiple-sprint performance through effects on the contractile machinery (interference with the effectiveness of Ca^{2+} activation at many sites in the excitation–contraction process (Fitts, 1994)) and its potential role in reducing the rate of glycolytic ATP production (via negative effects on key glycolytic enzymes such as phosphofructokinase (PFK) and glycogen phosphorylase (Harris et al., 1977; Spriet et al., 1987; Spriet et al., 1989)) (Figure 8.3). Finally, a decline in muscle pH may contribute to the occurrence of central fatigue. Indeed, the typical association between pH and EMG (Kent-Braun, 1999) is consistent with the role of pH in feedback to the central nervous system and a subsequent alteration in central motor drive during the development of fatigue. The suggestion that H^+ accumulation may contribute to fatigue during multiple-sprint exercise is supported by studies demonstrating a correlation between repeated-sprint ability and both muscle buffer capacity and changes in blood pH (Bishop et al., 2003; Bishop et al., 2004b; Bishop and Edge, 2006). Furthermore, greater improvements in repeated-sprint ability following training have also been reported in subjects with greater improvements in muscle buffer capacity (Edge et al., 2006). These findings suggest that ergogenic aids that can reduce the accumulation of H^+ (e.g. alkalising agents) or increase muscle buffer capacity (e.g. β-alanine) have the potential to improve intermittent- and repeated-sprint performance.

POTASSIUM (K⁺)

In addition to increases in H^+ accumulation, sprint exercise also results in other important ionic perturbations that may contribute to fatigue. In particular, sprint exercise increases the extracellular potassium (K^+) ion concentration ($[K^+]$) far beyond the narrow limits seen in resting subjects. For example, there is a greater than 200% increase in plasma $[K^+]$ after a 1 min all-out exercise bout on a motor-driven treadmill. It has been suggested by some (Medbo and Sejersted, 1990; Lindinger and Heigenhauser, 1991; Lindinger et al., 1992; Juel et al., 2000; Mohr et al., 2004) that subsequent alterations in sarcolemma excitability contribute to muscle fatigue by preventing cell activation. Interestingly, it has recently been demonstrated that the ingestion of alkalising agents (either sodium bicarbonate or sodium citrate) can reduce the exercise-induced increase in extracellular K^+ (Street et al., 2005; Sostaric et al., 2006). Thus, multiple-sprint performance may be improved by ergogenic aids that can attenuate the increase in extracellular K^+ (e.g. alkalising agents) or increase $Na^+–K^+$ pump activity (e.g. caffeine).

REACTIVE OXYGEN SPECIES

During skeletal muscle contraction, cellular reactions occurring in the mitochondria and elsewhere result in the production of ROS, such as hydrogen peroxide (H_2O_2) and superoxide (O_2^-) (Sachdev and Davies, 2008). At low levels, ROS in skeletal muscle can produce positive effects on glucose transport, ATPase activity, calcium release and mitochondrial biogenesis (Sakamoto and Goodyear, 2002). However, higher levels of ROS have been implicated in the fatigue process (Moopanar and Allen, 2005). There are several enzymatic antioxidants (e.g. superoxide dismutase, catalase, glutathione peroxidase), as well as non-enzymatic antioxidants (e.g. reduced glutathione, β-carotene and vitamins E and C) within skeletal muscle that act to buffer exercise-induced increases in ROS (Reid, 2001). To date, there is very little evidence that vitamin E and/or C supplementation can increase endogenous enzymatic antioxidant levels. However, administration of N-acetylcysteine has been shown to increase non-enzymatic antioxidant levels in skeletal muscle, and is associated with reduced fatigue during muscle stimulation (Reid et al., 1994) and increased cycle time to fatigue in humans (McKenna et al., 2006). This suggests that intermittent- and repeated-sprint performance may be improved by ergogenic aids that increase antioxidant levels in skeletal muscle (e.g. N-acetylcysteine or vitamins E and C).

Central fatigue

Although not as extensively studied, changes in skeletal muscle recruitment may also contribute to performance decrement during multiple-sprint exercise. For example, while not a universal finding (Hautier et al., 2000; Billaut et al., 2005; Billaut and Basset, 2007), the EMG amplitude signal of the vastus lateralis has been reported to decline in parallel with sprint performance during repeated-sprint exercise (Billaut et al., 2005; Billaut et al., 2006; Mendez-Villanueva et al., 2007b; Racinais et al., 2007), despite the constant maximal effort of the participants for each sprint (Figure 8.4). Moreover, neural adjustments such as a reduction in the central nervous system's drive to the active musculature have also been previously reported during repeated-sprint exercise (Drust et al., 2005; Racinais et al., 2007). Such neural adjustments have been suggested to be caused by an accumulation of metabolites and a consequent decrease in muscle pH, and/or some form of neural control through reflex regulation of muscle force to prevent muscle damage (Linssen et al., 1990; Linnamo et al., 2000; St Clair Gibson et al., 2001). In addition, at least during prolonged exercise, it has been suggested that a fall in plasma branched-chain amino acid (BCAA) levels, and an increase in plasma tryptophan, may lead to increased serotonin levels in the brain and also contribute to central fatigue (Nybo and Secher, 2004). This raises the possibility that ergogenic aids that can attenuate the increase in plasma tryptophan may improve intermittent- and repeated-sprint performance. However, while the ingestion of BCAAs during exercise has been proposed as a strategy to maintain plasma BCAA levels and to reduce brain tryptophan uptake, this does not appear to be effective (van Hall et al., 1995). Instead, carbohydrate ingestion has been shown to attenuate the rise in the free tryptophan–BCAA ratio by lowering the free fatty acid levels during exercise (free fatty acids (FFA) compete with tryptophan for binding sites on plasma albumin) (Davis et al., 1992).

Summary

Fatigue during multiple-sprint exercise appears to be a multi-factorial process that can be attributed to both "central" and "peripheral" mechanisms. Although not mutually exclusive, these peripheral factors can be broadly categorised as an insufficient supply of energy or factors related to the accumulation of metabolic by-products. In particular, fatigue during multiple-sprint exercise has been associated with the depletion of muscle PCr and glycogen stores and the accumulation of NH_3, H^+, K^+, ROS and plasma tryptophan. Various ergogenic aids can be used to manipulate energy stores and metabolite accumulation so as to better assess the contribution of these various factors to fatigue during multiple-sprint exercise.

ERGOGENIC AIDS

Despite more than a century of research, there is still no clear explanation for the mechanisms underlying the transient, exercise-induced reduction in the maximal force capacity of the muscle that occurs during multiple-sprint exercise. While it must be remembered that no one research model will provide all the answers, the use of ergogenic aids provides a valuable means to experimentally assess and/or confirm possible determinants of fatigue during multiple-sprint tasks. In addition, of course, athletes who perform multiple-sprint sports will also be tempted to take various ergogenic aids in the hope that it may enhance performance. The strength of this temptation is underlined by a poll of 198 Olympic-level power athletes (Bamberger and Yaeger, 1997) who were asked to respond to the following scenario: you are offered a banned substance with two guarantees. First, you will not be caught, and, second, by taking the substance you will win. Of the athletes asked if they would take the substance, only three said they would not. Thus, the effectiveness of various ergogenic aids to minimise fatigue and enhance performance during multiple-sprint activities is a topic of great interest to both scientists and athletes.

The term ergogenic is derived from the Greek words "ergon" (work) and "gennan" (to produce). Hence, an ergogenic aid usually refers to something that enhances work. With such a broad definition, ergogenic aids could include nutritional aids (e.g. supplements), physiological aids (e.g. training or taper techniques), psychological aids (e.g. imagery) and biomechanical aids (e.g. factors that modify technique). For the purpose of this chapter, the discussion of ergogenic aids that may improve multiple-sprint performance will be limited to ingestible substances. Furthermore, this chapter will only discuss substances that are not currently banned by the International Olympic Committee (IOC). While the legal implications regarding the use of ergogenic aids are generally well defined, the moral/ethical implications are less clear. The following section is concerned with how ergogenic aids may help to better understand the mechanisms contributing to fatigue during multiple-sprint exercise and does not constitute an endorsement or recommendation of any of the ergogenic aids discussed.

Caffeine

Classification and usage

Caffeine (1,3,7-trimethylxanthine) is the most commonly consumed drug in the world and is found in coffee, tea, cola, chocolate and various "energy" drinks (see Table 8.1). The actions of caffeine throughout the body correlate positively with plasma caffeine levels, which are governed by absorption, metabolism and excretion (Sinclair and Geiger, 2000). Almost 100% of orally administered caffeine is absorbed and it begins to appear in the blood within 5 min of ingestion (George, 2000). Typical experimental doses of caffeine (4–6 mg·kg^{-1} of body mass) will produce peak plasma concentration of 6–8 µg·ml^{-1} within 15–120 min after ingestion; plasma half-life ranges from three to ten hours (Tarnopolsky, 1994). While it remains unclear what the effective minimal or maximal doses are, it appears that ingestion of 2–3 mg·kg^{-1} of body mass of caffeine is sufficient to produce an ergogenic effect (Wiles et al., 1992; Kovaks et al., 1998), whereas ingestion of doses greater than 6 mg·kg^{-1} do not seem to provide a further enhancement of performance (Graham and Spriet, 1991).

Table 8.1 Typical caffeine content in common substances

Substance	Caffeine content (mg)
1 can of cola drink	40
1 cup of tea	50
1 cup of brewed coffee	100
1 No Doz caffeine tablet	100
1 can Red Bull (250 ml)	80
Guarana (100 mg)	100

Possible mechanisms

The ergogenic effects of caffeine have been attributed to a number of possible mechanisms, including; adenosine receptor antagonism (Fredholm *et al.*, 1999), CNS facilitation (Williams, 1991), increased Na^+–K^+ ATPase activity (Lindinger *et al.*, 1993), mobilisation of intracellular calcium (Sinclair and Geiger, 2000), and increased plasma catecholamine concentration (Mazzeo, 1991). It now seems likely that adenosine receptor antagonism is the primary mechanism of action and contributes to improved performance via increases in neurotransmitter release, motor unit firing rates and dopaminergic transmission (Graham, 2001). However, there is some evidence to support each of these mechanisms and it is probable that all contribute to the wide range of physiological responses to caffeine that make it ergogenic. A more detailed discussion on the possible mechanisms underlying the ergogenic effects of caffeine can be found elsewhere (Sinclair *et al.*, 2000a, 2000b).

Effects on multiple-sprint performance

While many studies have demonstrated that caffeine is ergogenic for the performance of prolonged, endurance exercise (Spriet, 1995), there is limited research that has investigated the effects of caffeine on single- or multiple-sprint performance. Following caffeine ingestion, an ~7% increase in maximal anaerobic power has been reported during a single 6 s sprint on a cycle ergometer (Anselme *et al.*, 1992). This ergogenic effect appears to be maintained when sprints are repeated. For example, a similar 7% increase in mean power was reported when ten male team-sport athletes ($\dot{V}O_{2peak}$ $56.5 \pm 8.0\,ml{\cdot}kg^{-1}{\cdot}min^{-1}$) performed an intermittent-sprint test consisting of two 36 min "halves", each comprising eighteen 4 s sprints with 2 min of active recovery at 35% $\dot{V}O_{2peak}$ between each sprint (Schneiker *et al.*, 2006). These results are supported by another study that reported enhanced simulated, high-intensity team-sport performance in competitive male rugby players (Stuart *et al.*, 2005). In contrast, another study by the same research group reported a negligible effect of caffeine ingestion on repeated-sprint performance (ten 20 m sprints performed every 10 s). All three of the above studies used a caffeine dose of 6 mg·kg^{-1} of body mass.

As a result of the improvements in initial sprint performance (Anselme *et al.*, 1992), there is the risk that caffeine ingestion may be ergolytic as fatigue develops, possibly due to an increase in the by-products of anaerobic metabolism. Indeed, in one study examining the effects of a 6 mg·kg^{-1} dose of caffeine on peak and mean power during four 30 s Wingate tests, each separated by 4 min, there was a non-significant trend towards enhanced performance of the first Wingate test, and a significantly reduced performance by the fourth test (Greer *et al.*, 1998). However, contrary to the results of this study (Greer *et al.*, 1998), other researcher have shown that although caffeine was able to significantly enhance performance of intermittent sprints, resulting in increased plasma lactate concentrations, this did not affect the ability of participants to maintain work efforts in the latter stages of the exercise protocol (Schneiker *et al.*, 2006). Thus, while further research is certainly warranted, these results indicate that caffeine ingestion ($6\,mg \cdot kg^{-1}$ body mass) is likely to improve intermittent, but not repeated, sprint performance. Furthermore, there is no apparent increase in the rate of fatigue development attributable to initial improvements in work and power achieved during multiple-sprint tests as a consequence of caffeine ingestion.

Creatine

Classification and usage

Creatine (Cr) is a naturally occurring, non-essential compound. Creatine can be obtained in the diet (from animal-based foods such as fresh fish and meat) or synthesised from the amino acids glycine, arginine and methionine, primarily in the liver, pancreas and kidneys. Creatine exists in free and phosphorylated forms (i.e. PCr), and approximately 95% of the body's creatine is stored in skeletal muscle. Total creatine concentrations in skeletal muscle (i.e. Cr + PCr) average around $120\,mmol \cdot kg\,dm$, with a higher capacity for storage in type II muscle fibres. Following a creatine loading phase (typically $\sim 0.3\,g \cdot kg^{-1} \cdot d^{-1}$ or $20\,g \cdot d^{-1}$ for 5–7 days), muscle creatine levels increase approximately 25% to what appears to be a maximum of about $160\,mmol \cdot kg\,dm^{-1}$ (Harris *et al.*, 1992; Hultman *et al.*, 1996). Thus, athletes can begin multiple-sprint exercise with greater levels of muscle creatine available for energy production. There is however, considerable variability in the increase in muscle creatine following supplementation; some individuals are "non-responders" (little or no increase in muscle creatine), whereas others are "high responders" (>30% increase in muscle creatine) (Harris *et al.*, 1992). As the body breaks down about 1–2 g of creatine per day, it is normally recommended to follow a creatine loading phase with a maintenance phase of $3–5\,g \cdot d^{-1}$.

Possible mechanisms

It is well established that creatine supplementation can increase both total creatine (TCr) and PCr concentrations in the muscle (Harris *et al.*, 1992; Snow *et al.*, 1998; McKenna *et al.*, 1999). As single and multiple sprints produce a severe reduction in intramuscular PCr concentration (Figure 8.6), it has been proposed that increasing muscle PCr stores may improve multiple-sprint performance via a reduced ATP degradation and a faster resynthesis of PCr between multiple sprints (Yquel *et al.*, 2002). In addition

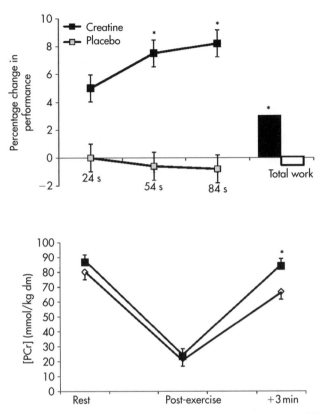

Figure 8.6 Changes in repeated-sprint performance and PCr resynthesis rate following short-term creatine supplementation (20 g·d for 5 days).

Note
* $P<0.05$

to increasing the contribution of PCr to ATP resynthesis, this could poten-tially decrease the reliance on anaerobic glycolysis during multiple-sprint exercise and thus reduce the accumulation of H^+ (Balsom *et al.*, 1993). There is also some evidence that increased muscle creatine levels may enhance oxygen uptake during high-intensity exercise (Rico-Sanz and Mendez-Marco, 2000), possibly via increased shuttling of high-energy phosphates between the cytosol and the mitochondria (i.e. the creatine-phosphate shuttle). While this could potentially contribute to a smaller performance decrement during mul-tiple-sprint exercise, this hypothesis remains to be tested.

Effects on multiple-sprint performance

While far from a universal finding, most research indicates that short-term creatine supplementation can improve multiple-sprint performance

(Kreider, 2003). For example, creatine supplementation ($15-20\,g\cdot d^{-1}$ for 5–6 days) has been reported to improve the performance of six 15 m running sprints interspersed with 30 s of recovery (Mujika *et al.*, 2000), five 10 s cycle sprints (interspersed with 60 s of recovery) (Wiroth *et al.*, 2001) and repeated 6 s cycle sprints with varying recovery intervals (Preen *et al.*, 2001). In the final study, the improved performance was associated with an improved PCr replenishment rate. Although not always explicitly reported, the improved repeated-sprint performance appears to be due to an attenuation of fatigue as creatine supplementation has been reported to have limited effects on single-sprint performance (Peyrebrune *et al.*, 1998).

It is important to note however, that not all studies have reported improvements in repeated-sprint ability following creatine supplementation. For example, one study reported that creatine supplementation ($30\,g\cdot d^{-1}$ for five days) did not affect the performance of five 10 s cycle sprints (interspersed with 180, 50 and 20 s rest periods), despite verifying a significant increase in TCr stores (McKenna *et al.*, 1999). While other studies have also reported no significant improvement in repeated "sprint" performance (Finn *et al.*, 2001), these studies have generally used exercise durations in excess of 10 s and are not considered as sprints for the purpose of this chapter. To date, no studies have investigated the effects of creatine supplementation on intermittent-sprint performance (i.e. maximal sprints efforts interspersed with recovery periods long enough to allow near complete recovery of sprint performance). Although one study did report that creatine supplementation did not affect the performance of two 10 s cycle sprints (interspersed with 180 s of rest), it also reported no effects of creatine supplementation on repeated-sprint performance (McKenna *et al.*, 1999). Thus, further research is required to investigate the effects of creatine supplementation on intermittent-sprint performance, especially as this type of performance is less likely to be influenced by PCr resynthesis rates.

Carbohydrates

Classification and usage

Carbohydrates are molecules containing carbon in a ratio to hydrogen and oxygen as found in water (i.e. CH_2O). Carbohydrates are found naturally in many foods and are typically subdivided into two categories – simple carbohydrates (e.g. "sugars" such glucose, fructose, sucrose and lactose) and complex carbohydrates made from long chains of glucose (e.g. "starches" found in grains and vegetables). Foods do not always occur exclusively as simple or complex carbohydrates, and some natural and processed foods may contain a mixture of carbohydrate types. In addition, there are many commercial carbohydrate supplements (e.g. gels, powders, sports bars and sports drinks). Carbohydrate ingestion to improve multiple-sprint performance normally has two goals:

1   To maximise carbohydrate stores before commencing exercise (i.e. "carbohydrate loading").
2   To maintain blood glucose levels and attenuate the decrease in muscle glycogen during exercise.

While the exact recommendation will depend on the athlete and the multiple-sprint task to be performed, the following general guidelines can be proposed:

1   To obtain an average of 8–10 g of carbohydrate per kilogram of body mass over the 24-hour period preceding exercise, while decreasing this consumption to:

    a 4 g·kg^{-1} body mass four hours before exercise; and
    b 1 g·kg^{-1} body mass one hour before exercise.

2   To drink about 250 ml of a 6% CHO solution (a typical sports drink) every 15 min during exercise.

Possible mechanisms

As the body's carbohydrate stores are limited, it has been suggested that maintaining muscle glycogen stores could potentially be important to attenuate fatigue during prolonged, multiple-sprint exercise (Balsom et al., 1999). As described above, insufficient muscle glycogen stores may contribute to fatigue during multiple-sprint exercise by decreasing the contribution of anaerobic glycogenolysis/glycolysis to energy supply during sprint exercise. In addition, as muscle glycogen levels fall, there will be an increased reliance on fat as an energy source and a consequent increase in blood FFA levels. As FFAs compete with tryptophan for binding sites on plasma albumin, an increase in the free tryptophan–BCAA ratio in the blood may facilitate the entry of tryptophan into the brain, causing increased brain serotonin formation and contributing to central fatigue (Nybo and Secher, 2004).

Effects on multiple-sprint performance

While many studies have demonstrated that carbohydrate ingestion is ergogenic for the performance of prolonged, endurance exercise (for review see Coggan and Coyle, 1991), there is limited research that has investigated the effects of carbohydrate ingestion on single- or multiple-sprint performance. However, in one of the few studies to date, it was reported that the raising of muscle glycogen stores resulted in better maintenance of power output over fifteen 6 s bouts of exercise and longer time to exhaustion when standardised 6 s bouts of high-intensity exercise were repeated until exhaustion (Balsom et al., 1999). In addition, following 45 min of simulated football activity, it has been demonstrated that football players were able to complete more distance (5.5%) during a repeated-sprint test, after carbohydrate loading, compared with their normal mixed diet (Bangsbo *et al.*, 1992b). These results suggest that carbohydrate loading can improve both repeated- and intermittent-sprint ability.

While there are some conflicting findings, carbohydrate ingestion (~200 ml of 6.9% glucose every 15 min) has been reported to reduce muscle glycogen utilisation (22%) during a set of six 15 min simulated team-sport activity (Nicholas *et al.*, 1999). It has also been shown that carbohydrate ingestion (0.5 l of 7% glucose 10 min before a football match and at half time) attenuated the decrease in muscle glycogen concentration compared with a group of players who consumed only water (Leatt and Jacobs, 1989). In theory, this better maintenance of muscle glycogen levels should permit players to better maintain multiple-sprint performance. In support of this, it has been reported that carbohydrate supplementation (~200 ml of 6% carbohydrate every 10 min) resulted in faster 15 m sprint times during a simulated rugby-league match (MacLaren and Close, 2000) and during an intermittent shuttle running test (Ali *et al.*, 2007) than consumption of a placebo. Carbohydrate ingestion (various volumes and concentrations) has also been reported to improve 20 m sprint times during the fourth quarter of a simulated team sport (Welsh *et al.*, 2002). Thus, the research is consistent in suggesting a role of carbohydrate ingestion in delaying fatigue and improving later sprint performance during prolonged multiple-sprint tests which simulate the activity patterns of team-sport athletes.

The beneficial effects of carbohydrate ingestion have generally been attributed to the better maintenance of muscle glycogen levels. However, it has been suggested that the beneficial effects of carbohydrate ingestion may actually be a consequence of the raised plasma glucose concentration which provides a source of CHO for the muscle and the central nervous system (Foskett *et al.*, 2008). In addition, compared with a placebo, consumption of a carbohydrate solution ($4.5 \, \text{ml·kg}^{-1}$ of 6.5% glucose every 15 min) has also been reported to maintain lower FFA levels in the blood during prolonged, intermittent, high-intensity running. Thus, as explained previously, this could provide an ergogenic effect during multiple-sprint exercise by attenuating brain serotonin formation and minimising central fatigue (Nybo and Secher, 2004).

Alkalising agents

Classification and usage

The bicarbonate system is present in both the intracellular and extracellular fluids, and operates to resist changes in H^+ concentration when a strong acid or base is added (Figure 8.7). When a strong acid is added to the fluid and H^+ ions are released by the acid, the bicarbonate ions (HCO_3^-), acting

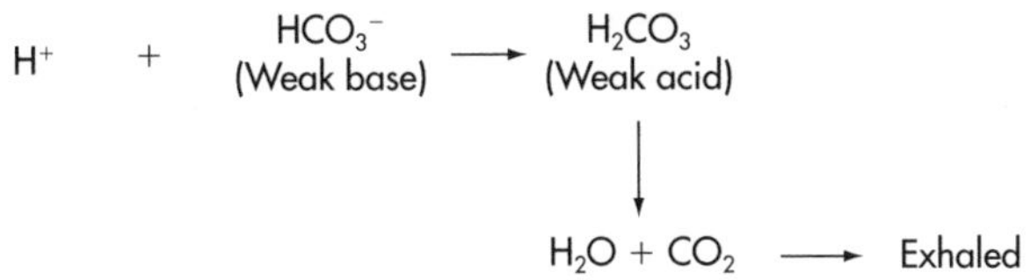

$$H^+ \quad + \quad \underset{\text{(Weak base)}}{HCO_3^-} \longrightarrow \underset{\text{(Weak acid)}}{H_2CO_3}$$

$$\downarrow$$

$$H_2O + CO_2 \longrightarrow \text{Exhaled}$$

Figure 8.7 The bicarbonate buffer system

as weak bases, combine with the H^+ ions, thus forming carbonic acid (H_2CO_3).

The [HCO_3^-] in the extracellular fluid is normally around $25\,mmol \cdot l^{-1}$ at rest, and this has been reported to increase by an average of $5.3\,mmol \cdot l^{-1}$ after the ingestion $0.3\,g \cdot kg^{-1}$ of body mass of sodium bicarbonate ($NaHCO_3$)(Matson and Vu Tran, 1993). While further research is required, $0.3\,g$ of $NaHCO_3$ per kilogram of body mass appears close to an optimal dose as it is generally accepted that $0.18\,g \cdot kg^{-1}$ is the threshold for induced alkalosis (Gao *et al.*, 1988; Horswill *et al.*, 1988) and that dosages higher than $0.3\,g \cdot kg^{-1}$ are likely to cause gastrointestinal discomfort in many subjects (Jones *et al.*, 1977). The most effective time to ingest $NaHCO_3$ has not been accurately determined. However, most authors agree that it should occur between 60 min and 90 min prior to exercise (McNaughton and Thompson, 2001; Bishop *et al.*, 2004a). In addition, it has been suggested that chronic $NaHCO_3$ administration over five days may be more applicable to athletes as they are less likely to suffer from gastrointestinal irritation, especially on the day of performance (McNaughton *et al.*, 1999). $NaHCO_3$ may be administered via the ingestion of capsules, in a solution or through intravenous injections. While future research is needed, the literature suggests that the optimal ingestion protocol would involve $0.2–0.3\,g \cdot kg^{-1}$ of $NaHCO_3$ taken 60–120 min prior to exercise. Instead of the more commonly used $NaHCO_3$, sodium citrate has also been used as the alkalising substance in some studies.

Possible mechanisms

The cell membrane is relatively impermeable to HCO_3^- (Mainwood and Worseley-Brown, 1975) and the ingestion of $NaHCO_3$ does not increase the resting intracellular pH or muscle buffer capacity (βm) (Bishop *et al.*, 2004a). Rather, the ingestion of alkalising agents (e.g. $NaHCO_3$) prior to exercise increases the extracellular buffer capacity and enhances the efflux of H^+ from the muscle into the blood, maintaining pH closer to normal levels during high-intensity exercise. As explained above, this should help to reduce the potential negative effects of H^+ accumulation on performance. While the ergogenic benefits of alkaline ingestion have been largely attributed to the enhanced extracellular buffer capacity, it has recently been demonstrated that the ingestion of alkalising agents (either sodium bicarbonate or citrate synthase) can also reduce the exercise-induced increase in extracellular K^+ (Street *et al.*, 2005; Sostaric *et al.*, 2006).

Effects on multiple-sprint performance

While many studies have investigated the effects of alkaline ingestion on high-intensity exercise performance, there is a paucity of studies that have investigated the effects of alkaline ingestion on multiple-sprint performance. One study has reported a significant improvement in power output during the third, fourth and fifth sprints of a repeated-sprint test (five 6s sprints performed every 30s) following the ingestion of $0.3g \cdot kg^{-1}$ of $NaHCO_3$ (Figure 8.8) (Bishop *et al.*, 2004a). In support of this finding, another study has also reported $NaHCO_3$ ingestion to be ergogenic for the performance of ten 10s cycle sprints with 50s of recovery between each sprint (Lavender and Bird, 1989). In contrast, $NaHCO_3$ ingestion produced only a small (~2%), non-significant improvement in the performance of ten 6s running sprints (on a non-motorised treadmill), separated by 30s recovery periods (Gaitanos *et al.*, 1991). While it is possible that these contrasting findings are due to differing effects of $NaHCO_3$ ingestion on running and cycling repeated-sprint performance, the more likely explanation is the relatively small change in blood pH reported in the final study (from 7.38 to 7.43) (possibly due to the greater time delay (150 min) between ingestion and exercise). Thus, while confirmatory research is required, it appears that alkaline ingestion leading to a large increase in pH (~0.1 of a pH unit) and $[HCO_3^-]$ (~5.0 mmol $\cdot l^{-1}$) is likely to improve repeated-sprint performance. Furthermore, while improved K^+ regulation may contribute, the greater production of lactate suggests that less inhibition of anaerobic glycolysis also plays a role (Figure 8.8).

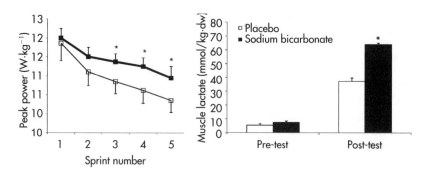

Figure 8.8 A: peak power output $(W \cdot kg^{-1})$ for each of the five sprints of the five 6s test of RSA, post-ingestion of sodium bicarbonate or placebo (NaCl); B: muscle lactate values before and after the RSA test. Values are mean ± SEM (N = 10).

Note
* denotes different to placebo ($P < 0.05$).

Alkaline ingestion may also improve intermittent-sprint performance, although the results are less convincing. It has been reported that $NaHCO_3$ ingestion significantly increased power output during a prolonged, inter- mittent-sprint test (ten 3 min blocks of 90 s at 40% $\dot{V}O_{2peak}$, 60 s at 60% $\dot{V}O_{2peak}$, a 14 s maximal sprint and 16 s of rest) (Price *et al.*, 2003). The results of this study are consistent with the findings of another study which investigated the effects of $NaHCO_3$ ingestion on the performance of an intermittent-sprint test involving shorter sprints (two 36 min "halves" of repeated 2 min blocks; all-out 4 s sprint, 100 s of active recovery at 35% $\dot{V}O_{2peak}$, 20 s rest), also performed on a cycle ergometer (Bishop and Clau- dius, 2005). It was reported that subjects performed significantly more work and achieved a higher peak power in almost half of the second-half sprints. Interestingly, the plasma $[HCO_3^-]$ peaked at 30.0 $mmol \cdot l^{-1}$ immedi- ately prior to the second half of the test (approximately 90 min after a second ingestion of 0.2 $g \cdot kg^{-1}$ $NaHCO_3$), and this may have contributed to why performance was improved in the second, but not the first, "half". The limited research to date therefore suggests that alkaline inges- tion leading to a large increase in $[HCO_3^-]$ (~5.0 $mmol \cdot l^{-1}$) is likely to improve intermittent-sprint performance, especially for longer duration sprints (~10 s).

CONCLUSION

As previously discussed, studies involving ergogenic aids can provide valu- able insights into the possible mechanisms of fatigue during multiple-sprint exercise. The studies involving creatine and carbohydrate supplementation provide good evidence that insufficient supply on energy (i.e. PCr and/or glycogen) is likely to contribute to fatigue during multiple-sprint tasks. It needs to be remembered however, that these supplements may exert their ergogenic effects via mechanisms other than better maintenance of energy stores. For example, it has been suggested that carbohydrate ingestion may improve performance by raising plasma glucose concentration or maintain- ing lower plasma FFA levels (and thus attenuating brain serotonin forma- tion). Studies involving ingestion of an alkaline substance (e.g. $NaHCO_3$) provide good evidence that the accumulation of certain metabolites (i.e. H^+) is likely to contribute to fatigue during multiple-sprint tasks. Once again however, more than one mechanism may be involved and the attenu- ation of exercise-induced increases in extracellular K^+ levels may also con- tribute to the ergogenic effects of alkaline ingestion. Finally, there is good evidence that caffeine ingestion can improve multiple-sprint performance. However, due to the many possible mechanisms of action of this ergogenic aid, it is difficult to draw strong conclusions regarding possible mechan- isms of fatigue during multiple-sprint exercise.

FIVE KEY PAPERS THAT SHAPED THE TOPIC AREA

Study 1. Spencer, M., Lawrence, S., Rechichi, C., Bishop, D., Dawson, B. and Goodman, C. (2004). Time–motion analysis of elite field-hockey: special reference to repeated-sprint activity. *Journal of Sports Sciences*, 22, 9, 843–850.

While previous time–motion analysis studies had reported the mean durations and frequencies of sprint activities during a game or a series of games, this was the first published data documenting the nature of repeated-sprint activity. It was clearly shown that, due to the unpredictable nature of team-sports, sprinting is not evenly distributed throughout a game, and intense periods of repeated-sprint activity do occur. By documenting the nature of repeated-sprints during a typical team sport, this paper demonstrates the importance of understanding the effects of different ergogenic aids on repeated-sprint ability.

Study 2. Schneiker, K.T., Bishop, D., Dawson, B. and Hackett, L.P. (2006). Effects of caffeine on prolonged intermittent-sprint ability in team-sport athletes. *Medicine and Science in Sports and Exercise*, 38, 578–585.

The purpose of this study was to determine the effects of acute caffeine ingestion on the performance of an intermittent-sprint exercise test designed to simulate the physiological demands specific to athletes participating in team sports. This study was unique as it was the first to utilise a test that incorporated intermittent sprints that were performed over a prolonged duration (~80 min). It was found that the total amount of sprint work performed during each half of the intermittent-sprint test, as well as the average peak power attained by participants during sprints in each half, were all significantly improved in the caffeine trial in comparison to placebo. In the caffeine trial, total sprint work was 8.5% greater during the first half and 7.6% greater in the second, when compared to placebo. Similarly, average peak power was 7.0% greater in the first half of the caffeine trial and 6.6% greater in the second, when compared to the placebo. The results of this study influenced many team sports to trial caffeine as an ergogenic aid.

Study 3. Mujika, I., Padilla, S., Ibanez, J., Izquierdo, M. and Gorostiaga, E. (2000). Creatine supplementation and sprint performance in soccer players. *Medicine and Science in Sports and Exercise*, 32 (2), 518–525.

This regularly cited study was one of the first to indicate that an acute creatine loading strategy that had repeatedly been shown to increase muscle creatine levels ($20 \, g \cdot d^{-1}$ for six days) has an ergogenic potential for highly trained soccer players. Creatine-supplemented players showed an improved performance during six 15 m sprints. These improvements could have a great impact on a player's performance level during actual competitive soccer match-play. The measured blood ammonia and blood lactate concentration values suggested that the observed ergogenic effects were associated with a better preservation of muscle ATP levels, and not an increased reliance on anaerobic glycolysis.

Study 4. Bangsbo, J., Norregaard, L. and Thorsoe, F. (1992). The effect of carbohydrate diet on intermittent exercise performance. *International Journal of Sports Medicine*, 13, 152–157.

While many previous studies have demonstrated the positive effect of a high carbohydrate diet on endurance performance, this was one of the first to show that performance during intermittent running could also be enhanced following the ingestion of a CHO-enriched diet for two days. Seven professional soccer players were tested twice, following a diet containing either 39% (C-diet) or 65% carbohydrate (CHO diet) during the two days prior to each test. The order of the diets was assigned randomly. The standardised test consisted initially of a field part (6,856 m) followed by treadmill running to exhaustion. The total mean running distance after the CHO-diet was 17.1 km, which was 0.9 km longer ($P < 0.05$) than after the C-diet. It is important to note however, that not all players benefited from the CHO-diet.

Study 5. Bishop, D., Edge, J., Davis, C. and Goodman, C. (2004). Induced metabolic alkalosis affects muscle metabolism and repeated-sprint ability. *Medicine and Science in Sports and Exercise*, 36, 807–813.

This study was the first to examine the effects of sodium bicarbonate ingestion on the metabolic responses of skeletal muscle to repeated high-intensity sprint efforts (five 6 s all-out sprints every 30 s). Despite no significant difference in post-test muscle pH between conditions, the sodium bicarbonate trial resulted in a greater post-test muscle lactate concentration and improved repeated-sprint ability. As sodium bicarbonate ingestion did not increase resting muscle pH or ßm, it is likely that the improved performance was a result of the greater extracellular buffer concentration increasing H^+ efflux from the muscles into the blood. The significant increase in post-test muscle lactate concentration in the sodium bicarbonate trial suggested that an increased anaerobic energy contribution was one mechanism by which sodium bicarbonate ingestion improved repeated-sprint ability.

GLOSSARY OF TERMS

ADP	adenosine diphosphate
AMP	adenosine monophosphate
ATP	adenosine triphosphate
BCAA	branched-chain amino acid
Ca^{2+}	calcium ion
CHO	carbohydrate
CNS	central nervous system
Cr	creatine
CV	coefficient of variation
EPO	erythropoietin
FFA	free fatty acid
FI	fatigue index
H^+	hydrogen ion
H_2O_2	hydrogen peroxide
HCO_3^-	bicarbonate ion
HO_2	hydrogen superoxide
IMP	inosine monophosphate
K^+	potassium ion
Mg^{2+}	magnesium ion
$NaHCO_3$	sodium bicarbonate
NH_3	ammonia
PCr	phosphocreatine
PFK	phosphofructokinase
P_i	inorganic phosphate
ROS	reactive oxygen species
S_{dec}	sprint score decrement
SR	sarcoplasmic reticulum
TCr	total creatine
W_{dec}	work decrement

REFERENCES

Ali, A., Williams, C., Nicholas, C.W. and Foskett, A. (2007). The influence of carbohydrate-electrolyte ingestion on soccer skill performance. *Medicine and Science in Sports and Exercise*, 39, 1969–1976.

Anselme, F., Collomp, K., Mercier, B., Ahmaiedi, S. and Prefaut, C. (1992). Caffeine increases maximal anaerobic power and blood lactate concentration. *European Journal of Applied Physiology and Occupational Physiology*, 65, 188–191.

Balsom, P., Ekblom, B. and Sjodin, B. (1994). Enhanced oxygen availablility during high intensity intermittent exercise decreases anaerobic metabolite concentration in blood. *Acta Physiologica Scandinavica*, 150, 455–456.

Balsom, P.D., Ekblom, B., Soderlund, K., Sjodin, B. and Hultman, E. (1993). Creatine supplementation and dynamic high-intensity intermittent exercise. *Scandinavian Journal of Medicine and Science in Sports*, 3, 143–149.

Balsom, P., Gaitanos, Soderlund, K. and Ekblom, B. (1999). High-intensity exercise and muscle glycogen availability in humans. *Acta Physiologica Scandinavica*, 165, 337–345.

Balsom, P.D., Seger, J.Y., Sjodin, B. and Ekblom, B. (1992a). Maximal-intensity intermittent exercise: effect of recovery duration. *International Journal of Sports Medicine*, 13, 528–533.

Balsom, P.D., Seger, J.Y., Sjodin, B. and Ekblom, B. (1992b). Physiological responses to maximal intensity intermittent exercise. *European Journal of Applied Physiology and Occupational Physiology*, 65, 144–149.

Bamberger, M. and Yaeger, D. (1997). Over the edge: special report. *Sports Illustrated*.

Bangsbo, J., Graham, T.E., Kiens, B. and Saltin, B. (1992a). Elevated muscle glycogen and anaerobic energy production during exhaustive exercise in man. *Journal of Physiology*, 451, 205–227.

Bangsbo, J., Norregaard, L. and Thorsoe, F. (1992b). The effect of carbohydrate diet on intermittent exercise performance. *International Journal of Sports Medicine*, 13, 152–157.

Barry, B.K. and Enoka, R.M. (2007). The neurobiology of muscle fatigue: 15 years later. *American Journal of Physiology: Integrative and Comparative Biology*, 47, 465–473.

Bigland-Ritchie, B., Furbush, F. and Woods, J.J. (1986). Fatigue of intermittent submaximal voluntary contractions: central and peripheral factors. *Journal of Applied Physiology*, 61, 421–429.

Billaut, F. and Basset, F.A. (2007). Effect of different recovery patterns on repeated sprint ability and neuromuscular responses. *Journal of Sports Sciences*, 25, 905–913.

Billaut, F. and Bishop, D. (2009). Muscle fatigue in males and females during multiple sprint exercise. *Sports Medicine*, 39 (4), 257–278.

Billaut, F., Basset, F.A. and Falgairette, G. (2005). Muscle coordination changes during intermittent cycling sprints. *Neuroscience Letters*, 380, 265–269.

Billaut, F., Basset, F.A., Giacomoni, M., Lemaître, F., Tricot, V. and Falgairette, G. (2006). Effect of high-intensity intermittent cycling sprints on neuromuscular activity. *International Journal of Sports Medicine*, 27, 25–30.

Bishop, D. and Claudius, B. (2005). Effects of induced metabolic alkalosis on prolonged intermittent-sprint performance. *Medicine and Science in Sports and Exercise*, 37, 759–767.

Bishop, D. and Edge, J. (2006). Determinants of repeated-sprint ability in females matched for single-sprint performance. *European Journal of Applied Physiology*, 97, 373–379.

Bishop, D. and Spencer, M. (2004). Determinants of repeated-sprint ability in well trained team-sport athletes and endurance-trained athletes. *Journal of Sports Medicine and Physical Fitness*, 44, 1–7.

Bishop, D., Edge, J., Davis, C. and Goodman, C. (2004a). Induced metabolic alkalosis affects muscle metabolism and repeated-sprint ability. *Medicine and Science in Sports and Exercise*, 36, 807–813.

Bishop, D., Edge, J. and Goodman, C. (2004b). Muscle buffer capacity and aerobic fitness are associated with repeated-sprint ability in women. *European Journal of Applied Physiology*, 92, 540–547.

Bishop, D., Lawrence, S. and Spencer, M. (2003). Predictors of repeated-sprint ability in elite female hockey players. *Journal of Science and Medicine in Sport*, 6, 199–209.

Bishop, D., Spencer, M., Duffield, R. and Lawrence, S. (2001). The validity of a repeated sprint ability test. *Journal of Science and Medicine in Sport*, 4, 19–29.

Bogdanis, G.C., Nevill, M.E., Boobis, L.H. and Lakomy, H.K. (1996). Contribution of phosphocreatine and aerobic metabolism to energy supply during repeated sprint exercise. *Journal of Applied Physiology*, 80 (3), 876–884.

Bogdanis, G.C., Nevill, M.E., Boobis, L.H., Lakomy, H.K. and Nevill, A.M. (1995). Recovery of power output and muscle metabolites following 30 s of maximal sprint cycling in man. *Journal of Physiology*, 482 (Pt 2), 467–480.

Boobis, L.H. (1987). Metabolic aspects of fatigue during sprinting. In D. Macleod, R.J. Maughan, M.A. Nimmo, T. Reilly and C. Williams (eds), *Exercise: Benefits, Limitations and Adaptations*. London: E & FN Spon.

Boobis, L.H., Williams, C. and Wootton, S. (1982). Human muscle metabolism during brief maximal exercise. *Journal of Physiology (Lond)*, 338, 22.

Boobis, L.H., Williams, C. and Wootton, S.A. (1983). Influence of sprint training on muscle metabolism during brief maximal exercise in man. *Journal of Physiology*, 342, 36–37.

Capriotti, P.V., Sherman, W.M. and Lamb, D.R. (1999) Reliability of power output during intermittent high-intensity cycling. *Medicine and Science in Sports and Exercise*, 31 (6), 913–915.

Casey, A., Constantin-Teodosiu, D., Howell, S., Hultman, E. and Greenhaff, P.L. (1996). Metabolic responses of type I and II muscle fibres during repeated bouts of maximal exercise in humans. *American Journal of Physiology*, 271, E38–43.

Chasiotis, D., Hultman, E. and Sahlin, K. (1982) Acidotic depression of cyclic AMP accumulation and phosphorylase b to a transformation in skeletal muscle of man. *Journal of Physiology*, 335, 197–204.

Cheetham, M.E., Boobis, L.H., Brooks, S. and Williams, C. (1986). Human muscle metabolism during sprint running. *Journal of Applied Physiology*, 61, 54–60.

Coggan, A.R. and Coyle, E.F. (1991). Carbohydrate ingestion during prolonged exercise: effects on metabolism and performance. *Exercise and Sport Sciences Reviews*, 19, 1–40.

Crowther, G.J., Carey, M.F., Kemper, W.F. and Conley, K.E. (2002). Control of glycolysis in contracting muscle. 1. Turning it on. *American Journal of Physiology*, 282, E67–73.

Davis, J.M., Bailey, S.P., Woods, J.A., Galiano, F.J., Hamilton, M.T. and Bartoli, W.P. (1992). Effects of carbohydrate feedings on plasma free tryptophan and branched-chain amino acids during prolonged cycling. *European Journal of Applied Physiology and Occupational Physiology*, 65, 513–519.

Dawson, B., Goodman, C., Lawrence, S., Preen, D., Polglaze, T., Fitzsimons, M. and Fournier, P. (1997). Muscle phosphocreatine repletion following single and repeated short sprint efforts. *Scandinavian Journal of Medicine and Science in Sports*, 7, 206–213.

Drust, B., Rasmussen, P., Mohr, M., Nielsen, B. and Nybo, L. (2005). Elevations in core and muscle temperature impairs repeated sprint performance. *Acta Physiologica Scandinavica*, 183, 181–190.

Dutka, T.L. and Lamb, G.D. (2004). Effect of low cytoplasmic [ATP] on excitation contraction coupling in fast-twitch muscle fibres of the rat. *Journal of Physiology*, 560, 451–468.

Edge, J., Bishop, D. and Goodman, C. (2005). Effects of high- and moderate-intensity training on metabolism and repeated sprints. *Medicine and Science in Sports and Exercise*, 37, 1975–1982.

Edge, J., Bishop, D., Hill-Haas, S., Dawson, B. and Goodman, C. (2006). Comparison of muscle buffer capacity and repeated-sprint ability of untrained, endurance-trained and team-sport athletes. *European Journal of Applied Physiology*, 96, 225–234.

Enoka, R.M. and Stuart, D.G. (1992). Neurobiology of muscle fatigue. *Journal of Applied Physiology*, 72, 1631–1648.

Finn, J.P., Ebert, T.R., Withers, R.T., Carey, M.F., Mackay, M., Philips, J.W. and Febbraio, M.A. (2001). Effect of creatine supplementation on metabolism and performance in humans during intermittent sprint cycling. *European Journal of Applied Physiology*, 84, 238–243.

Fitts, R.H. (1994). Cellular mechanisms of muscle fatigue. *Physiological Reviews*, 74, 49–94.

Foskett, A., Williams, C., Boobis, L. and Tsintzas, K. (2008). Carbohydrate availability and muscle energy metabolism during intermittent running. *Medicine and Science in Sports and Exercise*, 40, 96–103.

Fredholm, B.B., Bättig, K., Holmén, J., Nehlig, A. and Zvartau, E.E. (1999). Actions of caffeine in the brain with special reference to factors that contribute to its widespread use. *Pharmacological Reviews*, 51, 83–133.

Fuglevand, A., Zackowski, K., Huey, K. and Enoka, R. (1993). Impairment of neuromuscular propagation during human fatiguing contractions at submaximal forces. *Journal of Physiology*, 460, 549–572.

Gaitanos, G.C., Nevill, M.E., Brooks, S. and Williams, C. (1991). Repeated bouts of sprint running after induced alkalosis. *Journal of Sports Sciences*, 9, 355–369.

Gaitanos, G.C., Williams, C., Boobis, L.H. and Brooks, S. (1993). Human muscle metabolism during intermittent maximal exercise. *Journal of Applied Physiology*, 75 (2), 712–719.

Gandevia, S.C. (2001). Spinal and supraspinal factors in human muscle fatigue. *Physiological Reviews*, 81, 1725–1789.

Gao, J., Costill, D.L., Horswill, C.A. and Park, S.H. (1988). Sodium bicarbonate ingestion improves performance in interval swimming. *European Journal of Applied Physiology and Occupational Physiology*, 58, 171–174.

George, A.J. (2000). Central nervous system stimulants. Baillieres Best Practice and Research. *Clinical Endocrinology and Metabolism*, 14, 79–88.

Gibson, H. and Edwards, R.H.T. (1985). Muscular exercise and fatigue. *Sports Medicine*, 2, 120–132.

Glaister, M. (2005). Multiple sprint work: physiological responses, mechanisms of fatigue and the influence of aerobic fitness. *Sports Medicine*, 35, 757–777.

Graham, T.E. (2001). Caffeine and exercise: metabolism, endurance and performance. *Sports Medicine*, 31, 785–807.

Graham, T.E. and Spriet, L.L. (1991). Performance and metabolic responses to a high caffeine dose during prolonged exercise. *Journal of Applied Physiology*, 71, 2292–2298.

Greer, F., McLean, C. and Graham, T.E. (1998). Caffeine, performance, and metabolism during repeated Wingate tests. *Journal of Applied Physiology*, 85, 1504–1508.

Harris, R., Sahlin, K. and Hultman, E. (1977). Phosphagen and lactate contents of m. quadriceps femoris of man after exercise. *Journal of Applied Physiology*, 43, 852–857.

Harris, R.C., Soderlund, K. and Hultman, E. (1992). Elevation of creatine in resting and exercised muscle of normal subjects by creatine supplementation. *Clinical Science (Lond)*, 83, 367–374.

Hautier, C., Arsac, L., Deghdegh, K., Souquet, J., Belli, A. and Lacour, J. (2000). Influence of fatigue on EMG/force ratio and cocontraction in cycling. *Medicine and Science in Sports and Exercise*, 32, 839–843.

Holmyard, D.J., Cheetham, M.E., Lakomy, H.K.A. and Williams, C. (1987). Effects of recovery duration on performance during multiple treadmill sprints. In T. Reilly, A. Lees, K. Davids and W.J. Murphy (eds), *Science and Football*. London: E & F.N. Spon.

Horswill, C.A., Costill, D.L., Fink, W.J., Flynn, M.G., Kirwan, J.P., Mitchell, J.B. and Houmard, J.A. (1988). Influence of sodium bicarbonate on sprint performance: relationship to dosage. *Medicine and Science in Sports and Exercise*, 20, 566–569.

Hultman, E. and Sjoholm, H. (1983). Energy metabolism and contraction force of human skeletal muscle in situ during electrical stimulation. *Journal of Physiology*, 345, 525–532.

Hultman, E., Soderlund, K., Timmons, J.A., Cederblad, G. and Greenhaff, P.L. (1996). Muscle creatine loading in men. *Journal of Applied Physiology*, 81, 232–237.

Jones, N.L., McCartney, N., Graham, T., Spriet, L.L., Kowalchuk, J.M., Heigenhauser, G.J. and Sutton, J.R. (1985). Muscle performance and metabolism in maximal isokinetic cycling at slow and fast speeds. *Journal of Applied Physiology*, 59, 132–136.

Jones, N.L., Sutton, J.R., Taylor, R. and Toews, C.J. (1977). Effect of pH on cardiorespiratory and metabolic responses to exercise. *Journal of Applied Physiology*, 43 (6), 959–964.

Juel, C., Pilegaard, H., Nielsen, J.J. and Bangsbo, J. (2000). Interstitial K(+) in human skeletal muscle during and after dynamic graded exercise determined by microdialysis. *American Journal of Physiology: Regulatory, Integrative and Comparative Physiology*, 278, R400–406.

Karatzaferi, C., De Haan, A., Ferguson, R.A., Van Mechelen, W. and Sargeant, A.J. (2001). Phosphocreatine and ATP content in human single muscle fibres before and after maximum dynamic exercise. *European Journal of Physiology*, 442, 467–474.

Kent-Braun, J.A. (1999). Central and peripheral contributions to muscle fatigue in humans during sustained maximal effort. *European Journal of Applied Physiology*, 80, 57–63.

Kovaks, E., Stegan, J. and Brouns, F. (1998). Effect of caffeinated drinks on substrate metabolism, caffeine excretion, and performance. *Journal of Applied Physiology*, 85, 709–715.

Kreider, R. (2003). Effects of creatine supplementation on performance and training adaptations. *Molecular and Cellular Biochemistry*, 244, 89–94.

Krustrup, P., Mohr, M., Steensberg, A., Bencke, J., Kjaer, M. and Bangsbo, J.

(2006). Muscle and blood metabolites during a soccer game: implications for sprint performance. *Medicine and Science in Sports and Exercise*, 38, 1165–1174.

Lavender, G. and Bird, S.R. (1989). Effect of sodium bicarbonate ingestion upon repeated sprints. *British Journal of Sports Medicine*, 23 (1), 41–45.

Leatt, P.B. and Jacobs, I. (1989). Effect of glucose polymer ingestion on glycogen depletion during a soccer match. *Canadian Journal of Sport Sciences*, 14, 112–116.

Lindinger, M.I. and Heigenhauser, G.J. (1991). The roles of ion fluxes in skeletal muscle fatigue. *Canadian Journal of Physiology and Pharmacology*, 69, 246–253.

Lindinger, M.I., Graham, T.E. and Spriet, L.L. (1993). Caffeine attenuates the exercise induced increase in plasma [K+] in humans. *Journal of Applied Physiology*, 74, 1149–1155.

Lindinger, M.I., Heigenhauser, G.J., Mckelvie, R.S. and Jones, N.L. (1992). Blood ion regulation during repeated maximal exercise and recovery in humans. *American Journal of Physiology*, 262, R126–136.

Linnamo, V., Bottas, R. and Komi, P.V. (2000). Force and EMG power spectrum during and after eccentric and concentric fatigue. *Journal of Electromyography and Kinesiology*, 10, 293–300.

Linssen, W., Jacobs, M., Stegeman, D., Joosten, E. and Moleman, J. (1990). Muscle fatigue in McArdle's disease. Muscle fibre conduction velocity and surface EMG frequency spectrum during ischaemic exercise. *Brain*, 113, 1779–1793.

MacDonald, W.A. and Stephenson, D.G. (2004). Effects of ADP on action potential induced force responses in mechanically skinned rat fast-twitch fibres. *Journal of Physiology*, 559, 433–447.

McGawley, K. and Bishop, D. (2006). Reliability of a 5×6-s maximal cycling repeated-sprint test in trained female team-sport athletes. *European Journal of Applied Physiology*, 98, 383–393.

McGawley, K. and Bishop, D. (2008). Anaerobic and aerobic contribution to two, 5×6 s repeated-sprint bouts. *Coaching and Sport Science Journal*, 3, 52.

McKenna, M.J., Medved, I., Goodman, C.A., Brown, M.J., Bjorksten, A.R., Murphy, K.T., Petersen, A.C., Sostaric, S. and Gong, X. (2006). N-acetylcysteine attenuates the decline in muscle Na+, K+-pump activity and delays fatigue during prolonged exercise in humans. *Journal of Physiology (Lond)*, 576, 279–288.

McKenna, M.J., Morton, J., Selig, S.E. and Snow, R.J. (1999). Creatine supplementation increases muscle total creatine but not maximal intermittent exercise performance. *Journal of Applied Physiology*, 87, 2244–2252.

Maclaren, D. and Close, G.L. (2000). Effect of carbohydrate supplementation on simulated exercise of rugby league referees. *Ergonomics*, 43, 1528–1537.

McNaughton L. and Thompson, D. (2001). Acute versus chronic sodium bicarbonate ingestion and anaerobic work and power output. *Journal of Sports Medicine and Physical Fitness*, 41, 456–462.

McNaughton, L., Backx, K., Palmer, G. and Strange, N. (1999). Effects of chronic bicarbonate ingestion on the performance of high-intensity work. *European Journal of Applied Physiology and Occupational Physiology*, 80, 333–336.

Mainwood, G.W. and Worseley-Brown, P. (1975). The effect of extracellular pH and buffer concentration on the efflux of lactate from frog sartorius muscle. *Journal of Physiology*, 250, 1–22.

Matson, L.G. and Vu Tran, Z. (1993). Effects of sodium bicarbonate ingestion on anaerobic performance: a meta-analytic review. *International Journal of Sport Nutrition*, 3, 2–28.

Mazzeo, R.S. (1991). Catecholamine responses to acute and chronic exercise. *Medicine and Science in Sports and Exercise*, 23, 839–845.

Medbo, J.I. and Sejersted, O.M. (1990). Plasma potassium changes with high intensity exercise. *Journal of Physiology*, 421, 105–122.

Mendez-Villanueva, A., Bishop, D.J. and Hamer, P. (2007a). Reproducibility of a 6-s maximal cycling sprint test. *Journal of Science and Medicine in Sport*, 10, 323–326.

Mendez-Villanueva, A., Hamer, P. and Bishop, D. (2007b) Physical fitness and performance: fatigue responses during repeated sprints matched for initial mechanical output. *Medicine and Science in Sports and Exercise*, 39, 2219–2225.

Mendez-Villanueva, A., Hamer, P. and Bishop, D. (2008). Fatigue in repeated-sprint exercise is related to muscle power factors and reduced neuromuscular activity. *European Journal of Applied Physiology*, 103 (4), 411–419.

Merton, P. (1954). Voluntary strength and fatigue. *Journal of Physiology*, 123, 553–564.

Mohr, M., Nordsborg, N., Nielsen, J.J., Pedersen, L.D., Fischer, C., Krustrup, P. and Bangsbo, J. (2004). Potassium kinetics in human muscle interstitium during repeated intense exercise in relation to fatigue. *Pflugers Archives*, 448, 452–456.

Moopanar, T.R. and Allen, D.G. (2005). Reactive oxygen species reduce myofibrillar Ca^{2+} sensitivity in fatiguing mouse skeletal muscle at 37 degrees C. *Journal of Physiology*, 564, 189–199.

Mujika, I., Padilla, S., Ibanez, J., Izquierdo, M. and Gorostiaga, E. (2000). Creatine supplementation and sprint performance in soccer players. *Medicine and Science in Sports and Exercise*, 32, 518–525.

Nicholas, C.W., Tsintzas, K., Boobis, L. and Williams, C. (1999). Carbohydrate electrolyte ingestion during intermittent high-intensity running. *Medicine and Science in Sports and Exercise*, 31, 1280–1286.

Nybo, L. and Secher, N.H. (2004). Cerebral perturbations provoked by prolonged exercise. *Progress in Neurobiology* 72, 223–261.

Oliver, J.L. (2009). Is a fatigue index a worthwhile measure of repeated sprint ability. *Journal of Science and Medicine in Sport* 12 (1), 20–23.

Parolin, M.L., Chesley, A., Matsos, M.P., Spriet, L.L., Jones, N.L. and Heigenhauser, G.J. (1999). Regulation of skeletal muscle glycogen phosphorylase and PDH during maximal intermittent exercise. *American Journal of Physiology*, 277, E890–900.

Parra, J., Cadefau, J., Rodas, G., Amigo, N. and Cusso, R. (2000). The distribution of rest periods affects performance and adaptations of energy metabolism induced by high intensity training in human muscle. *Acta Physiologica Scandinavica*, 169, 157–165.

Pedersen, T.H., Nielsen, O.B., Lamb, G.D. and Stephenson, D.G. (2004). Intracellular acidosis enhances the excitability of working muscle. *Science*, 305, 1144–1147.

Peyrebrune, M.C., Nevill, M.E., Donaldson, F.J. and Cosford, D.J. (1998). The effects of oral creatine supplementation on performance in single and repeated sprint swimming. *Journal of Sports Sciences*, 16, 271–279.

Preen, D., Dawson, B., Goodman, C., Lawrence, S., Beilby, J. and Ching, S. (2001).

The effect of oral creatine supplementation on 80 minutes of repeated-sprint exercise. *Medicine and Science in Sports and Exercise*, 33, 814–825.

Price, M., Moss, P. and Rance, S. (2003). Effects of sodium bicarbonate ingestion on prolonged intermittent exercise. *Medicine and Science in Sports and Exercise*, 35 (8), 1303–1308.

Racinais, S., Bishop, D., Denis, R., Lattier, G., Mendez-Villaneuva, A. and Perrey, S. (2007). Muscle deoxygenation and neural drive to the muscle during repeated sprint cycling. *Medicine and Science in Sports and Exercise*, 39, 268–274.

Reid, M.B. (2001). Invited Review: redox modulation of skeletal muscle contraction: what we know and what we don't. *Journal of Applied Physiology*, 90, 724–731.

Reid, M.B., Stokic, D.S., Koch, S.M., Khawli, F.A. and Leis, A.A. (1994). N-acetyl-cysteine inhibits muscle fatigue in humans. *Journal of Clinical Investigations*, 94, 2468–2474.

Rico-Sanz, J. and Mendez-Marco, M.T. (2000). Creatine enhances oxygen uptake and performance during alternating intensity exercise. *Medicine and Science in Sports and Exercise*, 32 (2), 379–385.

Rico-Sanz, J., Zehnder, M., Buchli, R., Dambach, M. and Boutellier, U. (1999). Muscle glycogen degradation during simulation of a fatiguing soccer match in elite soccer players examined non-invasively by 13C-MRS. *Medicine and Science in Sports and Exercise*, 31, 1587–1593.

Sachdev, S. and Davies, K.J. (2008). Production, detection, and adaptive responses to free radicals in exercise. *Free Radical Biology and Medicine*, 44, 215–223.

Sahlin, K., Tonkonogi, M. and Soderlund, K. (1998). Energy supply and muscle fatigue in humans. *Acta Physiologica Scandinavica*, 162, 261–266.

Sakamoto, K. and Goodyear, L.J. (2002). Invited review: intracellular signaling in contracting skeletal muscle. *Journal of Applied Physiology*, 93, 369–383.

Schneiker, K.T., Bishop, D., Dawson, B. and Hackett, L.P. (2006). Effects of caffeine on prolonged intermittent-sprint ability in team-sport athletes. *Medicine and Science in Sports and Exercise*, 38, 578–585.

Sharp, R.L., Costill, D.L., Fink, W.J. and King, D.S. (1986). Effects of eights weeks of bicycle ergometer sprint training on human muscle buffer capacity. *International Journal of Sports Medicine*, 7, 13–17.

Sinclair, C.J. and Geiger, J.D. (2000). Caffeine use in sports: a pharmacological review. *Journal of Sports Medicine and Physical Fitness*, 40, 71–79.

Snow, R.J., Carey, M.F., Stathis, C.G., Febbraio, M.A. and Hargreaves, M. (2000). Effect of carbohydrate ingestion on ammonia metabolism during exercise in humans. *Journal of Applied Physiology*, 88, 1576–1580.

Snow, R.J., McKenna, M.J., Selig, S.E., Kemp, J., Stathis, C.G. and Zhao, S. (1998). Effect of creatine supplementation on sprint exercise performance and muscle metabolism. *Journal of Applied Physiology*, 84, 1667–1673.

Sostaric, S.M., Skinner, S.L., Brown, M.J., Sangkabutra, T., Medved, I., Medley, T., Selig, S.E., Fairweather, I., Rutar, D. and McKenna, M.J. (2006). Alkalosis increases muscle K+ release, but lowers plasma [K+] and delays fatigue during dynamic forearm exercise. *Journal of Physiology*, 570, 185–205.

Spencer, M., Bishop, D., Dawson, B. and Goodman, C. (2005). Physiological and metabolic responses of repeated-sprint activities: specific to field-based team sports. *Sports Medicine*, 35, 1025–1044.

Spencer, M., Bishop, D., Dawson, B., Goodman, C. and Duffield, R. (2006a).

Metabolism and performance in repeated cycle sprints: active versus passive recovery. *Medicine and Science in Sports and Exercise*, 38, 1492–1499.

Spencer, M., Fitzsimons, M., Dawson, B., Bishop, D. and Goodman, C. (2006b). Reliability of a repeated-sprint test for field-hockey. *Journal of Science and Medicine in Sport*, 9, 181–184.

Spencer, M., Lawrence, S., Rechichi, C., Bishop, D., Dawson, B. and Goodman, C. (2004). Time–motion analysis of elite field-hockey: special reference to repeated-sprint activity. *Journal of Sports Sciences*, 22 (9), 843–850.

Spriet, L. (1995). Caffeine and performance. *International Journal of Sport Nutrition*, 5, S84–99.

Spriet, L., Lindinger, M., McKelvie, R., Heigenhauser, G. and Jones, N. (1989). Muscle glycogenolysis and H^+ concentration during maximal intermittent cycling. *Journal of Applied Physiology*, 66, 8–13.

Spriet, L., Söderlund, K., Bergström, M. and Hultman, E. (1987). Skeletal muscle glycogenolysis, glycolysis, and pH during electrical stimulation in men. *Journal of Applied Physiology*, 62, 616–621.

St Clair Gibson, A., Lambert, M.I. and Noakes, T.D. (2001). Neural control of force output during maximal and submaximal exercise. *Sports Medicine*, 31, 637–650.

Stathis, C.G., Zhao, S., Carey, M.F. and Snow, R.J. (1999). Purine loss after repeated sprint bouts in humans. *Journal of Applied Physiology*, 87, 2037–2042.

Street, D., Nielsen, J.-J., Bangsbo, J. and Juel, C. (2005). Metabolic alkalosis reduces exercise-induced acidosis and potassium accumulation in human skeletal muscle interstitium. *Journal of Physiology*, 566, 481–489.

Stuart, G.R., Hopkins, W.G., Cook, C. and Cairns, S.P. (2005). Multiple effects of caffeine on simulated high-intensity team-sport performance. *Medicine and Science in Sports and Exercise*, 37, 1998–2005.

Tarnopolsky, M.A. (1994). Caffeine and endurance performance. *Sports Medicine*, 18, 109–125.

Taylor, J.L., Allen, G.M., Butler, J.E. and Gandevia, S.C. (2000). Supraspinal fatigue during intermittent maximal voluntary contractions of the human elbow flexors. *Journal of Applied Physiology*, 89, 305–311.

van Hall, G., Raaymakers, J.S., Saris, W.H. and Wagenmakers, A.J. (1995). Ingestion of branched-chain amino acids and tryptophan during sustained exercise in man: failure to affect performance. *Journal of Physiology*, 486 (Pt 3), 789–794.

Vollestad, N.K., Vaage, O. and Hermansen, L. (1984). Muscle glycogen depletion patterns in type I and subgroups of type II fibres during prolonged exercise in man. *Acta Physiologica Scandinavica*, 122, 433–441.

Welsh, R.S., Davis, J.M., Burke, J.R. and Williams, H.G. (2002). Carbohydrates and physical/mental performance during intermittent exercise to fatigue. *Medicine and Science in Sports and Exercise*, 34, 723–731.

Wiles, J.D., Bird, S.R., Hopkins, J. and Riley, M. (1992). Effect of caffeinated coffee on running speed, respiratory factors, blood lactate and perceived exertion during 1500-m treadmill running. *British Journal of Sports Medicine*, 26, 116–120.

Williams, J.H. (1991). Caffeine, neuromuscular function and high-intensity exercise performance. *Journal of Sports Medicine and Physical Fitness*, 31, 481–489.

Wiroth, J., Bermon, S., Andrei, S., Dalloz, E., Hebuterne, X. and Dolisi, C. (2001). Effects of oral creatine supplementation on maximal pedalling performance in older adults. *European Journal of Applied Physiology and Occupational Physiology*, 84, 533–539.

Wragg, C.B., Maxwell, N.S. and Doust, J.H. (2000). Evaluation of the reliability and validity of a soccer-specific field test of repeated sprint ability. *European Journal of Applied Physiology and Occupational Physiology*, 83, 77–83.

Yanagiya, T., Kanehisa, H., Kouzaki, M., Kawakami, Y. and Fukunaga, T. (2003). Effect of gender on mechanical power output during repeated bouts of maximal running in trained teenagers. *International Journal of Sports Medicine*, 24, 304–310.

Yquel, R.J., Arsac, L.M., Thiaudiere, E., Canioni, P. and Manier, G. (2002). Effect of creatine supplementation on phosphocreatine resynthesis, inorganic phosphate accumulation and pH during intermittent maximal exercise. *Journal of Sports Sciences*, 20, 427–437.

PART III

FATIGUE AND PATHOPHYSIOLOGY

CHAPTER 9

FATIGUE AND NEUROMUSCULAR DISEASES

Christine K. Thomas, Marine Dididze and Inge Zijdewind

OBJECTIVES

The objectives of this chapter are to:

- describe how to assess the potential sites of fatigue in relation to the neuromuscular changes that occur with impairment of the nervous system;
- review the magnitude of fatigue (force decline) and sites of fatigue in people with multiple sclerosis, post-polio syndrome, myasthenia gravis and spinal cord injury;
- examine whether current exercise treatments relieve fatigue in each disorder;
- summarise key papers that describe the relationship between weakness and fatigue in each disorder;
- use a case of spinal cord injury to illustrate how excessive fatigue of paralysed muscle reflects increased fatigability of the muscle fibres, not failure of neuromuscular transmission.

INTRODUCTION

Individuals with neuromuscular disorders often report that their daily activities are limited by fatigue. In many cases, these people are describing the need to exert more than the usual amount of effort to accomplish a task. However, in this chapter, we define fatigue as an exercise-induced reduction in the force-generating capacity of the neuromuscular system (Bigland-Ritchie and Woods, 1984; Enoka and Stuart, 1992; Gandevia, 2001). Thus, the mechanical (and electrical) output of the muscles and their activation are measured rather than relying on subjective report. The underlying cause(s) of fatigue (or the force deficit) can reflect failure at one or more levels in the central nervous system (CNS; central fatigue), peripheral nervous system (peripheral fatigue) or both (see the section, Sites of fatigue).

In this chapter we first describe how the integrity of each potential site of fatigue can be measured in humans, outlining issues that need consideration when the neuromuscular system is impaired. This is followed by a review of human studies that have evaluated fatigue in individuals who have one of four different disorders: multiple sclerosis, a disease that results in demyelination of central nervous system axons (Figure 9.1;

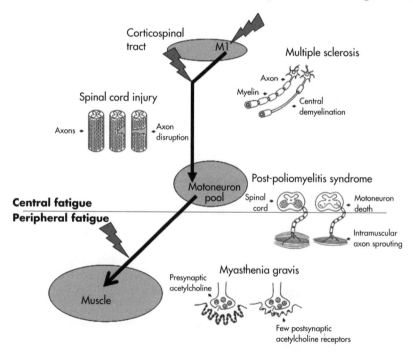

Figure 9.1 Stimulation of M1, the corticospinal tract and the peripheral nerve is delivered to separate fatigue that arises from central and/or peripheral sources. Each disorder involves different impairments. Multiple sclerosis results in central demyelination. Descending and ascending axons are disrupted with spinal cord injury. Motoneurons die when exposed to the poliomyelitis virus, and again many years later. Intact intramuscular axons sprout and inner-vate more muscle fibres after motoneuron death, but usually some fibres remain chronically denervated. In myasthenia gravis, acetylcholine release from the presynaptic membrane occurs as usual, but there are reduced numbers of acetylcholine receptors in the postsynaptic membrane of the neuromuscular junction

Compston and Coles, 2002); post-poliomyelitis syndrome, a disorder that involves additional muscle weakness, fatigue and pain many years after the poliomyelitis virus-induced death of spinal and supraspinal neurons (Trojan and Cashman, 2005); myasthenia gravis, an immunological condition that is characterised by rapid turnover of acetylcholine receptors in the postsynaptic membrane of the neuromuscular junction, resulting in decreased receptor availability (Keesey, 2004); and spinal cord injury, which usually results in paralysis (elimination of voluntary control) and paresis (weakness) of muscles innervated from spinal segments below the lesion (Thomas and Zijdewind, 2006). Exercise treatments currently used to reduce fatigue in each disorder are then examined to assess whether they are effective rehabilitation strategies.

During exercise of an intact neuromuscular system, fatigue-related changes occur simultaneously at various levels of the force-generating system. The primary site of fatigue is task-dependent (Bigland-Ritchie *et al.*, 1995; Enoka and Duchateau, 2008), but in many situations most of the force decline arises from failure of muscle processes (Allen *et al.*, 2008a, 2008b). We have chosen multiple sclerosis, post-poliomyelitis syndrome, myasthenia gravis and spinal cord injury to illustrate how the primary site of fatigue differs across disorders. Central fatigue is prevalent in multiple sclerosis. The chronic muscle weakness that can accompany death of motoneurons is an issue in poliomyelitis. Failure of neuromuscular transmission is problematic in myasthenia gravis. Increases in the intrinsic fatigability of paralysed muscle fibres largely limit force-generating capacity after spinal cord injury.

SITES OF FATIGUE

Exercise-induced changes at or above the level of the motoneuron are commonly classified as central fatigue, whereas changes at the neuromuscular junction or in the muscle are labelled as peripheral fatigue (Figure 9.2). The "twitch interpolation technique" is generally used to distinguish between central and peripheral fatigue. The amount of voluntary (central) drive to the muscle is evaluated by electrical stimulation of the appropriate peripheral nerve or muscle during maximal voluntary contractions. If the voluntary activation of the muscle is maximal, the electrical stimulation does not evoke any additional force. The presence of additional evoked force indicates incomplete (central) activation in an intact nervous system (Merton, 1954; Belanger and McComas, 1981; Taylor and Gandevia, 2008).

However, when a disorder or trauma physically disrupts descending inputs to spinal motoneurons or when there is a central conduction block, this will introduce a mismatch between the number of motoneurons that

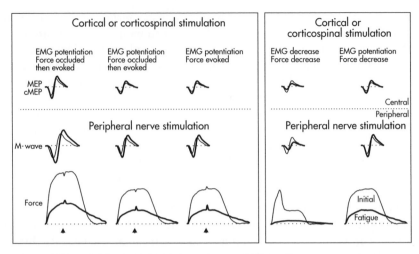

Figure 9.2 EMG and force evoked by stimulation of M1 or the corticospinal tract is compared to that recorded in response to peripheral nerve stimulation to separate central fatigue from peripheral fatigue. Left panel: Force occlusion shows complete muscle activation by voluntary drive. Evoked force demonstrates incomplete muscle activation (central fatigue). When a disorder disrupts central conduction, force evoked by the peripheral nerve system may represent central fatigue and/or activation of muscle that is no longer under voluntary control. Right panel: EMG area and force decrements with peripheral nerve stimulation show failure of neuromuscular transmission and muscle processes (peripheral fatigue). When only force declines with peripheral nerve stimulation and there is maintenance or potentiation of EMG area, only muscle processes are impaired (peripheral fatigue)

can be activated voluntarily and the number that can be excited by peripheral nerve stimulation. Thus, above-lesion stimulation (e.g. magnetic cortical stimulation), rather than below-lesion stimulation (peripheral nerve stimulation), is needed to delineate central fatigue from peripheral failure. Any force evoked by above-lesion stimulation shows submaximal muscle activation because the evoked and voluntary forces both arise from intact pathways (Figure 9.2). With peripheral nerve stimulation (below-lesion stimulation), it is impossible to distinguish whether evoked force represents submaximal voluntary drive or excitation of muscle fibres that are no longer accessible by voluntary effort (Thomas *et al.*, 1997). Changes in muscle stiffness and the handling of calcium in muscle may also alter twitch forces after fatigue, which if present, will lead to an overestimate of central fatigue (Carpentier *et al.*, 2001; Place *et al.*, 2008).

Central components of fatigue

During a voluntary contraction, electromyographic activity (EMG) gives a general measure of muscle activation. Although the voluntary EMG largely reflects the number of recruited motor units and the rates at which they fire, signal cancellation increases with contraction strength (Day and

Hulliger, 2001; Keenan et al., 2005). Thus, EMG measures underestimate the input to the muscle at higher force levels. In contrast, motor unit potentials also potentiate with repeated activity due to enhanced activation of the Na^+–K^+ pump (Hicks and McComas, 1989; Zijdewind and Kernell, 1994). Thus, surface EMG signals are difficult to interpret because changes in the magnitude of the signal with fatigue can arise from various sources. Nevertheless, force decline will occur with unit derecruitment, reductions in firing rate or conduction block, a sign of submaximal muscle activation which may limit task performance, particularly in muscles already weakened due to impairments in the neuromuscular system.

The EMG evoked by evaluations such as electrical or magnetic stimulation of the motor cortex or corticospinal tract can be compared to examine whether failure of central processes arises within the motor cortex, upstream of the motor cortex or in the spinal motoneuron pool. In a healthy nervous system, the amplitude of the EMG recorded in response to stimulation of the motor cortex (motor-evoked potential, MEP) can be compared to stimulation of the peripheral nerve (M wave) to provide an indication of the excitability of the motor cortex and corticospinal tract. Comparison of the response to stimulation of the corticospinal tract (cMEP) to the M wave gives an indication of the excitability of the spinal motoneurons and the efficacy of the synaptic input to these motoneurons (Taylor and Gandevia, 2008). Stimulation of the corticospinal tract is painful, however, and therefore is not often used in experiments that involve patients. When the nervous system is impaired, the amplitude of the evoked signals may decrease, either due to tissue damage, signal dispersion, conduction block or muscle atrophy. Thus, in all cases changes in MEPs (and cMEPs) during and after fatigue must be interpreted in relation to the M wave and under the same conditions. Voluntary drive increases system excitability compared to rest, which increases the magnitude of MEPs and cMEPs.

Relatively new techniques such as functional magnetic resonance imaging (fMRI), which shows changes in the blood flow in different brain regions, as well as the amount of input and neural processing in these areas, are also used to study fatigue-related changes in brain activation in patient groups and healthy subjects (Van Duinen et al., 2007; Post et al., 2009). These magnetic resonance imaging data are divided into two main categories: (1) anatomical scans, which provide a detailed image of the brain; and (2) functional scans, where the subject performs a predefined task during the brain scans. The activity evoked in brain areas can then be correlated with the timing of the task. Brain areas that show a significant correlation with task performance or changes in signal intensity can be compared between fatigued and less-fatigued individuals with a disorder, or between patients and control subjects.

Peripheral components of fatigue

Studies exploring peripheral fatigue typically examine the integrity of transmission across the neuromuscular junction and whether electrical, mechanical and metabolic processes fail within the muscle. Effective transmission of signals across the neuromuscular junction and electrical excitation of the muscle membrane is shown by maintenance (or potentiation) of M wave area as force declines. A decline in M wave amplitude does not necessarily indicate signal failure. With higher frequency stimulation, action potentials overlap and M wave amplitude declines from signal cancellation (Fuglevand, 1995). The conduction velocity of muscle fibres also slows with fatigue, resulting in dispersion of motor unit potentials, amplitude changes in the compound response, and an increase in M wave duration (Bigland-Ritchie *et al.*, 1982). Only when declines in M wave area are accompanied by force decrements is impulse conduction impaired (Figure 9.2).

In humans it is difficult to examine sites of fatigue beyond excitation of the muscle membrane. Muscle metabolism and biochemical changes have been examined by assessing changes in oxidative and glycolytic enzymes, lactate, phosphocreatine (PCr) and adenosine triphosphate (ATP) from muscle biopsies taken throughout a fatigue protocol (Vollestad *et al.*, 1988), use of nuclear magnetic resonance spectroscopy (Cady *et al.*, 1989) or from estimates of interstitial potassium levels (Sjogaard and McComas, 1995). Alterations in calcium handling are usually inferred by comparing changes in twitch and tetanic forces, and rates of contraction and relaxation, whereas failure of excitation–contraction coupling has largely been inferred by exclusion of the other possible sites that can be assessed. The importance of muscle membrane excitability, extracellular potassium accumulation, metabolic changes, reactive oxygen species and calcium handling in the failure of muscle processes (sites that underlie much of the force decline seen during exercise of a healthy nervous system) have largely been understood from studies in reduced preparations and single or skinned muscle fibres, as reviewed by Allen *et al.* (2008a, 2008b).

Similarly, the importance of afferent inputs to maintenance of muscle force has been explored little in human studies because it is difficult to measure these signals. Even so, sensory information makes substantial contributions to muscle activation and control of movement, and is likely impaired at peripheral, spinal and/or supraspinal levels after disease and trauma. For example: slowing of muscle relaxation matches the reduction in motor unit firing rates seen during some fatigue tasks (Bigland-Ritchie *et al.*, 1983, 1986; Bigland-Ritchie and Woods, 1984); motor unit firing rates are reduced when afferent inputs are blocked by anaesthesia of the peripheral nerve (Macefield *et al.*, 1993); firing of muscle spindle afferents facilitate muscle activity but is reduced during sustained contractions (Macefield *et al.*, 1991; Griffin *et al.*, 2001); golgi tendon organs respond during both weak and strong contractions (Crago *et al.*, 1982); cutaneous afferents are critical for control of grip forces (Johansson and Westling, 1988); and group III and IV afferents respond to chemical and mechanical stimuli produced during fatiguing muscle contractions (Mense, 1977; Kniffki *et al.*, 1979; Sinoway *et al.*, 1993).

MULTIPLE SCLEROSIS

Multiple sclerosis (MS) is the most common disease of the CNS that chronically disables young adults. The oligodendrocytes that surround the axons in the CNS atrophy and die, resulting in demyelination of the axons, glial scars and axonal degeneration. Without the myelin insulation around the axons, ions leak out of the axon and the amplitude of the action potentials travelling along the nerve fibres slowly declines. When the depolarisation is too small to open voltage-gated sodium channels at the nodes of Ranvier, the propagation of the action potentials is hindered and the flow of information stops, resulting in the symptoms of MS.

People with MS describe a wide range of symptoms. Changes in sensory input are common, including both diminished sensory input (numbness) and enhanced input (parasthesia, tingling and hyperesthesia); diminished motor output (e.g. a decline in muscle force and motor coordination); and a decline in cognitive function (e.g. lack of attention and concentration). Besides these symptoms, fatigue is frequently reported to interfere with a person's daily functioning (Krupp *et al.*, 1988; Bakshi, 2003). The Multiple Sclerosis Council for Clinical Practical Practice Guidelines (1998) published a consensus definition of fatigue for individuals with multiple sclerosis: "A subjective lack of physical and/or mental energy that is perceived by the individual or caregiver to interfere with usual and desired activities". Although this definition is much broader than we use in the present chapter, it raises several important points: lack of physical energy; perceived by the individual; and interference with usual activities.

Fatigue in multiple sclerosis

In patients with MS the initial strength of a muscle can be similar (Sheean *et al.*, 1997; de Haan *et al.*, 2000; Petajan and White, 2000; Perretti *et al.*, 2004) or diminished (Rice *et al.*, 1992; Kent-Braun *et al.*, 1994; Sharma *et al.*, 1995; de Ruiter *et al.*, 2001) compared to healthy individuals. In general, the strength and M waves of muscles in the upper body are better preserved than those of muscles in the lower body. Changes in skeletal muscle composition (fewer type I fibres, smaller fibres) and enzyme activity (lower aerobic enzyme activity) also suggest changes in muscle properties due to disuse (Kent-Braun *et al.*, 1997). However, most patients show a faster decline in force during sustained voluntary contractions even when the initial force is equal to control subjects (Sheean *et al.*, 1997; Petajan and White, 2000; cf. however, Thickbroom *et al.*, 2006).

Peripheral components of fatigue

Only a few studies have been conducted on the peripheral factors of fatigue in MS patients. In general, leg muscles are more prone to fatigue than arm or hand muscles. Two experiments in the upper limb have demonstrated similar peripheral fatigability in MS patients and healthy subjects when the initial muscle forces were comparable. During a fatigue test induced by electrical stimulation of the motor nerve, no significant difference in fatigability was seen in the hand muscle, adductor pollicis (de Ruiter *et al.*, 2001). This observation was strengthened by data that demonstrated a similar time-course for changes in PCr in MS patients and control subjects during a 3 min handgrip task (Petajan and White, 2000).

The data for the leg muscles are rather different, demonstrating a larger amount of peripheral fatigue in MS patients. Both force and resistance to fatigue were reduced in MS patients (quadriceps femoris, de Haan *et al.*, 2000; tibialis anterior, Sharma *et al.*, 1995). The larger force decline was accompanied by a larger decline in PCr levels and intracellular pH (Sharma *et al.*, 1995). The fact that the M wave was also smaller in the quadriceps of the MS patients suggests disuse atrophy or deconditioning. However, none of these changes correlated with levels of perceived fatigue (Sharma *et al.*, 1995).

Central components of fatigue

Data from Rice *et al.* (1992) showed that during a maximal voluntary contraction (MVC) of quadriceps, the motor unit firing frequencies were lower in MS patients than control subjects. Furthermore, experiments using superimposed twitches during MVCs showed lower central activation in MS patients during brief contractions (Sheean *et al.*, 1997; Ng *et al.*, 2004) and during a sustained maximal contraction (45 s) of the adductor pollicis muscle, resulting in larger force declines during the maximal contraction even though the initial force was equal to control subjects (Sheean *et al.*, 1997). However, no relation was found between the sense of fatigue during everyday situations and the amount of fatigue that was measured during these maximal contractions.

Changes in central excitability in MS patients have been examined further with cortical magnetic stimulation. The evoked motor response is generally smaller and arrives at a longer latency in MS patients compared to healthy control subjects (van der Kamp *et al.*, 1991; Perretti *et al.*, 2004; Thickbroom *et al.*, 2006) due to a decrease in conduction velocity in the corticospinal tract. During a fatiguing contraction MS patients show a larger increase in MEP (Perretti *et al.*, 2004; Thickbroom *et al.*, 2006). After the contraction, the MEP can be smaller (Liepert *et al.*, 1996; Thickbroom *et al.*, 2006) or similar (Sheean *et al.*, 1997; Petajan and White, 2000; Liepert *et al.*, 2005) compared to that seen in control subjects. However, no clear overall conclusion can be drawn from the cortical stimulation data at present.

In MS patients, anatomical scans obtained by MRI have been used to identify visible lesions. Most protocols concentrate on white-matter lesions because MS is thought to be a disease that affects the myelin. However, recent evidence also shows abnormalities in the grey matter (neurons). Comparisons of anatomical scans of fatigued versus less-fatigued MS patients has given only sparse evidence that the location of the white-matter lesion, lesion load (number and size of the lesions, Colombo *et al.*, 2000) or axonal damage (Tartaglia *et al.*, 2004) is important for the sense of fatigue. There is some evidence that glucose metabolism is reduced in prefrontal areas and in primary and secondary motor areas (Roelcke *et al.*, 1997). However, most studies did not find a strong relation between fatigue scores and MRI measurements (van der Werf *et al.*, 1998; Mainero *et al.*, 1999).

During functional scans, MS subjects had to perform a predefined task in an MR-scanner, and showed different brain activation patterns compared to that seen in control subjects (Lee *et al.*, 2000; Filippi *et al.*, 2002; Pantano *et al.*, 2005). Most experiments showed increased activation in the sensorimotor cortex ipsilateral to the limb that performed the task, as well as increased activation of secondary motor areas (e.g. supplementary motor area, premotor areas, cingulate motor area) in comparison to healthy subjects. This altered activation pattern could reflect adaptation in the usage of cortical areas. Ipsilateral sensorimotor activation is also seen more often in patients (e.g. after a stroke) and can reflect increased difficulty in performing a task. Experiments that compared MS patients who experience high or low levels of fatigue demonstrated that non-fatigued patients showed greater overall activation (ipsilateral cerebellum, rolandic operculum, precuneus, contralateral thalamus, contralateral middle frontal gyrus), whereas fatigued patients showed more activity in the contralateral cingulate motor area (Filippi *et al.*, 2002). Although the data set is still limited, the results suggest that the increased sense of fatigue experienced by people with MS could either be induced by changes in the cortical activation pattern or that alterations in activity itself result in an increased sense of fatigue.

Exercise in MS patients

The focus of most exercise programmes developed for people with MS is on aerobic fitness and isometric force. The overall conclusion is that aerobic exercise has a positive effect on general well-being, isometric strength, physical fitness and mobility-related daily life situations in patients suffering from MS (reviewed by Rietberg *et al.*, 2005; Motl and Gosney, 2008). However, the effects of this exercise on fatigue or the sense of fatigue are either small (Fragoso *et al.*, 2008; see for review Motl and Gosney, 2008) or absent (Petajan *et al.*, 1996; Rietberg *et al.*, 2005). Importantly, no negative effects were reported from the exercise.

Muscle weakness is a main determinant of increased fatigability (Thomas and Zijdewind, 2006). As weakness is indicated as a potential problem in MS, especially in leg muscles, exercise programmes that involve walking/stair climbing should be beneficial. Furthermore, the fact that MS-related fatigue is considered to involve more central than peripheral components suggests that exercise training programmes also have to be directed to central processes. Therefore, it is disappointing that the only study that focused on central effects of a (short-lasting) training programme did not find effects in brain activation, especially because there are some indications that differences in brain activation could be associated with increased levels of fatigue (Filippi *et al.*, 2002; see the section, Central components of fatigue). In the training study, patients and control subjects performed a thumb flexion-extension training task. In the pre-training condition MS patients showed an increased contralateral premotor activity. After the training, control subjects showed a decline in activity in contralateral sensorimotor cortex and inferior parietal lobule, whereas no decline in activation was seen in MS patients (Morgen *et al.*, 2004).

Since there is no means of preventing, treating or curing MS, optimising quality of life is extremely important. Interventions that reduce levels of fatigue are critical because fatigue is one of the most disabling symptoms of MS (Kurtzke, 1983). Presently, clinicians and health-givers are reluctant to stimulate MS patients to participate in exercise programmes. This is mainly caused by the knowledge that increased body temperature can substantiate motor disability in MS (Nelson and McDowell, 1959; Compston and Coles, 2002). However, exercise programmes also stimulate feelings of general well-being and can increase muscle strength, so promoting exercise programmes in MS patients may also reduce fatigue. Research concerning exercise programmes should also focus on protocols that induce changes in the central aspects of motor performance, and thereby diminish the central effects of fatigue in MS.

POST-POLIOMYELITIS SYNDROME

The poliomyelitis virus largely targets spinal motoneurons, with some death of other neurons in the cortex, cerebellum, brainstem and spinal cord, inducing various amounts of muscle paralysis, denervation and weakness (Dalakas, 1995). Provided some motoneurons to a muscle survive, muscle strength recovers to varying extents because spared axons sprout within the muscle and innervate nearby fibres that have no nerve supply (Trojan and Cashman, 2005). Voluntary contractions of these muscles are thus generated by stronger, but fewer, motor units. New muscle weakness occurs many years after the initial polio insult in 15–80% of people (Farbu et al., 2006), and is often accompanied by additional fatigue, pain, muscle atrophy and cold intolerance, all symptoms of the post-poliomyelitis syndrome (Halstead, 1991). Since the initial weakness was frequently greater in leg muscles compared to arm muscles (Trojan et al., 2001), the new symptoms often make walking and climbing stairs more difficult, and activities like shopping and cleaning harder in most individuals who experience post-poliomyelitis syndrome (PPS) (Farbu et al., 2006).

Muscle strength in PPS

Muscle weakness is a hallmark of poliomyelitis. In many cases, the muscle weakness seems to be more widespread and severe than documented or experienced initially. Leg and arm muscles rated as normal by manual evaluation only average 14–64% and 56–76% of the strength of age-matched controls, respectively (Beasley, 1961; Hildegunn et al., 2007). Motor unit counts are low in 87% of affected muscles, as expected, but also in 65% of muscles reported to be unaffected by the initial virus. On average, affected and unaffected muscles only have 41% and 61% of the usual complement of motor units, respectively (McComas et al., 1997). Large motoneurons also die preferentially (Hodes, 1949; Hodes et al., 1949), and with their high innervation ratios, chronic muscle denervation is likely. In support of this, imaging and biopsy studies show obvious muscle atrophy and intramuscular fat infiltration (Tollback et al., 1996; Beelen et al., 2003). Some muscles cannot be excited easily with electrical stimulation. Overused muscles have hypertrophied fibres, but are still weaker than usual (Tollback et al., 1992).

Chronic muscle weakness directly impacts fatigue when contractions have to be sustained or produced repeatedly. A weak muscle makes it more difficult to accomplish a given task. It has to work at a higher intensity relative to its maximal capacity to produce a given force, by recruiting more motor units and/or firing them faster, which can result in greater fatigue and/or limit the time the task can be performed. Thus, when new weakness arises many years after the initial polio insult, and ongoing strength declines range from 1–2% per year up to 18% in some older individuals (Grimby et al., 1998; Klein et al., 2000; Lord et al., 2002), it is not surprising that fatigue is repeatedly described as the most debilitating issue that limits daily activities.

Fatigue in PPS

Most people with PPS report that fatigue routinely limits their daily activities (83–89%, see Jubelt (2004)). They describe the fatigue as physical rather than mental, as a loss of strength during exercise, decreased exercise tolerance that worsens as the day progresses, tiredness, or as a lack of energy. Despite its prevalence, few studies have objectively examined fatigue in individuals with PPS. Some studies in people with PPS show force declines during exercise that exceed those recorded from age-matched control subjects (Sharma *et al.*, 1994; Allen *et al.*, 1994, 2004; Lupu *et al.*, 2008). Another indication of a reduced functional capacity is the inability of some people with PPS to complete an exercise protocol. Other studies report comparable force declines in people with PPS and those with an intact neuromuscular system (Agre and Rodriguez, 1990; Rodriquez and Agre, 1991; Rodriquez *et al.*, 1997; Grimby *et al.*, 1998; Samii *et al.*, 1998; Sunnerhagen *et al.*, 2000). Most of these studies have not examined the sites at which the signals fail and so the factors that contribute to the force decline.

Peripheral components of fatigue

Clinical examination of muscles affected by polio (and some muscles thought to be unaffected by polio) and the PPS usually show blocking of some EMG signals. There is also increased jitter in the electrical transmission of signals recorded from the muscle (Cashman *et al.*, 1987). However, at the whole-muscle level, the M wave is well-maintained when a train of pulses at 50 Hz is delivered to a peripheral nerve (Sharma *et al.*, 1994), as found in healthy individuals where most fatigue arises from failure of peripheral processes that are beyond transmission of impulses across the neuromuscular junction and excitation of the muscle membrane. Whether conduction of EMG signals is exacerbated during fatiguing contractions performed by people with PPS has not been evaluated. Given that the new muscle weakness in PPS likely reflects death of additional motoneurons and degeneration of axon terminals (Wiechers and Hubell, 1981; Weichers 1988), there will be new attempts to restore muscle innervation by intramuscular axon sprouting. The myelin encapsulating these axon branches will be thin initially; a feature that makes EMG failure at axon branch points a distinct possibility.

One study reports unremarkable metabolic changes when individuals with PPS perform submaximal intermittent contractions for a set time (Sharma *et al.*, 1994). If the exercise had continued, excitation–contraction coupling may have failed, as occurs in controls (Vollestad *et al.*, 1988). Rested muscles that are influenced by PPS have decreased oxidative and glycolytic enzyme activities, a low capillary density, and hybrid isoforms of myosin, presumably because the surviving motoneurons have not completely re-specified all of the muscle fibres that they now innervate (Borg *et al.*, 1988, 1989; Grimby *et al.*, 1996). Studies have not explored whether these factors contribute to the fatigue that people with PPS describe, although all seem likely to do so. Similarly, some studies show delays in the recovery of voluntary EMG and force after fatiguing exercise (Agre and Rodriquez, 1991; Sharma *et al.*, 1994; Grimby *et al.*, 1996; Sunnerhagen *et al.*, 2000), but the mechanisms responsible for this delayed recovery have not been explored.

Central components of fatigue

Voluntary drive to muscles weakened by polio and PPS is relatively high (>90%) during brief voluntary contractions when tested with the twitch interpolation technique (see Thomas and Zijdewind, 2006). When sustained or repeated contractions are needed, muscle activation has to increase. More motor units are recruited, or those units that are active can fire at higher rates, both of which will increase force. During fatiguing contractions performed by people with PPS there are increases in the surface EMG, motor unit recruitment and firing rates. MEPs show facilitation, and higher effort scores are reported (Allen *et al.*, 1994, 2004; Samii *et al.*, 1998; Sunnerhagen *et al.*, 2000), all indications that voluntary drive has increased. However, as exercise progresses, central drive usually declines. This reduction in muscle activation can be greater in some individuals with PPS versus control subjects (Allen *et al.*, 1994; Lupu *et al.*, 2008), indicating fatigue of central processes in people with PPS.

Exercise in PPS

To date, pharmacological interventions (e.g. pyridostigmine, amantadine, prenisolone) have been ineffective in changing the strength or the fatigability of muscles influenced by PPS (Farbu *et al.*, 2006). Thus, symptoms of fatigue are largely managed by exercise programmes that people with PPS find non-fatiguing, by lifestyle changes (use of assistive devices; mobility aids; self-pacing and monitoring exercise effect and intensity to avoid adverse effects; weight loss; cessation of smoking; naps, regular rests, more sleep; relaxation techniques; changes in a work station, seating or transportation; working at home or part time) (Jubelt, 2004; Trojan and Cashman, 2005). Since strength deficits differ across muscles and limbs in PPS, exercise is usually limited by the functional capability of the most-affected limb. Assisting that limb, by bracing a weak ankle for example, can be used to facilitate activities like walking (Halstead *et al.*, 1995).

In general, exercise has been used to prevent the effects of disuse. At the same time, the aim has been to prevent overuse or pain which can hinder further exercise. Numerous studies show that paced, submaximal exercise (strength training; non-swimming aquatic exercise; cardiopulmonary, aerobic exercise such as arm ergometry) that includes rest periods can improve the strength of the muscles least affected by PPS without adverse effects (see Agre, 1995; Jubelt, 2004; Farbu *et al.*, 2006). Although fatigue is usually reported on scales rather than measured, increases in work capacity have often been documented by progressive increases in the weight lifted, repetitions completed or work time. In contrast, there is little or no study of the weakest muscles in people with PPS, although passive movement may help maintain muscle (Dupont-Versteegden *et al.*, 1998) and flexibility, preventing contractures.

The long-term consequences of regular exercise on the strength and thus the fatigability of muscles weakened chronically by PPS are unknown. The factors that control the elimination of neuromuscular junctions in relation to activity are understood poorly, but the muscle fibre plays an important role in this process because decreases in acetylcholine receptors begin before the loss of presynaptic terminals (Balice-Gordon and Lichtman, 1994). Animal studies suggest that moderate levels of activity, rather than excessive activity or inactivity, best maintain or increase motor unit force in partially denervated muscles (Tam and Gordon, 2003). However, none of these studies are long-term, or involve aged animals with an early history of motoneuron death and muscle denervation. In PPS, muscle function is likely to be limited by the number of times that intact axons can sprout in an ageing nervous system and by the availability of nearby muscle fibres because axon sprouting is localised (Son and Thompson, 1995). Thus, exploring changes in fatigue in relation to motor unit strength in people with PPS and age-matched controls would be an interesting way to understand how axon sprouting limits muscle function in disorders that involve muscle denervation.

MYASTHENIA GRAVIS

Myasthenia gravis (MG) is a rare autoimmune neuromuscular junction disorder characterised by fluctuating weakness and fatigability in eye, facial, oropharyngeal, axial and/or limb muscles that increases with muscle use and improves with rest (Keesey, 2004). Most (85%) people with generalised myasthenia (oropharyngeal or limb muscle weakness) or the ocular form (50–60%) have antibodies against the nicotinic acetylcholine receptor on the postsynaptic membrane of the neuromuscular junction (Simpson, 1960; Lindstrom *et al.*, 1976; Thanvi and Lo, 2004), reducing the number of acetylcholine receptors (Fambrough *et al.*, 1973). Others with MG have antibodies that bind the muscle-specific protein kinase (MuSK), which mediates agrin-induced clustering of acetylcholine receptors during synapse formation (Glass *et al.*, 1996; Hoch *et al.*, 2001). This deficiency in acetylcholine receptors does not change the number of acetylcholine quanta released from the presynaptic terminal per nerve impulse, but it does reduce the amplitude of the endplate potentials and can result in blocking of nerve impulses (Elmqvist *et al.*, 1964; Engel *et al.*, 1976; Lambert *et al.*, 1976). In this chapter we focus on generalised MG. When people with this disorder perform voluntary contractions, there is often an increase in the firing variability (jitter) of different muscle fibres within a motor unit, a sign that the margin for safe transmission of impulses across the neuromuscular junction is reduced. When this jitter results in blocking of nerve impulses, a process that can be exacerbated during stronger contractions or repetitive nerve stimulation, there is acute muscle weakness (Ingram *et al.*, 1985; Trontelj and Stalberg, 1995). As a consequence, most (82%) people with MG complain of fatigue, and describe more impairment of physical function than cognitive function (Paul *et al.*, 2000). Both the acute muscle weakness and the fatigue make it more difficult to complete tasks like climbing stairs and overhead work (Grohar-Murray *et al.*, 1998; Juel and Massey, 2007).

Weakness in myasthenia gravis

Many people with generalised MG generate weak forces during shoulder abduction, elbow flexion, hand grip, knee extension, forward and backward flexion of the neck, bite, inspiration and expiration compared to healthy controls. Average forces across people and muscle groups varied from 34% to 87% of control values (Ringqvist, 1971; Secher and Petersen, 1984; Nicklin *et al.*, 1987; Mier-Jedrzejowicz *et al.*, 1988; Keenan *et al.*, 1995). The long-term reductions in muscle strength are explained in part by significant atrophy of muscle fibres, particularly type II fibres (Brooke and Engel, 1969; Ringqvist, 1971). However, muscle weakness also varies acutely and across muscles, by the hour, with general increases as the day progresses (Keesey *et al.*, 1981). Single fibre EMG has shown that this acute weakness relates to blocking of neuromuscular transmission in some fibres of some motor units during sustained weak voluntary contractions and at higher firing rates. Impulse blocking is exacerbated by increases in contraction duration, ischemia and temperature (Desmedt, 1957; Stalberg, 1980; Trontelj *et al.*, 1986, Rutkove *et al.*, 1998; Lo *et al.*, 2004; Sener and Yaman, 2008) as hydrolysis of acetylcholine by acetylcholine esterase becomes more rapid, and the shorter channel open time on the postsynaptic membrane results in endplate potentials of reduced amplitude (Lass and Fischbach, 1976; Hofmann *et al.*, 1966). In some cases, acute conduction block and decrements in single fibre EMG, compound action potentials and force are relieved by rest, cooling or acetylcholine esterase inhibitors, although the recovery is not necessarily sustained or related to the initial muscle strength (Ringqvist, 1971; Ingram *et al.*, 1985; Trontelj and Stalberg, 1991, 1995; Gilchrist *et al.*, 1994; Mermier *et al.*, 2006).

Fatigue in myasthenia gravis

During voluntary contractions, the blocking of nerve impulses in people with generalised MG is partially counteracted by increased mobilisation of acetylcholine from the presynaptic nerve terminal (Magleby and Zengel, 1976), but this potentiation is short-lived. It is followed by pronounced depression of endplate potentials, decrements in the compound muscle action potentials in response to repetitive nerve stimulation, and muscle force. These decrements often worsen several minutes after exercise and it takes 10–15 min to recover. These have been termed post-activation exhaustion (Desmedt, 1957; Lo *et al.*, 2004). Not surprisingly, people with MG describe fatigue as their most prominent and troubling symptom (Grohar-Murray *et al.*, 1998). This fatigue has largely been measured with visual analogue scales and questionnaires. Subjective reports of fatigue severity correlate best with restriction of activity and years since diagnosis (Kittiwatanapaisan *et al.*, 2003).

The few studies that have assessed the magnitude of the force declines during voluntary contractions show that it usually exceeds that measured in muscles of healthy subjects. After only ten brief maximal voluntary abductions of the shoulder, force declined by more than 50% in most people with MG (83%), but by only 6% in control subjects (Nicklin *et al.*, 1987). Repeated MVCs of thumb muscles resulted in significantly greater force declines (65%) in individuals with MG than in muscles of healthy subjects (21%), as well as within train EMG and force declines (Secher and Petersen, 1984). Fatigue can also depend on whether the muscles respond to acetylcholine esterase inhibitors. Repeated or sustained elbow flexor contractions resulted in similar force declines for subjects with generalised MG who did not respond to acetylcholine esterase inhibitors and control subjects (presumably less-affected muscles), whereas the force declines were typically greater than usual in those with MG who did respond to the inhibitors (Ringqvist, 1971). Thus, the inhibitors provided little functional benefit in these individuals. Similarly, in some participants reducing core body temperature improved endurance during repeated contractions of wrist extensors (Mermier *et al.*, 2006). None of these studies have examined the sites at which processes fail systematically, although certain observations suggest that the fatigue arises from both central and peripheral sites.

Peripheral factors in fatigue

Rests between contractions are particularly important in individuals with generalised MG because during sustained maximal contractions of the thumb the force declined abruptly in the first second, then was followed by a steady decline in force (Secher and Petersen, 1984). The initial sudden force decrement in people with MG, but not in controls, probably reflects that component related to blocking of signal transmission across the neuromuscular junction. The slower force declines in both groups of subjects likely reflect failure of other muscle processes. Consistent with this suggestion, spectra at rest (inorganic phosphate P_i/ATP, PCr/ATP, P_i/PCr ratios, muscle pH) were similar in individuals with mild or moderate-to-severe MG and controls. Repeated plantar flexion of the ankle at submaximal intensities every 2 s for 3 min also resulted in similar metabolic changes in those with mild MG and controls. However, those with more severe MG symptoms had higher P_i/ATP and P_i/PCr ratios at the end of the exercise and lower muscle pH, suggesting reduced oxidative metabolism and a shift towards glycolytic metabolism. Recovery of PCr levels was also poor in individuals with more severe MG (Ko *et al.*, 2008). Thus, although impairment at the neuromuscular junction is considered the main cause of peripheral fatigue in MG, studies also show failure of muscle processes.

Central factors in fatigue

A few observations show that the force declines seen in muscles influenced by generalised MG also involve fatigue of central processes, although quantitative assessments are lacking. For example, transdiaphragmatic pressure during maximal sniffs was reduced in eight individuals with mild or moderate generalised MG, but pressure during bilateral phrenic nerve stimulation was only reduced in three subjects, suggesting central fatigue (Mier-Jedrzejowicz *et al.*, 1988). Moreover, when additional efforts were made during sustained voluntary contractions of thumb muscles, there were transitory increases in muscle force in subjects with MG and healthy subjects, a sign of central fatigue (Secher and Petersen, 1984).

Exercise in MG

Although fatigue is common in people with MG, these individuals are often left to self-manage this symptom as treatment has focused on the disease process itself. Some people have been advised to refrain from any kind of exercise. Others are told to exercise judiciously and to learn their limits. Thus, there are no specific protocols to guide the management of fatigue. Most people with MG (90%) change their lifestyle. For example, they learn to pace themselves and not push to the limit, they avoid stress, they participate in low-impact physical activities during peak energy times, take rests and naps, organise their environment to conserve energy, and use assistive devices like a walking cane (Grohar-Murray *et al.*, 1998).

Since exercise can exacerbate the symptoms of MG, few individuals participate in it (7%; Mermier *et al.*, 2006). However, the few studies that have explored the acute effects of exercise training generally show positive results. Although most improvements have occurred in muscle strength, this may also facilitate reductions in fatigability. For example, in individuals with mild or moderate generalised MG, 91% (and eventually all subjects) were able to repeatedly extend the knee maximally every 2s for 3min, three times per week. Strength increased after ten weeks of training, but there was no change in endurance. Most people were unable to complete this same programme with arm muscles because they were unable to complete the repetitions or could not increase the load to be lifted. Thus, physical training with low loads is safe in mild MG and results in force improvements in certain muscles (Lohi *et al.*, 1993). Various kinds of respiratory training performed over 2–3 months (diaphragmatic breathing, interval-based inspiratory muscle training with progressive increases in load from low to moderate intensities, breathing with pursed lips, restrictive breathing throughout the respiratory cycle) all improve maximal inspiratory and expiratory pressures (respiratory muscle strength), maximal voluntary ventilation, chest wall mobility and respiratory pattern. The greatest improvements were seen in the weakest muscles (Gross and Meiner, 1993; Weiner *et al.*, 1998; Fregonezi *et al.*, 2005). Resistance exercise (lifting weights) with creatine supplementation in one case of MG resulted in increases in the strength of leg flexion and extension, although there was no control for the effects of creatine intake (Stout *et al.*, 2001). In another case of McArdle disease and MG, three months of aerobic exercise (walking progressively more until 60min of exercise was possible without rest) resulted in a return to independent living and increases in peak oxygen uptake (Lucia *et al.*, 2007). No adverse effects have been reported in these studies. Thus, it seems that people with MG do have the ability to perform low-intensity exercise and the potential to improve with training. Exploring ways to sustain exercise with less weakness and fatigue, particularly in muscles less affected by MG, and possibly in association with cooling of core body temperature, could enhance the health and quality of life of people with MG.

SPINAL CORD INJURY

Trauma can occur at any level of the spinal cord. The injury usually disrupts descending inputs to the cord, which results in paralysis of muscles (elimination of voluntary control) innervated from spinal segments below the lesion. Damage to ascending fibres induces sensory impairments. Near the lesion epicentre, it is common for motoneurons to die and for spinal roots to be damaged leading to muscle denervation. As occurs with poliomyelitis, intramuscular sprouting of intact axons will restore muscle innervation provided that the muscle denervation is not extensive or complete (see Thomas and Zijdewind, 2006). In this section, the focus is on paralysed muscles that retain their innervation. Although not under voluntary control, these muscles are often induced to contract artificially by electrical stimulation of the peripheral nerve or intramuscular nerve branches, either as a means of exercise and/or to restore functional behaviour. In either situation, it is typical for fatigue of the muscle fibres to limit work capacity and time.

Muscle strength

Weakness of paralysed muscles also limits performance because more muscle has to be used to generate a given force. Muscle fibres almost always atrophy following paralysis, often to half of their original size (Martin et al., 1992; Mohr et al., 1997; Castro et al., 1999). In most cases this atrophy results in significant muscle weakness (Peckham et al., 1976; Thomas et al., 1997; Gerrits et al., 1999). However, when contractions of paralysed muscles are evoked by supramaximal nerve stimulation, some muscles are as strong as those of uninjured subjects (Lieber et al., 1986; Stein et al., 1992; Gordon and Mao, 1994; Shields, 1995; Thomas, 1997b). This apparent muscle strength may reflect changes in specific tension (force/cross-sectional area), stiffness due to increases in connective tissue, more effective transmission of force to the tendon and/or alterations in muscle use (Lieber et al., 1986; cf. Cope et al., 1986; Maganaris et al., 2006). Although there are large reductions in the use of paralysed muscles (Stein et al., 1992; Tepavac et al., 1992), they are not inactive. Involuntary contractions (muscle spasms) begin a few weeks after injury (Curt and Dietz, 1999). Activation of motor units at high rates may facilitate maintenance of strength (Kernell et al., 1987; Thomas and Ross, 1997), particularly if the contractions are against resistance (Shields and Dudley-Javoroski, 2006; Kim et al., 2007). Chronic excitation of motor units at low rates can weaken paralysed muscles (Kernell et al., 1987; Zijdewind and Thomas, 2001), as can medication like baclofen, which is often given to reduce muscle spasms (Thomas et al., 2008).

Fatigue after spinal cord injury

When muscle paralysis is due to disruption of descending inputs to spinal motoneurons, any central contributions to fatigue are eliminated. Thus, force declines must reflect failure at the neuromuscular junction or processes within the muscle. Repeated stimulation of single paralysed motor units at high frequencies shows that the force declines are accompanied by maintenance or potentiation of the evoked EMG. Thus, transmission of the signals across the neuromuscular junction and excitation of the muscle membrane are effective. The fatigability of the muscle fibres themselves has increased, possibly due to changes in excitation–contraction coupling, calcium handling, metabolism, and/or vascularisation (Klein et al., 2006). Few studies on whole paralysed muscles have recorded both the evoked EMG and force in response to supramaximal stimulation because the stimuli are painful when sensation is intact, and strong muscle contractions may result in bone fracture after spinal cord injury (SCI) (Biering-Sorensen *et al.*, 1990; Garland *et al.*, 1992; Wilmet *et al.*, 1995). The data are also influenced by muscle ischemia. Even so, these studies confirm the idea that the excessive fatigue originates within the muscle fibres (Stein *et al.*, 1992; Shields, 1995; Thomas, 1997a).

Fatigue of paralysed muscles increases steadily for two years after injury, then at a much slower rate (Shields *et al.*, 2006a), and is accompanied by conversion of muscle fibres from type I (slow) to type II (fast); reductions in oxidative enzymes, muscle capillaries and mitochondria; reduced blood flow; and slower increases in blood flow during evoked contractions (Martin *et al.*, 1992; Rochester *et al.*, 1995a; Andersen *et al.*, 1996; Hopman *et al.*, 1996; Burnham *et al.*, 1997; Mohr *et al.*, 1997; Castro *et al.*, 1999; Gerrits *et al.*, 1999; Olive *et al.*, 2003). Fatigue resistance can be restored slowly, and to a large extent, by repeated muscle stimulation (Peckham *et al.*, 1976; Stein *et al.*, 1992; Shields and Dudley-Javoroski, 2006), suggesting that much of the increase in fatigue relates to chronic changes in muscle use.

Fatigue of paralysed muscles is also often attributed to the different ways that motor units are activated by electrical stimulation versus voluntary contractions (stimulation at constant versus variable frequencies; recruitment of strong, fatigable units before weak units; occlusion of blood flow during strong contractions; synchronous versus asynchronous activation of motor units; antidromic activation of motoneurons with stimulation). The present evidence suggests that these factors only make small contributions to the excessive declines in force.

Inclusion of one of two short intervals at the beginning of a train of pulses increases the evoked force, a finding that holds across muscles, motor units, species, and after muscle paralysis (Griffin *et al.*, 2002), but several groups have shown that use of variable frequency trains does not reduce fatigue in different paralysed muscles (Thomas *et al.*, 2003; Janssen *et al.*, 2004; Scott *et al.*, 2007). Stimulation frequency is more critical; higher frequencies generate more force but greater fusion, slowing relaxation and fatigue in a shorter time (Thomas *et al.*, 2003; Kebaetse *et al.*, 2005). Despite this, higher stimulus frequencies have to be used to maintain evoked force when it is monitored on-line because lower frequencies do not elicit sufficient force to reach the preset target (Shields *et al.*, 2006b).

With respect to recruitment order, electrical stimulation of both paralysed muscles and muscles of uninjured subjects has recruited motor units in no consistent order, weak then strong units, or fast-contracting units preferentially (McComas *et al.*, 1971; Bergmans, 1973; Brown *et al.*, 1981; Dengler *et al.*, 1988; Trimble and Enoka, 1991; Thomas *et al.*, 2002), suggesting that axon excitability is influenced by biophysical properties and the proximity of axons to the stimulus. Relative force declines were also smaller when stimuli excited only part of a paralysed muscle versus the whole muscle, even after correction for ischemia, suggesting that stimulation does not necessarily activate the most fatigable units first (Godfrey et al., 2002). All fibres show increases in oxidative enzymes, with training suggesting a lack of preferential activation of particular fibre types (Rochester et al., 1995a, 1995b). Moreover, any effects from recruiting fatigable units preferentially by electrical stimulation are minimised after chronic paralysis because all units become fatigable after SCI (Klein et al., 2006).

The increased fatigability of muscle fibres may also explain why ischemia has little influence on the declining force. When whole paralysed muscles and motor units were stimulated repeatedly at 40 Hz, the force decreased to 24% and 34% initial after 2 min, a difference of only 10% (Thomas et al., 2003; Klein et al., 2006). Contraction of a single motor unit will not involve ischemia, so the large differences in fatigue of paralysed motor units (force after 2 min is 34% initial) and that measured in uninjured subjects (78% initial) must reflect enhanced fatigue of the muscle fibres. Thus, occlusion of blood flow during evoked whole-muscle contractions only makes a small contribution to the overall fatigability. Consistent with this suggestion, fatigue of paralysed muscles is reduced somewhat when blood pressure is increased during stimulation of paralysed muscles (by voluntary contraction of other muscles (Butler et al., 2004)), but not when blood flow is increased prior to exercise (Olive et al., 2004).

With peripheral nerve stimulation, the stimulus travels along the axon to the muscle, resulting in a compound muscle action potential (orthodromic response, M wave), and the stimulus travels back to the spinal cord. The antidromic response to stimulation does excite some motoneurons to generate a second action potential (an F wave) at a longer latency after the direct muscle response (Magladgery and McDougal, 1950), additional muscle activity that may increase fatigue. However, F wave incidence and magnitude were comparable during fatigue of muscles in SCI and uninjured subjects, suggesting any changes in motoneuron excitability following SCI did not exacerbate fatigue (Butler and Thomas, 2003).

Finally, the most promising way to reduce acute fatigue seems to be to interleave stimulation of different parts of a muscle so some fibres can rest while others work (Wise et al., 2001). Possible ways to implement activation of only part of a paralysed human muscle may include high-frequency stimulation to block the action potentials in some axons (Bhadra and Kilgore, 2004), or placement of stimulating electrodes on different nerve branches or areas of large muscles like the diaphragm (diMarco et al., 2005).

Exercise after SCI

Paralysed muscles that remain innervated after SCI are often exercised for maintenance of general health and/or to perform functional behaviour (e.g.

walking) by electrical stimulation of the appropriate peripheral nerves or intramuscular nerve branches. Not only can training reverse chronic changes in muscle properties for the most part, but early intervention can largely prevent deterioration and may be more effective (Shields and Dudley-Javoroski, 2006, 2007). Training effects from electrical stimulation usually occur over weeks to months. As with voluntary contractions, progressive overload and increases in repetitions are needed for effects which also regress when exercise is stopped. Thus, compliance influences outcome. Changes include: (1) an increase in muscle strength due to an increase in the cross-sectional area of the muscle fibres and a reduction in perimysial tissue, particularly when work is against resistance or a load (Peckham *et al.*, 1976; Mohr *et al.*, 1997; Crameri *et al.*, 2004; Mahoney *et al.*, 2005; Shields and Dudley-Javoroski, 2006, 2007); (2) an increase in endurance which is associated with increases in oxidative enzymes, increases in the proportion of type IIA muscle fibres, slowing of muscle contraction and relaxation which may permit use of lower stimulation frequencies and thus less fatigue (Peckham *et al.*, 1976; Martin *et al.*, 1992; Stein *et al.*, 1992; Rochester *et al.*, 1995a, 1995b; Andersen *et al.*, 1996; Mohr *et al.*, 1997; Gerrits *et al.*, 2002; Shields and Dudley-Javoroski, 2006, 2007); and (3) other health benefits such as cardiovascular conditioning with more robust effects possible when injury level is lower, improved sensitivity to insulin and increases in bone mineral density when loading of bone occurs (Glaser, 1994; Creasey *et al.*, 2004; Nash, 2005; Shields and Dudley-Javoroski, 2006, 2007; Frotzler *et al.*, 2008).

The magnitude of the training effects also depends on the initial condition of the muscle, muscle length and the extent to which the muscles are excited by electrical stimulation. Weak and/or fatigable muscles may limit the amount of exercise that is possible initially. However, a severely atrophied muscle may show large training effects with time (Mohr *et al.*, 1997). Increasing fatigue resistance can thus magnify subsequent training effects. Muscles that have shortened are often weaker than muscles with a longer rest length (Gerrits *et al.*, 2005). Often only small amounts of muscle are stimulated, which can exacerbate fatigue and limit central haemodynamic responses (Glaser, 1994), so there is reason to activate more of a muscle, different parts of a muscle in sequence to permit rest and blood perfusion, and/or additional muscles.

For functional benefits, electrical stimulation of several paralysed muscles via surface, percutaneous or implanted electrodes is coordinated to generate behaviours such as walking, grasping, respiration or control of the bladder or bowel. The strength of the contractions is varied by modulating the stimulation parameters, usually pulse amplitude or duration, while frequency is kept constant (Creasey *et al.*, 2004; Peckham and Knutson, 2005). Since these systems are typically implemented in individuals with chronic injuries, the paralysed muscles are often weak, fatigable and in need of conditioning. For example, two weeks after implantation of electrodes near the motor points of the paralysed diaphragm in individuals who are only able to breath independently for up to two hours, stimulation at 20 Hz for 10–15 min·h^{-1} is gradually increased over 2–3 months or longer until the muscles have the strength and endurance to wean the person off a mechanical ventilator (diMarco et al., 2005). Success occurs in most cases, demonstrating the remarkable capacity of muscle to adapt to changes in use. But again, one of the limits to performing these activities is fatigue, but of various muscles. For example, fatigue of arm muscles can limit the ability to stand because they are used to support body weight. Activation of more muscle with epimysial electrodes versus a nerve cuff electrode extends standing time (Fischer et al., 2006). Thus, optimal ways to condition muscles to minimise fatigue, to deliver stimulation to multiple muscles for energy efficient movement and to sense whether or not the behaviour is appropriate or needs adjustment all need further exploration.

CASE STUDY

Fatigue is excessive in muscles paralysed chronically by SCI. Comparison of whole-muscle and motor unit data from the same SCI individual show that the large force declines measured in response to repeated electrical stimulation largely reflect an increase in the intrinsic fatigability of the muscle fibres. Force evoked from the whole thenar muscles declined to 21% of initial values after 2 min of supramaximal median nerve stimulation (13 pulses at 40 Hz each second) in the right hand of a 26-year-old man, nine years after he sustained a complete spinal cord injury at C4 (American Spinal Injury Association classification A, i.e. motor and sensory complete) in a diving accident. He took no medication to control increases in muscle tone or spasticity. This fatigue at the whole-muscle level must reflect failure within the muscle because the evoked EMG area potentiated (18% higher than initial after 2 min), a sign of effective transmission of electrical signals. Ischemia effects during whole-muscle stimulation were also small, because intraneural stimulation of three different thenar motor units in the same muscle resulted in forces of 16%, 22% and 40% initial after 2 min of the same stimulation protocol. These units produced weak, intermediate or strong maximal forces compared to the population results, suggesting these thenar unit data are a reasonable sample for these thenar muscles. The mean unit force after 2 min was 26% initial, only 5% higher than for the whole muscle. The similarity in the whole-muscle and motor unit force declines in the same muscle suggest that the intrinsic fatigability of the muscle fibres has increased significantly with chronic paralysis (Thomas et al., 2003; Klein et al., 2006), consistent with the increase in percentage of type II muscle fibres.

CONCLUSION

Fatigue routinely limits the daily lives of individuals with various neurological disorders, including MS, PPS, SCI and MG, even though the primary deficit differs across conditions. As a consequence, muscle atrophy and weakness from disuse is common in all of these disorders, particularly PPS and SCI. Chronic denervation in PPS and acute conduction block at the neuromuscular junction in MG exacerbate weakness further. Muscle weakness from both acute and chronic sources induces fatigue prematurely, simply because a larger fraction of a muscle has to be activated to maintain the force necessary to perform a given task (Thomas and Zijdewind, 2006). Thus, regular exercise directed towards strength improvement could facilitate fatigue resistance and functional capacity in all of these conditions.

Despite the importance of regular exercise for general health and well-being, and its potential to facilitate performance of daily activities without adverse effects, there are few training studies in those with an impaired nervous system. Considerable research is needed to provide some consensus on the type of exercise to use, its intensity, duration and frequency to provide functional impact. Only sensitive measures of muscle strength are useful, because numerous studies have shown deficits are often underestimated with manual evaluation. Similarly, both the magnitude and sites of fatigue require objective assessment because force declines do not necessarily correlate with the subjective sense of fatigue or effort, and thus, do not always exceed that measured in muscles of healthy subjects. Failure at various sites is likely to contribute to the force declines, but one is likely to predominate in each condition. Exercise programmes need to focus on that site. For example, exercise that taxes central processing may help people with MS reduce their routine sense of fatigue. Working in a cooler, temperature-controlled environment, with rests, self-pacing and monitoring may prolong work time and reduce recovery time in PPS and MG. Assistive devices often facilitate behaviour after SCI. Moreover, the large variability in symptom severity within a condition and across muscles makes it critical to assess each situation, to tailor the exercise programme to a level that each person can accomplish and to regularly monitor performance so that additional exercise challenges can be assigned as improvement occurs. Comparisons of data from affected and unaffected muscles may reveal differences that are not apparent from clinical assessments, so suitable data from healthy muscles are needed to adequately interpret data. The influence of medication on functional capacity is another under-explored area. Compliance also changes outcome, so is important to document. But most encouraging is the positive effect that regular controlled exercise can have on physical and psychological function. The challenge is to learn how to train the system to optimise functional changes.

FIVE KEY PAPERS THAT SHAPED THE TOPIC AREA

Study 1. Thomas, C.K. and Zijdewind, I. (2006). Fatigue of muscles weakened by death of motoneurons. *Muscle and Nerve*, 33, 21–41.

This study reviews how muscle weakness exacerbates fatigue in diseases and trauma where spinal motoneurons die (e.g. amyotrophic lateral sclerosis, poliomyelitis, SCI). Irrespective of the source of the chronic muscle weakness (disuse, chronic denervation from motoneuron death, both possibilities), it is imperative to distinguish between acute force decreases (fatigue) that result from chronic weakness of the muscle itself compared to deficits that relate to an increase in the intrinsic fatigability of the muscle fibres. With a weak muscle, a person has to work that muscle at a higher percentage of its maximal force-generating capacity to perform daily tasks. This can induce a decrement in work capacity in itself, without the muscle fibres necessarily having unusual fatigability.

Study 2. Sheean, G.L., Murray, N.M., Rothwell, J.C., Miller, D.H. and Thompson, A.J. (1997). An electrophysiological study of the mechanism of fatigue in multiple sclerosis. *Brain*, 120, 299–315.

Patients diagnosed with MS and control subjects performed maximal thumb adduction for 45 s. The authors measured the force, muscle activity (EMG) and central activation by comparing the size of twitches superimposed on the contraction to the size of the potentiated twitch at rest. These data show that maximal force and central activation did not differ between control subjects and MS patients in the non-fatigued situation. After the fatiguing contraction, maximal force was significantly lower in MS patients and there was a stronger decline in central activation. This reduction in central activation was clearly visible as an increase in the superimposed twitch. These data demonstrate that central fatigue contributes progressively to a decline in force in MS patients.

Study 3. Halstead, L.S., Gawne, A.C. and Pham, B.T. (1995). National rehabilitation hospital limb classification for exercise, research, and clinical trials in post-polio patients. *Annals of the New York Academy of Sciences*, 753, 343–353.

In PPS, the most-impaired limb largely defines the kinds of exercise that are feasible. The authors of this study use the history of acute muscle weakness, current muscle strength, atrophy and intramuscular EMG data to classify the potential function of each limb (five classes: from least-affected to severe muscle atrophy) and to prescribe exercise. For example, in a case of left leg impairment, right leg involvement was expected but absent. Exercise was thus expanded from arm ergometry to aerobic exercise (walking with the left leg braced; a cane to improve stability and to reduce weight-bearing on the left leg). Use of the same classification system can allow improvements to be monitored and data comparisons to be made across studies.

Study 4. Ringqvist, I. (1971). Muscle strength in myasthenia gravis: effects of exhaustion and anticholinesterase related to muscle fibre size. *Acta Neurologica Scandinavica*, 47, 619–641.

Ringqvist shows that many muscles are weak in people with MG, in large part due to significant atrophy of type II muscle fibres. Furthermore, force only increased in 50% of elbow flexor muscles with acetylcholine esterase inhibitors. However, the enhancement in strength from the inhibitors had limited functional effects because these muscles fatigued more than non-responsive muscles and muscles of healthy subjects during either intermittent or sustained voluntary contractions. Neither the muscle strength nor the fatigability related to the presence of neuromuscular junction block. This is one of the few studies that quantifies muscle function in MG and how muscle strength and fatigability change in response to a common treatment.

Study 5. Shields, R.K. and Dudley-Javoroski, S. (2006). Musculoskeletal plasticity after acute spinal cord injury: effects of long-term neuromuscular electrical stimulation training. *Journal of Neurophysiology*, 95, 2380–2390.

The authors of this study initiated long-term stimulation of paralysed soleus muscles within six weeks of SCI. Of importance, the tibial nerve was stimulated long-term, supramaximally against resistance, at a high frequency for soleus (15 Hz for 667 ms every 2 s; four bouts of 125 trains, five days each week for two years), and objective monitoring showed high subject compliance. The greatest increases in soleus torque occurred within the first six months, peaking at 1.5 years. Although muscle fatigability increased progressively with time post-injury, chronic stimulation prevented some of this increase in fatigue. Thus, muscle activity can preserve the strength and fatigue resistance of paralysed muscles to some extent when exercise is implemented early and delivered consistently over time.

GLOSSARY OF TERMS

ATP	adenosine triphosphate
cMEP	corticospinal tract motor-evoked potential
CNS	central nervous system
EMG	electromyographic activity
fMRI	functional magnetic resonance imaging
M wave	compound muscle action potential

MEP	motor-evoked potential
MG	myasthenia gravis
MS	multiple sclerosis
MuSK	muscle-specific protein kinase
MVC	maximal voluntary contraction
PCr	phosphocreatine
P_i	inorganic phosphate
PPS	post-poliomyelitis syndrome
SCI	spinal cord injury

ACKNOWLEDGEMENTS

The authors thank Dr A. Kutubidze for help with Figure 9.1. Funded by USPHS grant NS-30226, The Miami Project to Cure Paralysis, and the University Medical Center Groningen.

REFERENCES

Agre, J.C. (1995). The role of exercise in the patient with post-polio syndrome. *Annals of the New York Academy of Sciences*, 753, 321–334.

Agre, J.C. and Rodriquez, A.A. (1990). Neuromuscular function: comparison of symptomatic and asymptomatic polio subjects to control subjects. *Archives of Physical Medicine and Rehabilitation*, 71, 545–551.

Agre, J.C. and Rodriquez, A.A. (1991). Intermittent isometric activity: its effect on muscle fatigue in postpolio subjects. *Archives of Physical Medicine and Rehabilitation*, 72, 971–975.

Allen D.G., Lamb, G.D. and Westerblad, H. (2008a). Skeletal muscle fatigue: cellular mechanisms. *Physiological Reviews*, 88, 287–332.

Allen, D.G., Lamb, G.D. and Westerblad, H. (2008b). Impaired calcium release during fatigue. *Journal of Applied Physiology*, 104, 296–305.

Allen, G.M., Gandevia, S.C., Neering, I.R., Hickie, I., Jones, R. and Middleton, J. (1994). Muscle performance, voluntary activation and perceived effort in normal subjects and patients with prior poliomyelitis. *Brain*, 117, 661–670.

Allen, G.M., Middleton, J., Katrak, P.H., Lord, S.R. and Gandevia, S.C. (2004). Prediction of voluntary activation, strength and endurance of elbow flexors in postpolio patients. *Muscle and Nerve*, 30, 172–181.

Andersen, J.L., Mohr, T., Biering-Sørensen, F., Galbo, H. and Kjaer, M. (1996). Myosin heavy chain isoform transformation in single fibers from m. vastus lateralis in spinal cord injured individuals: effects of long term functional electrical stimulation (FES). *Pflügers Archives: European Journal of Physiology*, 431, 513–518.

Bakshi, R. (2003). Fatigue associated with multiple sclerosis: diagnosis, impact and management. *Multiple Sclerosis*, 9, 219–227.

Balice-Gordon, R.J. and Lichtman, J.W. (1994). Long-term synapse loss induced by focal blockade of postsynaptic receptors. *Nature*, 372, 519–524.

Beasley, W.C. (1961). Quantitative muscle testing: principles and applications to research and clinical services. Archives *of Physical Medicine and Rehabilitation*, 42, 398–425.

Beelen, A., Nollet, F., de Visser, M., de Jong, B.A., Lankhorst, G.J. and Sargeant, A.J. (2003). Quadriceps muscle strength and voluntary activation after polio. *Muscle and Nerve*, 28, 218–226.

Belanger, A.Y. and McComas, A.J. (1981). Extent of motor unit activation during effort. *Journal of Applied Physiology*, 51, 1131–1135.

Bergmans, J. (1973). Physiological observations on single human nerve fibres. In J.E. Desmedt (ed.), *New Developments in Electromyography and Clinical Neurophysiology*. Basel: Karger, pp. 89–127.

Bhadra, N. and Kilgore, K.L. (2004). Direct current electrical conduction block of peripheral nerve. *IEEE Transactions on Neural Systems and Rehabilitation Engineering*, 12, 313–324.

Biering-Sorensen, F., Bohr, H.H. and Schaadt, O.P. (1990). Longitudinal study of bone mineral content in the lumbar spine, the forearm and the lower extremities after spinal cord injury. *European Journal of Clinical Investigations*, 20, 330–335.

Bigland-Ritchie, B. and Woods, J.J. (1984). Changes in muscle contractile properties and neural control during human muscular fatigue. *Muscle and Nerve*, 7, 691–699.

Bigland-Ritchie, B., Dawson, N.J., Johansson, R.S. and Lippold, O.C. (1986). Reflex origin for the slowing of motoneurone firing rates in fatigue of human voluntary contractions. *Journal of Physiology*, 379, 451–459.

Bigland-Ritchie, B., Johansson, R., Lippold, O.C., Smith, S. and Woods, J.J. (1983). Changes in motoneurone firing rates during sustained maximal voluntary contractions. *Journal of Physiology*, 340, 335–346.

Bigland-Ritchie, B., Kukulka, C.G., Lippold, O.C. and Woods, J.J. (1982). The absence of neuromuscular transmission failure in sustained maximal voluntary contractions. *Journal of Physiology*, 330, 265–278.

Bigland-Ritchie, B., Rice, C.L., Garland, S.J. and Walsh, M.L. (1995). Task-dependent factors in fatigue of human voluntary contractions. *Advances in Experimental Medicine and Biology*, 384, 361–380.

Borg, K., Borg, J., Dhoot, G., Edstrom, L., Grimby, L. and Thornell, L.E. (1989). Motoneuron firing and isomyosin type of muscle fibres in prior polio. *Journal of Neurology, Neurosurgery and Psychiatry*, 52, 1141–1148.

Borg, K., Borg, J., Edström, L. and Grimby, L. (1988). Effects of excessive use of remaining muscle fibers in prior polio and LV lesion. *Muscle and Nerve*, 11, 1219–1230.

Brooke, M.H. and Engel, W.K. (1969). The histographic analysis of human muscle biopsies with regard to fiber types 3: myotonias, myasthenia gravis, and hypokalemic periodic paralysis. *Neurology*, 19, 469–477.

Brown, W.F., Kadrie, H.A. and Milner-Brown, H.S. (1981). Rank order of recruitment of motor units with graded electrical stimulation of median or ulnar nerves in normal subjects and in patients with entrapment neuropathies. In J.E. Desmedt (ed.), *Motor Unit Types, Recruitment and Plasticity in Health and Disease*. Basel: Karger, pp. 319–330.

Burnham, R., Martin, T., Stein, R., Bell, G., Maclean, I. and Steadward, R. (1997). Skeletal muscle fibre type transformation following spinal cord injury. *Spinal Cord*, 35, 86–91.

Butler, J.E. and Thomas, C.K. (2003). Effects of sustained stimulation on the excitability of motoneurons innervating paralyzed and control muscles. *Journal of Applied Physiology*, 94, 567–575.

Butler, J.E., Ribot-Ciscar, E., Zijdewind, I. and Thomas, C.K. (2004). Increased blood pressure can reduce fatigue of thenar muscles paralyzed after spinal cord injury. *Muscle and Nerve*, 29, 575–584.

Cady, E.B., Jones, D.A., Lynn, J. and Newham, D.J. (1989). Changes in force and intracellular metabolites during fatigue of human skeletal muscle. *Journal of Physiology*, 418, 311–325.

Carpentier, A., Duchateau, J. and Hainaut, K. (2001). Motor unit behaviour and contractile changes during fatigue in the human first dorsal interosseus. *Journal of Physiology*, 534, 903–912.

Cashman, N.R., Maselli, R., Wollmann, R.L., Roos, R., Simon, R. and Antel, J.P. (1987). Late denervation in patients with antecedent paralytic poliomyelitis. *The New England Journal of Medicine*, 317, 7–12.

Castro, M.J., Apple Jr, D.F., Hillegass, E.A and Dudley, G.A. (1999). Influence of complete spinal cord injury on skeletal muscle cross-sectional area within the first 6 months of injury. *European Journal of Applied Physiology and Occupational Physiology*, 80, 373–378.

Colombo, B., Martinelli, B.F., Rossi, P., Rovaris, M., Maderna, L., Filippi, M. and Comi, G. (2000). MRI and motor evoked potential findings in nondisabled multiple sclerosis patients with and without symptoms of fatigue. *Journal of Neurology*, 247, 506–509.

Compston, A. and Coles, A. (2002). Multiple sclerosis. *Lancet*, 359, 1221–1231.

Cope, T.C., Bodine, S.C., Fournier, M. and Edgerton, V.R. (1986). Soleus motor units in chronic spinal transected cats: physiological and morphological alterations. *Journal of Neurophysiology*, 55, 1202–1220.

Crago, P.E., Houk, J.C. and Rymer, W.Z. (1982). Sampling of total muscle force by tendon organs. *Journal of Neurophysiology*, 47, 1069–1083.

Crameri, R.M., Cooper, P., Sinclair, P.J., Bryant, G. and Weston, A. (2004). Effect of load during electrical stimulation training in spinal cord injury. *Muscle and Nerve*, 29, 104–111.

Creasey, G.H., Ho, C.H., Triolo, R.J., Gater, D.R., DiMarco, A.F., Bogie, K.M. and Keith, M.W. (2004). Clinical applications of electrical stimulation after spinal cord injury. *Journal of Spinal Cord Medicine*, 27, 365–375.

Curt, A. and Dietz, V. (1999). Neurologic recovery in SCI. *Archives of Physical Medicine and Rehabilitation*, 80, 607–608.

Dalakas, M.C. (1995). Pathogenetic mechanisms of post-polio syndrome: morphological, electrophysiological, virological, and immunological correlations. *Annals of the New York Academy of Sciences*, 753, 167–185.

Day, S.J. and Hulliger, M. (2001). Experimental simulation of cat electromyogram: evidence for algebraic summation of motor-unit action-potential trains. *Journal of Neurophysiology*, 86, 2144–2158.

De Haan, A., de Ruiter, C.J., Van der Woude, L.H. and Jongen. P.J. (2000). Contractile properties and fatigue of quadriceps muscles in multiple sclerosis. *Muscle and Nerve*, 23, 1534–1541.

de Ruiter, C.J., Jongen, P.J., Van der Woude, L.H. and de Haan, A. (2001). Contractile speed and fatigue of adductor pollicis muscle in multiple sclerosis. *Muscle and Nerve*, 24, 1173–1180.

Dengler, R., Stein, R.B. and Thomas, C.K. (1988). Axonal conduction velocity and force of single human motor units. *Muscle and Nerve*, 11, 136–145.

Desmedt, J.E. (1957). Nature of the defect of neuromuscular transmission in myasthenic patients: post-tetanic exhaustion. *Nature*, 179, 156–157.

DiMarco, A.F., Onders, R.P., Ignagni, A., Kowalski, K.E. and Mortimer, J.T. (2005). Phrenic nerve pacing via intramuscular diaphragm electrodes in tetraplegic subjects. *Chest*, 127, 671–678.

Dupont-Versteegden, E.E., Houle, J.D., Gurley, C.M. and Peterson, C.A. (1998). Early changes in muscle fiber size and gene expression in response to spinal cord transection and exercise. *American Journal of Physiology*, 275, C1124–1133.

Elmqvist, D., Hofmann, W.W., Kugelberg, J. and Quastel, D.M. (1964). An electrophysiological investigation of neuromuscular transmission in myasthenia gravis. *Journal of Physiology*, 174, 417–434.

Engel, A.G., Tsujihata, M., Lambert, E.H., Lindstrom, J.M. and Lennon, V.A. (1976). Experimental autoimmune myasthenia gravis: a sequential and quantitative study of the neuromuscular junction ultrastructure and electrophysiologic correlations. *Journal of Neuropathology and Experimental Neurology*, 35, 569–587.

Enoka, R.M. and Duchateau, J. (2008). Muscle fatigue: what, why and how it influences muscle function. *Journal of Physiology*, 586, 11–23.

Enoka, R.M. and Stuart, D.G. (1992). Neurobiology of muscle fatigue. *Journal of Applied Physiology*, 72, 1631–1648.

Fambrough, D.M., Drachman, D.B. and Satyamurti, S. (1973). Neuromuscular junction in myasthenia gravis: decreased acetylcholine receptors. *Science*, 182, 293–295.

Farbu, E., Gilhus, N.E., Barnes, M.P., Borg, K., de Visser, M., Driessen, A., Howard, R., Nollet, F., Opara, J. and Stalberg, E. (2006). EFNS guideline on diagnosis and management of post-polio syndrome: report of an EFNS task force. *European Journal of Neurology*, 13, 795–801.

Filippi, M., Rocca, M.A., Colombo, B., Falini, A., Codella, M., Scotti, G. and Comi, G. (2002). Functional magnetic resonance imaging correlates of fatigue in multiple sclerosis. *Neuroimage*, 15, 559–567.

Fisher, L.E., Miller, M.E., Nogan, S.J., Davis, J.A., Anderson, J.S., Murray, L.M., Tyler D.J. and Triolo, R.J. (2006). Preliminary evaluation of a neural prosthesis for standing after spinal cord injury with four contact nerve-cuff electrodes for quadriceps stimulation. *Conference Proceedings of the IEEE in Engineering in Medicine and Biology Society*, 1, 3592–3595.

Fragoso, Y.D., Santana, D.L. and Pinto, R.C. (2008). The positive effects of a physical activity program for multiple sclerosis patients with fatigue. *NeuroRehabilitation*, 23, 153–157.

Fregonezi, G.A., Resqueti, V.R., Guell, R., Pradas, J. and Casan, P. (2005). Effects of 8-week, interval-based inspiratory muscle training and breathing retraining in patients with generalized myasthenia gravis. *Chest*, 128, 1524–1530.

Frotzler, A., Coupaud, S., Perret, C., Kakebeeke, T.H., Hunt, K.J., Donaldson, N.N. and Eser, P. (2008). High-volume FES-cycling partially reverses bone loss in people with chronic spinal cord injury. *Bone*, 43, 169–176.

Fuglevand, A.J. (1995). The role of the sarcolemma action potential in fatigue. *Advances in Experimental Medicine and Biology*, 384, 101–108.

Gandevia, S.C. (2001). Spinal and supraspinal factors in human muscle fatigue. *Physiological Reviews*, 81, 1725–1789.

Garland, D.E., Stewart, C.A., Adkins, R.H., Hu, S.S., Rosen, C., Liotta, F.J. and Weinstein, D.A. (1992). Osteoporosis after spinal cord injury. *Journal of Orthopaedic Research*, 10, 371–378.

Gerrits, H.L., de Haan, A., Hopman, M.T., Der Woude, L.H., Jones, D.A. and Sargeant, A.J. (1999). Contractile properties of the quadriceps muscle in individuals with spinal cord injury. *Muscle and Nerve*, 22, 1249–1256.

Gerrits, H.L., Hopman, M.T., Sargeant, A.J., Jones, D.A. and de Haan, A. (2002). Effects of training on contractile properties of paralyzed quadriceps muscle. *Muscle and Nerve*, 25, 559–567.

Gerrits, K.H., Maganaris, C.N., Reeves, N.D., Sargeant, A.J., Jones, D.A. and de Haan, A. (2005). Influence of knee joint angle on muscle properties of paralyzed and nonparalyzed human knee extensors. *Muscle and Nerve*, 32, 73–80.

Gilchrist, J.M., Massey, J.M. and Sanders, D.B. (1994). Single fiber EMG and repetitive stimulation of the same muscle in myasthenia gravis. *Muscle and Nerve*, 17, 171–175.

Glaser, R.M. (1994). Functional neuromuscular stimulation: exercise conditioning of spinal cord injured patients. *International Journal of Sports Medicine*, 15, 142–148.

Glass, D.J., Bowen, D.C., Stitt, T.N., Radziejewski, C., Bruno, J., Ryan, T.E., Gies, D.R., Shah, S., Mattsson, K., Burden, S.J., DiStefano, P.S., Valenzuela, D.M., DeChiara, T.M. and Yancopoulos, G.D. (1996). Agrin acts via a MuSK receptor complex. *Cell*, 85, 513–523.

Godfrey, S., Butler, J.E., Griffin, L. and Thomas, C.K. (2002). Differential fatigue of paralyzed thenar muscles by stimuli of different intensities. *Muscle and Nerve*, 26, 122–131.

Gordon, T. and Mao, J. (1994). Muscle atrophy and procedures for training after spinal cord injury. *Physical Therapy*, 74, 50–60.

Griffin, L., Garland, S.J., Ivanova, T. and Gossen, E.R. (2001). Muscle vibration sustains motor unit firing rate during submaximal isometric fatigue in humans. *Journal of Physiology*, 535, 929–936.

Griffin, L., Godfrey, S. and Thomas, C.K. (2002). Stimulation pattern that maximizes force in paralyzed and control whole thenar muscles. *Journal of Neurophysiology*, 87, 2271–2278.

Grimby, G., Stålberg, E., Sandberg, A. and Stibrant, S.K. (1998). An 8-year longitudinal study of muscle strength, muscle fiber size, and dynamic electromyogram in individuals with late polio. *Muscle and Nerve*, 21, 1428–1437.

Grimby, L., Tollbäck, A., Müller, U. and Larsson L. (1996). Fatigue of chronically overused motor units in prior polio patients. *Muscle and Nerve*, 19, 728–737.

Grohar-Murray, M.E., Becker, A., Reilly, S. and Ricci, M. (1998). Self-care actions to manage fatigue among myasthenia gravis patients. *Journal of Neuroscience Nursing*, 30, 191–199.

Gross, D. and Meiner Z. (1993). The effect of ventilatory muscle training on respiratory function and capacity in ambulatory and bed-ridden patients with neuromuscular disease. *Monaldi Archives for Chest Disease*, 48, 322–326.

Halstead, L.S. (1991). Assessment and differential diagnosis for post-polio syndrome. *Orthopedics*, 14, 1209–1217.

Halstead, L.S., Gawne, A.C. and Pham, B.T. (1995). National rehabilitation hospital limb classification for exercise, research, and clinical trials in post-polio patients. *Annals of the New York Academy of Sciences*, 753, 343–353.

Hicks, A. and McComas, A.J. (1989). Increased sodium pump activity following repetitive stimulation of rat soleus muscles. *Journal of Physiology*, 414, 337–349.

Hildegunn, L., Jones, K., Grenstad, T., Dreyer, V., Farbu, E. and Rekand, T. (2007). Perceived disability, fatigue, pain and measured isometric muscle strength in patients with post-polio symptoms. *Physiotherapy Research International*, 12, 39–49.

Hoch, W., McConville, J., Helms, S., Newsom-Davis, J., Melms, A. and Vincent, A. (2001). Auto-antibodies to the receptor tyrosine kinase MuSK in patients with myasthenia gravis without acetylcholine receptor antibodies. *Nature Medicine*, 7, 365–368.

Hodes, R. (1949). Selective destruction of large motoneurons by poliomyelitis virus: I: conduction velocity of motor nerve fibers of chronic poliomyelitis patients. *Journal of Neurophysiology*, 12, 257–266.

Hodes, R., Peacock Jr, S.M. and Bodian, D. (1949). Selective destruction of large motoneurons by poliomyelitis virus: II: size of motoneurons in the spinal cord of rhesus monkeys. *Journal of Neuropathology and Experimental Neurology*, 8, 400–410.

Hofmann, W.W., Parsons, R.L. and Feigen, G.A. (1966). Effects of temperature and drugs on mammalian motor nerve terminals. *American Journal of Physiology*, 211, 135–140.

Hopman, M.T., van Asten, W.N. and Oeseburg, B. (1996). Changes in blood flow in the common femoral artery related to inactivity and muscle atrophy in individuals with long-standing paraplegia. *Advances in Experimental Medicine and Biology*, 388, 379–383.

Ingram, D.A., Davis, G.R., Schwartz, M.S. and Swash, M. (1985). The effect of continuous voluntary activation on neuromuscular transmission: a SFEMG study of myasthenia gravis and anterior horn cell disorders. *Electroencephalography and Clinical Neurophysiology*, 60, 207–213.

Janssen, T.W., Bakker, M., Wyngaert, A., Gerrits, K.H. and De Haan, A. (2004). Effects of stimulation pattern on electrical stimulation-induced leg cycling performance. *Journal of Rehabilitation Research and Development*, 41, 787–796.

Johansson, R.S. and Westling, G. (1988). Coordinated isometric muscle commands adequately and erroneously programmed for the weight during lifting task with precision grip. *Experimental Brain Research*, 71, 59–71.

Jubelt, B. (2004). Post-polio syndrome. *Current Treatment Options in Neurology*, 6, 87–93.

Juel, V.C. and Massey, J.M. (2007). Myasthenia gravis. *Orphanet Journal of Rare Diseases*, 2 (44), 1–13.

Kebatse, M.B., Lee, S.C., Johnston, T.E. and Binder-Macleod, S.A. (2005). Strategies that improve paralyzed human quadriceps femoris muscle performance during repetitive, nonisometric contractions. *Archives in Physical Medicine and Rehabilitation*, 86, 2157–2164.

Keenan, K.G., Farina, D., Maluf, K.S., Merletti, R. and Enoka, R.M. (2005). Influence of amplitude cancellation on the simulated surface electromyogram. *Journal of Applied Physiology*, 98, 120–131.

Keenan, S.P., Alexander, D., Road, J.D., Ryan, C.F., Oger, J. and Wilcox, P.G. (1995). Ventilatory muscle strength and endurance in myasthenia gravis. *European Respiratory Journal*, 8, 1130–1135.

Keesey, J.C. (2004). Clinical evaluation and management of myasthenia gravis. *Muscle and Nerve*, 29, 484–505.

Keesey, J.C, Buffkin, D., Kebo, D., Ho, W. and Herrmann Jr, C. (1981). Plasma exchange alone as therapy for myasthenia gravis. *Annals of the New York Academy of Sciences*, 377, 729–743.

Kent-Braun, J.A., Ng, A.V., Castro, M., Weiner, M.W., Gelinas, D., Dudley, G.A. and Miller, R.G. (1997). Strength, skeletal muscle composition, and enzyme activity in multiple sclerosis. *Journal of Applied Physiology*, 83, 1998–2004.

Kent-Braun, J.A., Sharma, K.R., Weiner, M.W. and Miller, R.G. (1994). Effects of exercise on muscle activation and metabolism in multiple sclerosis. *Muscle and Nerve*, 17, 1162–1169.

Kernell, D., Eerbeek, O., Verhey, B.A. and Donselaar, Y. (1987). Effects of physiological amounts of high- and low-rate chronic stimulation on fast-twitch muscle of the cat hindlimb; I: speed- and force-related properties. *Journal of Neurophysiology*, 58, 598–613.

Kim, S.J., Roy, R.R., Zhong, H., Suzuki, H., Ambartsumyan, L., Haddad, F., Baldwin, K.M. and Edgerton, V.R. (2007). Electromechanical stimulation ameliorates inactivity-induced adaptations in the medial gastrocnemius of adult rats. *Journal of Applied Physiology*, 103, 195–205.

Kittiwatanapaisan, W., Gauthier, D.K., Williams, A.M. and Oh, S.J. (2003). Fatigue in myasthenia gravis patients. *Journal of Neuroscience Nursing*, 35, 87–93.

Klein, C.S., Hager-Ross, C.K. and Thomas, C.K. (2006). Fatigue properties of human thenar motor units paralysed by chronic spinal cord injury. *Journal of Physiology*, 573, 161–171.

Klein, M.G., Whyte, J., Keenan, M.A., Esquenazi, A. and Polansky, M. (2000). Changes in strength over time among polio survivors. *Archives in Physical Medicine and Rehabilitation*, 81, 1059–1064.

Kniffki, K.D., Schomburg, E.D. and Steffens, H. (1979). Synaptic responses of lumbar alpha-motoneurones to chemical algesic stimulation of skeletal muscle in spinal cats. *Brain Research*, 160, 549–552.

Ko, S.F., Huang, C.C., Hsieh, M.J., Ng, S.H., Lee, C.C., Lin, T.K., Chen, M.C. and Lee, L. (2008). 31P MR spectroscopic assessment of muscle in patients with myasthenia gravis before and after thymectomy: initial experience. *Radiology*, 247, 162–169.

Krupp, L.B., Alvarez, L.A., LaRocca, N.G. and Scheinberg, L.C. (1988). Fatigue in multiple sclerosis. *Archives of Neurology*, 45, 435–437.

Kurtzke, J.F. (1983). Rating neurologic impairment in multiple sclerosis: an expanded disability status scale (EDSS). *Neurology*, 33, 1444–1452.

Lambert, E.H., Lindstrom, J.M. and Lennon, V.A. (1976). End-plate potentials in experimental autoimmune myasthenia gravis in rats. *Annals of the New York Academy of Sciences*, 274, 300–318.

Lass, Y. and Fischbach, G.D. (1976). A discontinuous relationship between the acetylcholine-activated channel conductance and temperature. *Nature*, 263, 150–151.

Lee, M., Reddy, H., Johansen-Berg, H., Pendlebury, S., Jenkinson, M., Smith,

S., Palace, J. and Matthews, P.M. (2000). The motor cortex shows adaptive functional changes to brain injury from multiple sclerosis. *Annals of Neurology*, 47, 606–613.

Lieber, R.L., Friden, J.O., Hargens, A.R. and Feringa, E.R. (1986). Long-term effects of spinal cord transection on fast and slow rat skeletal muscle: II: morphometric properties. *Experimental Neurology*, 91, 435–448.

Liepert, J., Kotterba, S., Tegenthoff, M. and Malin, J.P. (1996). Central fatigue assessed by transcranial magnetic stimulation. *Muscle and Nerve*, 19, 1429–1434.

Liepert, J., Mingers, D., Heesen, C., Baumer, T. and Weiller, C. (2005). Motor cortex excitability and fatigue in multiple sclerosis: a transcranial magnetic stimulation study. *Multiple Sclerosis*, 11, 316–321.

Lindstrom, J.M., Seybold, M.E., Lennon, V.A., Whittingham, S. and Duane, D.D. (1976). Antibody to acetylcholine receptor in myasthenia gravis. Prevalence, clinical correlates, and diagnostic value. *Neurology*, 26, 1054–1059.

Lo, Y.L., Dan, Y.F., Leoh, T.H., Tan, Y.E., Nurjannah, S. and Ratnagopal, P. (2004). Effect of exercise on repetitive nerve stimulation studies: new appraisal of an old technique. *Journal of Clinical Neurophysiology*, 21, 110–113.

Lohi, E.L., Lindberg, C. and Andersen, O. (1993). Physical training effects in myasthenia gravis. *Archives in Physical Medicine and Rehabilitation*, 74, 1178–1180.

Lord, S.R., Allen, G.M., Williams, P. and Gandevia, S.C. (2002). Risk of falling: predictors based on reduced strength in persons previously affected by polio. *Archives in Physical Medicine and Rehabilitation*, 83, 757–763.

Lucia, A., Mate-Munoz, J.L., Perez, M., Foster, C., Gutierrez-Rivas, E. and Arenas, J. (2007). Double trouble (McArdle's disease and myasthenia gravis): how can exercise help? *Muscle and Nerve*, 35, 125–128.

Lupu, V.D., Danielian, L., Johnsen, J.A., Vasconcelos, O.M., Prokhorenko, O.A., Jabbari, B., Campbell, W.W. and Floeter, M.K. (2008). Physiology of the motor cortex in polio survivors. *Muscle and Nerve*, 37, 177–182.

McComas, A.J., Fawcett, P.R., Campbell, M.J. and Sica, R.E. (1971). Electrophysiological estimation of the number of motor units within a human muscle. *Journal of Neurology and Neurosurgical Psychiatry*, 34, 121–131.

McComas, A.J., Quartly, C. and Griggs, R.C. (1997). Early and late losses of motor units after poliomyelitis. *Brain*, 120, 1415–1421.

Macefield, G., Hagbarth, K.E., Gorman, R., Gandevia, S.C. and Burke, D. (1991). Decline in spindle support to alpha-motoneurones during sustained voluntary contractions. *Journal of Physiology*, 440, 497–512.

Macefield, V.G., Gandevia, S.C., Bigland-Ritchie, B., Gorman, R.B. and Burke, D. (1993). The firing rates of human motoneurones voluntarily activated in the absence of muscle afferent feedback. *Journal of Physiology*, 471, 429–443.

Maganaris, C.N., Reeves, N.D., Rittweger, J., Sargeant, A.J., Jones, D.A., Gerrits, K. and De Haan, A. (2006). Adaptive response of human tendon to paralysis. *Muscle and Nerve*, 33, 85–92.

Magladery, J.W. and McDougal Jr, D.B. (1950). Electrophysiological studies of nerve and reflex activity in normal man: I: identification of certain reflexes in the electromyogram and the conduction velocity of peripheral nerve fibres. *Bulletin John Hopkins Hospital*, 86, 265–290.

Magleby, K.L. and Zengel, J.E. (1976). Long term changes in augmentation, potentiation, and depression of transmitter release as a function of repeated synaptic activity at the frog neuromuscular junction. *Journal of Physiology*, 257, 471–494.

Mahoney, E.T., Bickel, C.S., Elder, C., Black, C., Slade, J.M., Apple Jr, D. and Dudley, G.A. (2005). Changes in skeletal muscle size and glucose tolerance with electrically stimulated resistance training in subjects with chronic spinal cord injury. *Archives in Physical Medicine and Rehabilitation*, 86, 1502–1504.

Mainero, C., Faroni, J., Gasperini, C., Filippi, M., Giugni, E., Ciccarelli, O., Rovaris, M., Bastianello, S., Comi, G. and Pozzilli, C. (1999). Fatigue and magnetic resonance imaging activity in multiple sclerosis. *Journal of Neurology*, 246, 454–458.

Martin, T.P., Stein, R.B., Hoeppner, P.H. and Reid D.C. (1992). Influence of electrical stimulation on the morphological and metabolic properties of paralyzed muscle. *Journal of Applied Physiology*, 72, 1401–1406.

Mense, S. (1977). Nervous outflow from skeletal muscle following chemical noxious stimulation. *Journal of Physiology*, 267, 75–88.

Mermier, C.M., Schneider, S.M., Gurney, A.B., Weingart, H.M. and Wilmerding, M.V. (2006). Preliminary results: effect of whole-body cooling in patients with myasthenia gravis. *Medicine and Science in Sports and Exercise*, 38, 13–20.

Merton, P.A. (1954). Voluntary strength and fatigue. *Journal of Physiology*, 123, 553–564.

Mier-Jedrzejowicz, A.K., Brophy, C. and Green, M. (1988). Respiratory muscle function in myasthenia gravis. *American Review of Respiratory Disease*, 138, 867–873.

Mohr, T., Andersen, J.L., Biering-Sørensen, F., Galbo, H., Bangsbo, J., Wagner, A. and Kjaer, M. (1997). Long-term adaptation to electrically induced cycle training in severe spinal cord injured individuals. *Spinal Cord*, 35, 1–16.

Morgen, K., Kadom, N., Sawaki, L., Tessitore, A., Ohayon, J., McFarland, H., Frank, J., Martin, R. and Cohen, L.G. (2004). Training-dependent plasticity in patients with multiple sclerosis. *Brain*, 127, 2506–2517.

Motl, R.W. and Gosney, J.L. (2008). Effect of exercise training on quality of life in multiple sclerosis: a meta-analysis. *Multiple Sclerosis*, 14, 129–135.

Multiple Sclerosis Council for Clinical Practical Guidelines (1998). *Fatigue and Multiple Sclerosis: Evidence-based Management Strategies for Fatigue in Multiple Sclerosis*. Washington: Paralyzed Veterans of America, p. 2.

Nash, M.S. (2005). Exercise as a health-promoting activity following spinal cord injury. *Journal of Neurological Physical Therapy*, 29, 87–103.

Nelson D.A. and McDowell, F. (1959). The effects of induced hyperthermia on patients with multiple sclerosis. *Journal of Neurology, Neurosurgery and Psychiatry*, 22, 113–116.

Ng, A.V., Miller, R.G., Gelinas, D. and Kent-Braun, J.A. (2004). Functional relationships of central and peripheral muscle alterations in multiple sclerosis. *Muscle and Nerve*, 29, 843–852.

Nicklin, J., Karni, Y. and Wiles, C.M. (1987). Shoulder abduction fatiguability. *Journal of Neurology and Neurosurgical Psychiatry*, 50, 423–427.

Olive, J.L., Slade, J.M., Bickel, C.S., Dudley, G.A. and McCully, K.K. (2004). Increasing blood flow before exercise in spinal cord-injured individuals does not alter muscle fatigue. *Journal of Applied Physiology*, 96, 477–482.

Olive, J.L., Slade, J.M., Dudley, G.A. and McCully, K.K. (2003). Blood flow and muscle fatigue in SCI individuals during electrical stimulation. *Journal of Applied Physiology*, 94, 701–708.

Pantano, P., Mainero, C., Lenzi, D., Caramia, F., Iannetti, G.D., Piattella, M.C., Pestalozza, I., Di, L.S., Bozzao, L. and Pozzilli, C. (2005). A longitudinal fMRI study on motor activity in patients with multiple sclerosis. *Brain*, 128, 2146–2153.

Paul, R.H., Cohen, R.A., Goldstein, J.M. and Gilchrist, J.M. (2000). Fatigue and its impact on patients with myasthenia gravis. *Muscle and Nerve*, 23, 1402–1406.

Peckham, P.H. and Knutson, J.S. (2005). Functional electrical stimulation for neuromuscular applications. *Annual Reviews in Biomedical Engineering*, 7, 327–360.

Peckham, P.H., Mortimer, J.T. and Marsolais, E.B. (1976). Alteration in the force and fatigability of skeletal muscle in quadriplegic humans following exercise induced by chronic electrical stimulation. *Clinical Orthopaedics*, 326–333.

Perretti, A., Balbi, P., Orefice, G., Trojano, L., Marcantonio, L., Brescia-Morra, V., Ascione, S., Manganelli, F., Conte, G. and Santoro, L. (2004). Post-exercise facilitation and depression of motor evoked potentials to transcranial magnetic stimulation: a study in multiple sclerosis. *Clinical Neurophysiology*, 115, 2128–2133.

Petajan, J.H. and White, A.T. (2000). Motor-evoked potentials in response to fatiguing grip exercise in multiple sclerosis patients. *Clinical Neurophysiology*, 111, 2188–2195.

Petajan, J.H., Gappmaier, E., White, A.T., Spencer, M.K., Mino, L. and Hicks, R.W. (1996). Impact of aerobic training on fitness and quality of life in multiple sclerosis. *Annals of Neurology*, 39, 432–441.

Place, N., Yamada, T., Bruton, J.D. and Westerblad, H. (2008). Interpolated twitches in fatiguing single mouse muscle fibres: implications for the assessment of central fatigue. *Journal of Physiology*, 586, 2799–2805.

Post, M., Steens, A., Renken, R., Maurits, N.M. and Zijdewind, I. (2009). Voluntary activation and cortical activity during a sustained maximal contraction: an fMRI study. *Human Brain Mapping*, 30 (3), 1014–1027.

Rice, C.L., Vollmer, T.L. and Bigland-Ritchie, B. (1992). Neuromuscular responses of patients with multiple sclerosis. *Muscle and Nerve*, 15, 1123–1132.

Rietberg, M.B., Brooks, D., Uitdehaag, B.M. and Kwakkel, G. (2005). Exercise therapy for multiple sclerosis. *Cochrane Database Systematic Review*, CD003980.

Ringqvist, I. (1971). Muscle strength in myasthenia gravis: effects of exhaustion and anticholinesterase related to muscle fibre size. *Acta Neurologica Scandinavica*, 47, 619–641.

Rochester, L., Barron, M.J., Chandler, C.S., Sutton, R.A., Miller, S. and Johnson, M.A. (1995a). Influence of electrical stimulation of the tibialis anterior muscle in paraplegic subjects: 2: morphological and histochemical properties. *Paraplegia*, 33, 514–522.

Rochester, L., Chandler, C.S., Johnson, M.A., Sutton, R.A. and Miller, S. (1995b). Influence of electrical stimulation of the tibialis anterior muscle in paraplegic subjects: 1: contractile properties. *Paraplegia*, 33, 437–449.

Rodriquez, A.A. and Agre, J.C. (1991). Electrophysiologic study of the quadriceps muscles during fatiguing exercise and recovery: a comparison of symptomatic and asymptomatic postpolio patients and controls. *Archives in Physical Medicine and Rehabilitation*, 72, 993–997.

Rodriquez, A.A., Agre, J.C. and Franke, T.M. (1997). Electromyographic and neuromuscular variables in unstable postpolio subjects, stable postpolio subjects, and control subjects. *Archives in Physical Medicine and Rehabilitation*, 78, 986–991.

Roelcke, U., Kappos, L., Lechner-Scott, J., Brunnschweiler, H., Huber, S., Ammann, W., Plohmann, A., Dellas, S., Maguire, R.P., Missimer, J., Radu, E.W., Steck, A. and Leenders, K.L. (1997). Reduced glucose metabolism in the frontal cortex and basal ganglia of multiple sclerosis patients with fatigue: a 18F-fluorodeoxyglucose positron emission tomography study. *Neurology*, 48, 1566–1571.

Rutkove, S.B., Shefner, J.M., Wang, A.K., Ronthal, M. and Raynor, E.M. (1998). High-temperature repetitive nerve stimulation in myasthenia gravis. *Muscle and Nerve*, 21, 1414–1418.

Samii, A., Lopez-Devine, J., Wasserman, E.M., Dalakas, M.C., Clark, K., Grafman, J. and Hallett, M. (1998). Normal postexercise facilitation and depression of motor evoked potentials in postpolio patients. *Muscle and Nerve*, 21, 948–950.

Scott, W.B., Lee, S.C., Johnston, T.E., Binkley, J. and Binder-Macleod, S.A. (2007). Effect of electrical stimulation pattern on the force responses of paralyzed human quadriceps muscles. *Muscle and Nerve*, 35, 471–478.

Secher, N.H. and Petersen, S. (1984). Fatigue of voluntary contractions in normal and myasthenic human subjects. *Acta Physiologica Scandinavica*, 122, 243–248.

Sener, H.O. and Yaman, A. (2008). Effect of high temperature on neuromuscular jitter in myasthenia gravis. *European Neurology*, 59, 179–182.

Sharma, K.R., Kent-Braun, J., Mynhier, M.A., Weiner, M.W. and Miller, R.G. (1994). Excessive muscular fatigue in the postpoliomyelitis syndrome. *Neurology*, 44, 642–646.

Sharma, K.R., Kent-Braun, J., Mynhier, M.A., Weiner, M.W. and Miller, R.G. (1995). Evidence of an abnormal intramuscular component of fatigue in multiple sclerosis. *Muscle and Nerve*, 18, 1403–1411.

Sheean, G.L., Murray, N.M., Rothwell, J.C., Miller, D.H. and Thompson, A.J. (1997). An electrophysiological study of the mechanism of fatigue in multiple sclerosis. *Brain*, 120, 299–315.

Shields, R.K. (1995). Fatigability, relaxation properties, and electromyographic responses of the human paralyzed soleus muscle. *Journal of Neurophysiology*, 73, 2195–2206.

Shields, R.K. and Dudley-Javoroski, S. (2006). Musculoskeletal plasticity after acute spinal cord injury: effects of long-term neuromuscular electrical stimulation training. *Journal of Neurophysiology*, 95, 2380–2390.

Shields, R.K. and Dudley-Javoroski, S. (2007). Musculoskeletal adaptations in chronic spinal cord injury: effects of long-term soleus electrical stimulation training. *Neurorehabilitation and Neural Repair*, 21, 169–179.

Shields, R.K., Chang, Y.J., Dudley-Javoroski, S. and Lin, C.H. (2006a). Predictive model of muscle fatigue after spinal cord injury in humans. *Muscle and Nerve*, 34, 84–91.

Shields, R.K., Dudley-Javoroski, S. and Cole, K.R. (2006b). Feedback-controlled stimulation enhances human paralyzed muscle performance. *Journal of Applied Physiology*, 101, 1312–1319.

Simpson, J.A. (1960). Myasthenia gravis: a new hypothesis. *Scottish Medical Journal*, 5, 419–436.

Sinoway, L.I., Hill, J.M., Pickar, J.G. and Kaufman, M.P. (1993). Effects of contraction and lactic acid on the discharge of group III muscle afferents in cats. *Journal of Neurophysiology*, 69, 1053–1059.

Sjogaard, G. and McComas, A.J. (1995). Role of interstitial potassium. *Advances in Experimental Medicine and Biology*, 384, 69–80.

Son, Y.J. and Thompson, W.J. (1995). Schwann cell processes guide regeneration of peripheral axons. *Neuron*, 14, 125–132.

Stalberg, E. (1980). Clinical electrophysiology in myasthenia gravis. *Journal of Neurology and Neurosurgical Psychiatry*, 43, 622–633.

Stein, R.B., Gordon, T., Jefferson, J., Sharfenberger, A., Yang, J.F., Tötösy de Zepetnek, J. and Belanger, M. (1992). Optimal stimulation of paralyzed muscle after human spinal cord injury. *Journal of Applied Physiology*, 72, 1393–1400.

Stout, J.R., Eckerson, J.M., May, E., Coulter, C. and Bradley-Popovich, G.E. (2001). Effects of resistance exercise and creatine supplementation on myasthenia gravis: a case study. *Medicine and Science in Sports and Exercise*, 33, 869–872.

Sunnerhagen, K.S., Carlsson, U., Sandberg, A., Stålberg, E., Hedberg, M. and Grimby, G. (2000). Electrophysiologic evaluation of muscle fatigue development and recovery in late polio. *Archives of Physical Medicine and Rehabilitation*, 81, 770–776.

Tam, S.L. and Gordon, T. (2003). Mechanisms controlling axonal sprouting at the neuromuscular junction. *Journal of Neurocytology*, 32, 961–974.

Tartaglia, M.C., Narayanan, S., Francis, S.J., Santos, A.C., De Stefano, N., Lapierre, Y. and Arnold, D.L. (2004). The relationship between diffuse axonal damage and fatigue in multiple sclerosis. *Archives in Neurology*, 61, 201–207.

Taylor, J.L. and Gandevia, S.C. (2008). A comparison of central aspects of fatigue in submaximal and maximal voluntary contractions. *Journal of Applied Physiology*, 104, 542–550.

Tepavac, D., Swenson, J.R., Stenehjem, J., Sarjanovic, I. and Popovic, D. (1992). Microcomputer-based portable long-term spasticity recording system. *IEEE Transactions on Bio-medical Engineering* 39, 426–431.

Thanvi, B.R. and Lo, T.C. (2004). Update on myasthenia gravis. *Postgraduate Medical Journal*, 80, 690–700.

Thickbroom, G.W., Sacco, P., Kermode, A.G., Archer, S.A., Byrnes, M.L., Guilfoyle, A. and Mastaglia, F.L. (2006). Central motor drive and perception of effort during fatigue in multiple sclerosis. *Journal of Neurology*, 253, 1048–1053.

Thomas, C.K. (1997a). Fatigue in human thenar muscle paralysed by spinal cord injury. *Journal of Electromyography and Kinesiology*, 7, 15–26.

Thomas, C.K. (1997b). Contractile properties of human thenar muscles paralyzed by spinal cord injury. *Muscle and Nerve*, 20, 788–799.

Thomas, C.K. and Ross, B.H. (1997). Distinct patterns of motor unit behavior during muscle spasms in spinal cord injured subjects. *Journal of Neurophysiology*, 77, 2847–2850.

Thomas, C.K. and Zijdewind, I. (2006). Fatigue of muscles weakened by death of motoneurons. *Muscle and Nerve*, 33, 21–41.

Thomas, C.K., Nelson, G., Than, L. and Zijdewind, I. (2002). Motor unit activation order during electrically evoked contractions of paralyzed or partially para-lyzed muscles. *Muscle and Nerve*, 25, 797–804.

Thomas, C.K., Griffin, L., Godfrey, S., Ribot-Ciscar, E. and Butler, J.E. (2003). Fatigue of paralyzed and control thenar muscles induced by variable or constant frequency stimulation. *Journal of Neurophysiology*, 89, 2055–2064.

Thomas, C.K., Häger-Ross, C.K. and Klein, C.S. (2008). Baclofen weakens motor units paralyzed by spinal cord injury. *Society for Neuroscience*, 859, 10.

Thomas, C.K., Zaidner, E.Y., Calancie, B., Broton, J.G. and Bigland-Ritchie, B. (1997). Muscle weakness, paralysis, and atrophy after human cervical spinal cord injury. *Experimental Neurology*, 148, 414–423.

Tollback, A., Knutsson, E., Borg, J., Borg, K. and Jakobsson, F. (1992). Torque-velocity relation and muscle fibre characteristics of foot dorsiflexors after long-term overuse of residual muscle fibres due to prior polio or L5 root lesion. *Scandinavian Journal of Rehabilitative Medicine*, 24, 151–156.

Tollback, A., Soderlund, V., Jakobsson, F., Fransson, A., Borg, K. and Borg, J. (1996). Magnetic resonance imaging of lower extremity muscles and isokinetic strength in foot dorsiflexors in patients with prior polio. *Scandinavian Journal of Rehabilitation Medicine*, 28, 115–123.

Trimble, M.H. and Enoka, R.M. (1991). Mechanisms underlying the training effects associated with neuromuscular electrical stimulation. *Physical Therapy*, 71, 273–280.

Trojan, D.A. and Cashman, N.R. (2005). Post-poliomyelitis syndrome. *Muscle and Nerve*, 31, 6–19.

Trojan, D.A., Collet, J., Pollak, M.N., Shapiro, S., Jubelt, B., Miller, R.G., Agre, J.C., Munsat, T.L., Hollander, D., Tandan, R., Robinson, A., Finch, L., Ducruet, T. and Cashman, N.R. (2001). Serum insulin-like growth factor-I (IGF-I) does not correlate positively with isometric strength, fatigue, and quality of life in post-polio syndrome. *Journal of the Neurological Sciences*, 182, 107–115.

Trontelj, J.V. and Stalberg, E. (1991). Single motor end-plates in myasthenia gravis and LEMS at different firing rates. *Muscle and Nerve*, 14, 226–232.

Trontelj, J.V. and Stalberg, E. (1995). Single fiber electromyography in studies of neuromuscular function. *Advances in Experimental Medicine and Biology*, 384, 109–119.

Trontelj, J.V., Mihelin, M., Fernandez, J.M. and Stålberg, E. (1986). Axonal stimulation for end-plate jitter studies. *Journal of Neurology and Neurosurgical Psychiatry*, 49, 677–685.

van der Kamp, W., Maertens de Noordhout, A., Thompson, P.D., Rothwell, J.C., Day, B.L. and Marsden, C.D. (1991). Correlation of phasic muscle strength and corticomotoneuron conduction time in multiple sclerosis. *Annals of Neurology*, 29, 6–12.

van der Werf, S.P., Jongen, P.J., Nijeholt, G.J., Barkhof, F., Hommes, O.R. and Bleijenberg, G. (1998). Fatigue in multiple sclerosis: interrelations between fatigue complaints, cerebral MRI abnormalities and neurological disability. *Journal of Neurological Sciences*, 160, 164–170.

van Duinen, H., Renken, R., Maurits, N. and Zijdewind, L. (2007). Effects of motor fatigue on human brain activity, and fMRI study. *Neuroimage*, 35, 1438–1449.

Vollestad, N.K., Sejersted, O.M., Bahr, R., Woods, J.J. and Bigland-Ritchie, B. (1988). Motor drive and metabolic responses during repeated submaximal contractions in humans. *Journal of Applied Physiology*, 64, 1421–1427.

Weiner, P., Gross, D., Meiner, Z., Ganem, R., Weiner, M., Zamir, D. and Rabner, M. (1998). Respiratory muscle training in patients with moderate to severe myasthenia gravis. *The Canadian Journal of Neurological Sciences*, 25, 236–241.

Wiechers, D.O. (1988). New concepts of the reinnervated motor unit revealed by vaccine-associated poliomyelitis. *Muscle and Nerve*, 11, 356–364.

Wiechers, D.O. and Hubbell, S.L. (1981). Late changes in the motor unit after acute poliomyelitis. *Muscle and Nerve*, 4, 524–528.

Wilmet, E., Ismail, A.A., Heilporn, A., Welraeds, D. and Bergmann, P. (1995). Longitudinal study of the bone mineral content and of soft tissue composition after spinal cord section. *Paraplegia*, 33, 674–677.

Wise, A.K., Morgan, D.L., Gregory, J.E. and Proske, U. (2001). Fatigue in mammalian skeletal muscle stimulated under computer control. *Journal of Applied Physiology*, 90, 189–197.

Zijdewind, I. and Kernell, D. (1994). Fatigue-associated EMG behaviour of the first dorsal interosseous and adductor pollicis muscles in different groups of subjects. *Muscle and Nerve*, 17, 1044–1054.

Zijdewind, I. and Thomas, C.K. (2001). Spontaneous motor unit behaviour in human thenar muscles after spinal cord injury. *Muscle and Nerve*, 24, 952–962.

CHAPTER 10

MUSCLE FATIGUE IN MUSCULAR DYSTROPHIES

N.B.M. Voet, A.C.H. Geurts, G. Bleijenberg, M.J. Zwarts, G.W. Padberg and B.G.M. van Engelen

OBJECTIVES

The aim of this chapter is to provide an overview on:

- the prevalence and assessment of fatigue in muscular dystrophies;
- the pathophysiological determinants of fatigue in muscular dystrophies;
- the possible treatment options of fatigue in muscular dystrophies.

INTRODUCTION

The aim of this chapter is to provide an update on the prevalence, relevance, causes and treatment of fatigue in muscular dystrophies. In a study by McDonald *et al.*, the three problems most frequently cited as "very significant" by patients with slowly progressive neuromuscular disease (n = 811) were muscle weakness (57%), difficulty exercising (43%) and fatigue (40%) (McDonald, 2002). In a study by Kalkman *et al.*, 61% of patients with facioscapulohumeral dystrophy (n = 139) and 74% of patients with myotonic dystrophy (n = 322) were "severely fatigued" (Kalkman *et al.*, 2005). The muscular dystrophies are an inherited group of more than 30 distinct progressive disorders resulting from defects in a number of genes required for normal muscle structure and function. They are characterised by progressive loss of muscle strength and integrity and they have a variable distribution and severity (Emery, 2002). We will, however, limit our review to those main types of diseases which are most frequent: Duchenne and Becker muscular dystrophy, myotonic dystrophy type 1, facioscapulohumeral muscular dystrophy and the limb girdle muscular dystrophies. A more extensive overview of muscular dystrophies can be found in Engel and Franzini-Armstrong (2004).

In this chapter, we will distinguish two main types of fatigue (Table 10.1). Physiological fatigue, or muscle fatigue, has been defined as a reduction in maximal voluntary muscle force (MVC) during exercise. Experienced fatigue, on the other hand, is the subjective feeling of fatigue. Muscle fatigue is not necessarily accompanied by experienced fatigue, or vice versa.

High-quality studies about fatigue in muscular dystrophy are scarce. Nevertheless, after a general introduction on muscular dystrophies, addressing both clinical and pathophysiological aspects, the prevalence of experienced fatigue in muscular dystrophies in the literature will be critically reviewed and the putative underlying pathophysiological mechanisms of muscle fatigue and experienced fatigue will be outlined. Finally, literature about treatment of muscle fatigue and experienced fatigue in muscular dystrophies will be reviewed and recommendations for future research will be made. Throughout the chapter, the scientific knowledge will be illustrated by a clinical case report that describes the experienced fatigue of a 59-year-old man with facioscapulohumeral dystrophy, Mr A.

Table 10.1 Dimensions and definitions of fatigue as used in this chapter

---

Muscle fatigue, or *physiological fatigue,* can be objectively assessed in a laboratory setting and is defined as the total amount of voluntary loss of force during a sustained maximal voluntary muscle force. It contains both peripheral and central components, a distinction that is based on whether the loss of capacity to generate force originates from the muscle system or the central nervous system.

- *Peripheral fatigue* can be determined as the loss of force during constant electrical stimulation, applied to the motor nerve or motor endplate during a sustained MVC.
- *Central fatigue* can be determined as an increase in central activation failure during exercise. *Central activation failure (CAF)* is defined as submaximal central activation. CAF can be measured by the "twitch interpolation technique". If the central activation is submaximal, the electrical stimulation, applied to the motor nerve or motor endplate during a sustained MVC, will result in an increased exertion of force compared to exertion of force without electrical stimulation, demonstrating CAF. CAF can be present already at the start of a sustained MVC (Schillings *et al.* 2007).

Experienced fatigue is by definition a subjective entity and can be assessed only by self-report, for example with the subscale fatigue of the Checklist Individual Strength, and does not necessarily correlate with physiological fatigue. Experienced fatigue has many dimensions: for example, activity limitations in daily life, physical inactivity, sleep disturbances, concentration problems, loss of social participation, psychological distress, lack of sense of control over fatigue, cognitions about the possible cause of fatigue, and pain.

DYSTROPHINOPATHIES

Duchenne muscular dystrophy (DMD) is the most common form of the human muscular dystrophies. Becker muscular dystrophy (BMD) is a less frequent and more benign form of the disease. The incidence of DMD is approximately one in 3,500 live male births. By comparison, BMD is found in one in 30,000 male births (Emery, 1991). Both are X-linked recessive disorders and are caused by a mutation in the DMD gene, which is located on chromosome Xp21 and encodes for the production of dystrophin. One-third of the cases are due to spontaneous mutations (Aartsma-Rus et al., 2006). The primary abnormality in DMD is the lack of dystrophin. In BMD, the protein is reduced in amount or abnormal in size. Dystrophin is a 427 kilodalton protein normally found at the cytoplasmic face of the muscle cell surface membrane, functioning as a component of a large, tightly associated glycoprotein complex (Monaco et al., 1988). In its absence, the glycoprotein complex is digested by proteases. This may initiate the degeneration of muscle fibres, resulting in muscle weakness and potential mechanical injury from tissue stress in rest and during exercise (Petrof et al., 1993; Muntoni et al., 2003; Lapidos et al., 2004).

Diagnosis is informed by characteristic clinical findings among which progressive symmetrical muscle weakness is the most important, affecting proximal limb muscles more than the distal muscles. Initially, only lower-limb muscles are affected, accompanied by pseudohypertrophy of the calf muscles. Common musculoskeletal complications are kyphoscoliosis and muscle contractures. Because dystrophin is also found in the heart, brain and the smooth muscles, frequent concomitant manifestations are cardiomyopathy and mental retardation (Muntoni, 2003).

In DMD, the clinical symptoms first present between three and five years of age, and patients generally lose ambulation at 7–12 years. In the past, death usually occurred from cardiac or respiratory causes in the late teens or early twenties. Recently, however, respiratory support has been able prolong survival into the fourth decade (Eagle et al., 2007). The diagnosis is supported by a family history suggestive of X-linked recessive inheritance (Engel and Franzini-Armstrong, 2004) or by dystrophin immunostaining of muscle tissue (Muntoni, 2001; Griggs and Bushby, 2005; Aartsma-Rus et al., 2006). Serum creatine phosphokinase (CK) level is generally increased to levels that are 50–100 times the reference range (i.e. as high as $20,000 \, mU \cdot ml^{-1}$). The diagnosis is confirmed by identifying abnormalities in the dystrophin gene by mutation analysis of DNA from peripheral blood leukocytes (Aartsma-Rus et al., 2006).

In BMD, the distribution of muscle wasting and weakness is closely similar to that in DMD, but the course of the disease is more benign and far less predictable, with first clinical symptoms presenting at around 12 years old. Many patients remain ambulatory into adult life (Emery, 2002; Aartsma-Rus et al., 2006).

No curative treatment is available for either disease, although first attempts are being made: the gene transfer technique by intramuscular injection of an anti-sense oligonucleotide is under development (van Deutekom *et al.*, 2007), but several hurdles still need to be overcome (Cossu and Sampaolesi, 2007). Therefore, emphasis is currently on respiratory care, treatment of cardiological complications and optimising the quality of life by symptomatic physiotherapeutic and medical treatments (Bogdanovich *et al.*, 2004). There is evidence that corticosteroid therapy in DMD can reduce speed of decline of muscle strength and function (De Groot, 2006; Manzur *et al.*, 2008).

MYOTONIC DYSTROPHY

Myotonic dystrophy (MD) is the second-most common muscular dystrophy. There are two major forms: MD1, also known as Steinert's disease, and MD2, a multisystem disease, also known as proximal myotonic myopathy (PROMM). In this chapter, we will limit the discussion to MD1, which is more frequent.

MD1 is divided into congenital, classical and minimal phenotypes according to the age of the symptom onset and disease severity. The congenital form of MD1 will not be further considered in this chapter (see Engel and Franzini-Armstrong (2004)). The prevalence of MD1 is approximately one in 8,000 in the general population (Machuca-Tzili *et al.*, 2005). MD1 is an autosomal-dominant disorder, of which the molecular basis is expansion of an unstable repeat sequence in a non-coding part of the dystrophia myotonica protein kinase (DMPK) gene on chromosome 19. The repeat expansion enlarges with each generation, which leads to earlier onset and increased severity of symptoms with each affected generation, a phenomenon which is known as "anticipation" (Howeler *et al.*, 1989). There is increasing support for the theory that disruption of RNA metabolism, which has effects on many other genes, explains the multisystemic nature of the disease (Schara and Schoser, 2006).

MD1 is clinically characterised by muscle weakness of the distal limbs, progressing to the proximal limbs with gradual occurrence of myotonia (delayed relaxation after muscle contraction). Weakness occurs most frequently in facial muscles, the distal muscles of the forearm, and the ankle dorsiflexors, with onset of symptoms in the second, third or fourth decade (Day *et al.*, 2003). Associated findings include muscle pain, cognitive and psychological changes, cataracts, cardiac conduction defects and endocrine disorders (George *et al.*, 2004; Machuca-Tzili *et al.*, 2005; Kalkman, 2006). Excessive daytime sleepiness is found in about one-third of patients (Culebras, 1996; Phillips *et al.*, 1999).

The diagnosis can be suspected clinically by a positive family history and by identifying the symptoms mentioned above. Specific genetic testing to demonstrate the presence of an expanded CTG repeat in the DMPK gene is the gold standard for the diagnosis of MD1 (Machuca-Tzili et al., 2005). Life expectancy is reduced for patients with MD1. Respiratory insufficiency and cardiac diseases are the most common causes of death (Reardon et al., 1993; de Die-Smulders et al., 1998; Mathieu et al., 1999). There is no disease-modifying therapy available for the treatment of MD1. Therefore, treatment is symptomatic (Harper et al., 2004).

FACIOSCAPULOHUMERAL DYSTROPHY

Facioscapulohumeral dystrophy (FSHD) is the third-most common muscular dystrophy. The estimated prevalence is one in 20,000 persons (Padberg, 1982). FSHD is an autosomal-dominant disease. It is associated with subtelomeric contraction of chromosome 4q, with loss of tandem repeat-units. In general, the disorder is more severe in a patient with a lower number of repeats. The pathogenetic mechanisms in FSHD are unknown. The presence of some extramuscular manifestations in FSHD suggests the involvement of a gene with pleiotropic effects or, alternatively, the involvement of multiple genes (Tawil and van der Maarel, 2006). FSHD derives its name from the muscle groups that are mainly affected first: facial and shoulder girdle muscles. During disease progression, humeral, abdominal, pelvic girdle and foot extensor muscles can become involved as well (Padberg, 1982). Lower abdominal muscles are weaker than the upper abdominal muscles, causing a "Beevor's sign", a physical finding specific to FSHD (Shahrizaila and Wills, 2005). The most commonly described extramuscular manifestations are the high-frequency hearing loss and retinal telangiectasias, occurring in 75% and 60% of affected individuals, respectively (Tawil and van der Maarel, 2006). The heart is not affected in most cases, though arrhythmias and conduction defects have been described (Emery, 2002). The median age of onset is around 17 years, but the onset of clinical symptoms varies from infancy to the seventh decade (Padberg, 1982).

Although the exact gene defect or genetic mechanism is not yet known, a DNA test is available for FSHD, which detects a specific deletion in chromosome 4q35. This diagnostic test is abnormal in 95–98% of typical FSHD cases (Kohler et al., 1999; Ricci et al., 1999; Vitelli et al., 1999).

The course of FSHD is usually slowly progressive, but the severity among patients is extremely variable, ranging from isolated facial weakness to severe generalised weakness, with approximately 20% of patients eventually becoming wheelchair-dependent (Padberg, 1982). Many patients report a relapsing course, with long periods of quiescence interrupted by periods of rapid deterioration involving a particular muscle group, often heralded by pain in the affected limb. Most of the patients have a normal life expectancy (Tawil and van der Maarel, 2006).

Currently, there is no genetic or pharmaceutical curative treatment available for FSHD. Only two randomised controlled trials have been published. Recent trials of albuterol, also known as salbutamol (van der Kooi *et al.*, 2004, 2007), folic acid and methionine (van der Kooi *et al.*, 2006) and creatine, a dietary supplement for building muscle (Walter *et al.*, 2000), did not confirm or refute a significant effect of either of these treatments (Rose and Tawil, 2004). The mainstay of management is, therefore, treatment of symptoms, prevention of secondary problems and improvement of functional abilities and quality of life (Tawil and van der Maarel, 2006).

LIMB GIRDLE MUSCULAR DYSTROPHIES

The limb girdle muscular dystrophies (LGMD) are a group of disorders that are historically grouped together because of the shared clinical feature of predominant involvement of the "limb girdle" (pelvic and shoulder) musculature. However, it is recognised that there is a broad heterogeneity of presentation and muscle involvement in the LGMD group (Bushby *et al.*, 2007). The overall frequency has been estimated to be one in 14,000 to one in 200,000 (Emery, 1991). Most cases of LGMD are inherited in an autosomal-recessive fashion (Zatz *et al.*, 2003). However, families with an autosomal-dominant pattern of inheritance have also been described, which probably account for about 10% of all LGMDs (Kirschner and Bonnemann, 2004). The emergence of a LGMD phenotype can result from mutations in any of at least 19 different genes (Daniele *et al.*, 2007). The discovery of genetically distinct subtypes has redefined the classification of LGMD and has led to a nomenclature designating the autosomal dominant form as LGMD1A, 1B, 1C, etc., and the autosomal recessive form as LGMD2A, 2B, 2C, etc. (Kirschner and Bonnemann, 2004). The proteins causing LGMD have a wide range of localisation across the muscle fibre, from sarcolemma to nuclear envelope, with various functions (Bushby *et al.*, 2007; Daniele *et al.*, 2007).

Weakness may affect proximal muscles of the shoulder girdle (scapulohumeral type), the pelvic girdle (pelvifemoral type), or both. Neck flexor and extensor muscles may be concurrently involved. Facial weakness, when present, is usually mild and, in most cases, totally absent. Even in mild cases, there is preferential weakness and atrophy of the biceps muscle. Distal muscle strength is usually preserved, even at the late stage of the disease. Selectivity of muscle involvement and clinical characteristics such as hypertrophy of the calves or tongue and late stage cardiac complications are associated more or less specifically with each of the different forms (Engel and Franzini-Armstrong, 2004).

The single constant biochemical abnormality in LGMD is the elevation of the CK level. In autosomal-recessive types of LGMD, serum CK is always increased, up to 200 times the normal range. DNA analysis to detect a mutation in the affected gene(s) is the gold standard of diagnosis (Norwood *et al.*, 2007). Reported age of onset of LGMDs varies among the different mutations and is between one and 50 years, although some patients may be asymptomatic. Compared with the autosomal-dominant type, autosomal-recessive LGMD is usually associated with earlier age of onset, more rapid progression and relatively high CK values. Morbidity and mortality rates vary, but with early onset the course is generally rapid (Engel and Franzini-Armstrong, 2004). Treatment is supportive and consists of physical therapy, assistive devices and monitoring of respiratory function and cardial complications. Treatment is generally aimed at prolonging survival and improving quality of life (Daniele *et al.*, 2007).

Clinical case: disease description

Mr A is a 59-year-old man who broke his clavicle in a football game when he was 18 years old. A year after the accident he went back to his general practitioner because symptoms of pain and decreased functioning of his shoulder did not disappear. He was referred to a neurologist, who clinically diagnosed a "muscle disease" when he was 19 years old. At that time, he knew that his mother, who was wheelchair-dependent, had a "muscle disease", but neither the diagnosis, nor the prognosis of her condition was known. Decades later, his neurologist told him he had a muscle disease which was known as "Landouzy–Dejerine", the former name of FSHD. The diagnosis FSHD was genetically confirmed 30 years later. At that time, an autosomal-dominant inheritance pattern could be recognised in his family. Many persons in every generation appeared to be affected by the disease.

Mr A experiences a relapsing course of FSHD with long stable periods followed by periods of clear deterioration. Currently, facial, shoulder girdle, humeral, abdominal, pelvic girdle and foot dorsiflexor muscles are involved. He is still ambulant, but his unaided walking distance is restricted to approximately 100 m. Outdoor he uses a rollator, which increases his walking distance to 250 m. He is very afraid of becoming wheelchair-dependent, just like his mother. Mr A lives together with his wife in an apartment at ground level. He works four days a week as an IT specialist and spends a lot of time in volunteer activities. He plays the saxophone in a band.

EXPERIENCED FATIGUE

Assessment of experienced fatigue

Distinguishing experienced fatigue from muscle weakness, the key feature in muscular dystrophy, may be difficult. Asking patients to describe their fatigue will lead to several descriptions, varying from sleepiness to weakness to exercise intolerance to exhaustion. Hence, experienced fatigue is, therefore, a multi-dimensional concept with possible contributions of, for example, physical, cognitive and motivational factors (Table 10.1). Although experienced fatigue is difficult to define, it still is a valuable concept which can be reliably measured by using questionnaires. An often-used questionnaire for the experience of fatigue and its behavioural consequences is the Checklist Individual Strength (CIS). The CIS consists of four subscales: one scale for experienced fatigue, so called "CIS-fatigue", and three scales for reduction in motivation, physical activity and concentration, respectively. Higher scores indicate higher levels of fatigue, more concentration problems, a greater decrease in motivation and lower levels of activities (Vercoulen *et al.*, 1994). The Abbreviated Fatigue Questionnaire (AFQ) is another short, reliable, and easy-to-use instrument to determine the intensity of a patient's experienced fatigue. It consists of four questions that have to be answered on a seven-point Likert scale. A lower total score indicates a higher degree of fatigue (Alberts *et al.*, 1997).

Prevalence and impact of experienced fatigue

Kalkman *et al.* measured the prevalence of "severe experienced fatigue" in 598 neuromuscular patients, among which were 139 patients with FSHD and 322 patients with MD. Both patient groups experienced high levels of fatigue (Kalkman *et al.*, 2005). The mean CIS-fatigue score in the FSHD group was 36.5 (SD ± 12.5) and in the MD group 40.4 (SD ± 11.8). In the FSHD group 61% of patients were "severely fatigued" (determined by a CIS-fatigue score equal or above 35). In the MD group, this was 74%. In both groups, age showed low but significant correlations with fatigue severity, indicating that, in general, older patients experienced somewhat greater fatigue. Severely fatigued patients scored lower on all Short Form-36 (SF-36) scales than the non-severely fatigued patients, suggesting a relation between experienced fatigue and activity limitations. There appeared to be several differences between MD and FSHD patients. Patients with MD had higher scores for experienced fatigue, reported greater problems with concentration and had more difficulties with initiative and planning than patients with FSHD. In FSHD patients and MD patients, social functioning was related to fatigue severity.

Irrespective of its cause, fatigue has a major impact on daily functioning and quality of life (Sharpe and Wilks, 2002; Gielissen *et al.*, 2007a). For example, in a study by van der Werf in patients with MD (n = 32) and FSHD or LGMD (n = 20), severe fatigue was associated with greater levels of psychological distress and more physical and psychosocial limitations, as measured with the Sickness Impact Profile (SIP), the Symptom Checklist-90 (SCL-90) and the Beck Depression Inventory Primary Care (BDI-PC (van der Werf, 2003)). In the study by Kalkman *et al.*, severely fatigued patients with FSHD or MD also had lower scores on all subscales of the SF-36, which monitors disease burden. This suggests a relation between experienced fatigue and the level of activity and social participation (Kalkman *et al.*, 2005). Apparently, fatigue is not only a frequent, but also a relevant problem in muscular dystrophy.

Clinical case: experienced fatigue

Mr A has suffered from fatigue since the age of 40. He considers his fatigue and muscle pain to be the most relevant and disabling consequences of his disease. He defines his fatigue as a lack of energy which restrains him from activities. After walking approximately 100 m, he has to stop because of severe fatigue and muscle pain. These symptoms are comparable with the exhaustion he felt after playing football in his younger years. That type of exhaustion, however, felt positive, in contrast to the negative feeling associated with the present fatigue. Fatigue has a significant and deleterious impact on his life that goes beyond the other symptoms of FSHD. It takes almost two hours to prepare himself for work every morning. After he has dressed, he often falls asleep due to exhaustion. He can only travel by car because other forms of transport are too strenuous. He describes himself as a "healthy mind in an aged body".

DETERMINANTS OF FATIGUE

As fatigue in muscular dystrophy is a multi-dimensional concept (see assessment of experienced fatigue and Table 10.1), it is important to understand factors that contribute to fatigue. Based on such an analysis, preventive and therapeutic interventions can be developed. The critical pathophysiological determinants of muscle fatigue and experienced fatigue will, therefore, be described in the next section.

Pathophysiological studies of muscle fatigue

Because of practical reasons, pathophysiological studies depend to a large extent on animal models. A review of Wineinger *et al.* summarises the literature regarding the physiological fatigue characteristics of skeletal muscles in animal models of muscular dystrophy (Wineinger *et al.*, 2002). Muscle fatigue in animal studies was expressed as a percentage of initial force, i.e. physiological fatigue. Force was measured by recording the action potential (AP) of muscles and muscle-evoked tension. Two rodent models (mdx mouse and dystrophic hamster) have been studied most extensively. The dystrophic hamster, lacking normal sarcoglycan, was used as a model for LGMD. The mdx mouse lacks dystrophin, and was therefore considered a model for DMD.

Significant variability has been observed before in studies of muscle fatigue in dystrophic animals, which may be due to different experimental conditions. Because of this variability, it is difficult to evaluate muscle fatigue in animal models of muscular dystrophy. Still, some trends can be recognised (Table 10.2).

The dystrophic soleus muscle fatigued more slowly or at the same rate as that of healthy animals. The soleus is largely composed of slow-twitch type I oxidative muscle fibres and is considered to be fatigue resistant. His-tological studies showed an increase in the proportion of type I muscle fibres in the dystrophic soleus muscle, which could explain the increased resistance to fatigue. The dystrophic extensor digitorum longus (EDL) was weaker than in healthy animals and generally more fatigable. The EDL muscle has a majority of type IIB fibres which are easily fatigable. Pagala *et al.* described that type IIB dystrophic muscle fibres are more susceptible to degeneration, in contrast to type I muscle fibres (Pagala *et al.*, 1991). No difference was found in fatigability between healthy and dystrophic diaphragm muscles. The diaphragm is composed of fast oxidative IIA muscle fibres, which are relatively fatigue resistant. It appears that differ-ence in fatigability of dystrophic animal muscles compared to muscles of healthy animals can largely be explained by differences in muscle fibre types (Table 10.2).

Apparently, type I, in contrast to type II muscle fibres of dystrophic animal muscles have the potential to regenerate. Because aerobic training increases the proportion of type I muscle fibres and, with that, fatigue-resistance of healthy muscles, aerobic training could be effective in decreas-ing fatigability of dystrophic muscles, as well through the same mechanism (Ingjer, 1979).

Increased muscle fatigue has often been attributed to a decrease in the metabolic potential of the individual muscle fibres. It is known that the levels of some energy metabolites like creatine are decreased in muscular dystrophies such as DMD (Griffin *et al.*, 2001; Sharma *et al.*, 2003b), which may aggravate muscle weakness and muscle fatigue. Interestingly, in some types of LGMD, in which muscles are less severely affected, creatine does not seem to be decreased, indicating that the level of creatine may serve as a biomarker for the severity of muscle weakness and muscle fatigue (Sharma *et al.*, 2003a). Furthermore, the decrease in concentrations

Table 10.2 The difference in fatigability of dystrophic animal muscles compared to healthy animal muscles can be explained by differences in muscle fibre types

Muscle fibre type	Fatigability of dystrophic animal muscles compared to healthy animal muscles
Type I (slow-twitch, oxidative)	↓/=
Type IIA (fast-twitch, oxidative)	=
Type IIB (fast-twitch, glycolytic)	↑

of other metabolites such as choline and lactate was less severe in LGMD compared to DMD, suggesting that these metabolites could also be potential biomarkers (Sharma *et al.*, 2003a).

In summary, these reports indicate that abnormal metabolite profiles could serve as specific biomarkers to characterise the severity of muscular dystrophies.

Peripheral versus central fatigue

Until recently, the emphasis in clinical research in muscular dystrophies was on peripheral fatigue. However, not only peripheral impairments, but also changes within the central nervous system could be responsible for increased fatigue. Schillings *et al.* first investigated central aspects of physiological fatigue in patients with muscular dystrophy (Schillings *et al.*, 2007). Both peripheral and central aspects of fatigue were determined during a sustained MVC of elbow flexion in patients with FSHD (n = 65) and MD (n = 79) (Figure 10.1).

Unexpectedly, overall physiological fatigue and peripheral fatigue were smaller in neuromuscular patients compared with healthy controls. Moreover, in patients with FSHD and MD, physiological fatigue did not correlate with the level of experienced fatigue. In contrast, Schulte-Mattler *et al.* described excessive peripheral fatigue in a mixed group of neuromuscular disorders, among which are FSHD and MD (Schulte-Mattler *et al.*, 2003). This discrepancy may be explained by a difference in the exercises. The

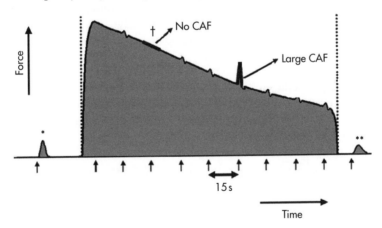

Figure 10.1 Schematic representation of peripheral and central fatigue. The figure shows the decline over time (within 2 min) of the maximum voluntary force (on the Y-axis), which is peripheral fatigue. The arrows indicate the moments of superimposed electrical endplate stimulation. The twitch interpolation may induce increments in muscle force with examples of a negligible (†) and a large CAF (#). A (near) absent response indicates a full voluntary activation of the muscle. The "at-rest twitches" are visible before (*) and following (**) the contraction, with the post-experimental twitch being clearly lower, indicative of peripheral fatigue

Source: Adapted from Zwarts *et al.*, 2008. This figure is not the registration of an individual patient.

type of exercise in the study by Schillings *et al.*, i.e. isometric contraction at maximal force level, is hardly ever required in daily life and may also decrease blood supply. Schulte-Mattler *et al.* elicited fatigue by intermittent and non-tetanic contractions to avoid blood vessel occlusion. This type of exercise may be clinically more relevant and valid for measuring physiological fatigue. CAF and central fatigue in the study by Schillings *et al.* were measured by the twitch interpolation technique (see Chapter 2; Gandevia, 2001). Central fatigue was minimal in all groups and did not differ between groups. Nor did it have any relation with experienced fatigue. Remarkably, CAF at the start of sustained MVC was enlarged in patients compared to controls. CAF in patients was related to the level of experienced fatigue. An increased CAF further decreases the maximal voluntary force in patients with muscular dystrophy. The cause of this decreased central activation cannot be determined by currently available techniques. It could be that the activation pattern of the central nervous system is not able to compensate for the peripheral problems in muscular dystrophies. The increased CAF could also be considered a beneficial adaptation, which prevents the affected muscles from excessive fatigue.

Vicious circle of physical inactivity

Fatigue may result in patients altering their lifestyles to avoid activities. Low physical activity levels may lead to even greater weakness and atrophy of skeletal muscles, which causes a vicious circle of disuse and weakness. Physical inactivity in turn can lead to chronic cardiovascular and muscle deconditioning and increased cardiovascular health risks (McDonald, 2002). For example, the maximal oxygen uptake ($\dot{V}O_2$max) is abnormally low in patients with muscular dystrophy (Lewis and Haller, 1989). Body-composition measurements in muscular dystrophy patients by various methods indicate reduced fat-free mass (FFM) and increased adiposity relative to able-bodied control subjects of comparable ages and body weights (Kilmer *et al.*, 1995; McDonald *et al.*, 1995; Johnson *et al.*, 1995). The excess body fat of muscular dystrophy patients additionally impairs mobility and further increases the risk of cardiovascular disease.

In a study by McCrory *et al.*, resting energy expenditure (REE) and total daily energy expenditure (TEE) were measured by indirect calorimetry and heart rate monitoring, respectively (McCrory *et al.*, 1998). Relatively active muscular dystrophy patients (FSHD, LGMD, MD and BMD) did not differ in REE, but had a lower estimated TEE. They also had a higher energy cost of physical activity than able-bodied subjects of the same gender who were similar in age and weight, even after adjustment for FFM differences. It is possible that the lower amount of time spent in physical activity by muscular dystrophy patients can be attributed to the higher energy cost. An alternative explanation is that persons with muscular dystrophy avoid physical activity because of the widespread belief that too much strain on the muscles will accelerate the disease process (overwork weakness). Fear of physical activity, or fear of damaging the muscle, may also contribute to the reduced central activation in patients with muscular dystrophy as described above. Irrespective of its cause, physical inactivity should be discouraged in muscular dystrophy patients because of an increasing risk of cardiovascular disease and muscle deconditioning.

Perpetuating factors of experienced fatigue

Experienced fatigue can be regarded as a multimodal concept, with a wide variety of contributing factors in patients with muscular dystrophy. These factors can be categorised into predisposing, precipitating and perpetuating factors. Predisposing factors include the presence of muscular dystrophy, whereas precipitating factors include acute physical stresses such as a concomitant disease or a period of relative deterioration of muscle function. These factors cannot be treated, in contrast to perpetuating factors, which contribute to the continuation of experienced fatigue. Kalkman *et al.* used a longitudinal design to investigate the perpetuating factors of experienced fatigue in patients with FSHD (n = 60) and MD (n = 70) (Kalkman *et al.*, 2007b). Structural equation techniques, also referred to as "causal modelling" were used. Based on longitudinal data, separate models for FSHD and MD were developed. The model of perpetuating factors of experienced fatigue in FSHD differed from the model for MD, the main difference being physical (in)activity and pain. The model fit was best for FSHD (Figure 10.2).

In FSHD, the level of physical (in)activity has a central place in the model. Lower levels of physical activity contribute to higher levels of experienced fatigue and, through that, to a lower health status. The level of physical activity is directly and negatively influenced by loss of muscle strength. In addition, pain complaints influence levels of experienced

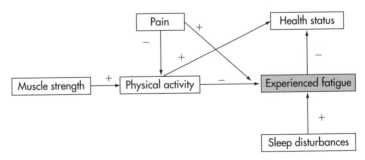

Figure 10.2 Adjusted model of perpetuating factors of experienced fatigue in patients with FSHD (n = 60). –: negative relationships; +: positive relationships. Severe muscle strength, (self-reported) physical inactivity, pain and sleep disturbances were significantly associated with the level of experienced fatigue.

Source: Adapted from Kalkman *et al.*, 2007b.

fatigue both directly and indirectly by decreasing physical activity. In contrast, in MD, physical activity and pain did not differ between patients with and without severe experienced fatigue and, therefore, did not significantly contribute to experienced fatigue.

Yet, sleep disturbances lead to higher levels of experienced fatigue in both FSHD and MD patients. The observed patterns of perpetuating factors are unique for FSHD and MD, and are different from the model of experienced fatigue in chronic fatigue syndrome (Vercoulen et al., 1994). They can be used as a basis to develop evidence-based interventions to reduce fatigue. Specific attention should be paid to sleep disturbances in both patient groups. Specifically in FSHD, treatment of fatigue should also be directed at increasing physical activity and reducing pain complaints.

Experienced fatigue and psychiatric disorders

Fatigue is a characteristic of a number of affective disorders, and an association between experienced fatigue and psychiatric symptoms has been reported in a number of central nervous system disorders, including Parkinson's disease and multiple sclerosis (Surtees et al., 2003; Schrag, 2004). In this perspective, it is relevant to know whether psychiatric comorbidity is associated with fatigue severity in muscular dystrophies. Although in the study by Kalkman et al. (see the section on prevalence and impact of experienced fatigue), severe experienced fatigue was related to higher levels of psychological distress in both patients with FSHD and MD, most of the severely fatigued patients did not fulfil the operational criteria of depression (Kalkman et al., 2005). The authors argued that severe experienced fatigue can, therefore, not be seen as merely a sign of depression. In a later study by Kalkman et al. using the Structured Clinical Interview for DSM-IV, axis 1 disorders and BDI lifetime and current psychiatric disorders (mood disorders, anxiety disorders and substance-related disorders) were equally prevalent in a large cohort of MD and FSHD patients, and were equally or even less present than these disorders in the general (Dutch) population (De Graaf et al., 2000; Kalkman et al., 2007a). The most common psychiatric disorders were depression and phobias. Psychiatric comorbidity was not associated with fatigue severity or muscle strength in the various neuromuscular disorders. In conclusion, psychiatric comorbidity is not an explanation for experienced fatigue in FSHD and MD.

Clinical case: perpetuating factors of fatigue

Mr A experiences fatigue during activities of daily life, work and leisure, but also when reading a book due to concentration problems. He easily falls asleep by day. At night, his sleep is often disturbed by muscle pain. In the past, when he was less physically disabled, he exercised at a low intensity. In his youth, he played football. In adult life, he practised swimming once a week and, later on, physical fitness, but avoided excessive training because of fear of overuse. Nevertheless, many years ago he altered his lifestyle. He stopped swimming and playing football. Currently, he is physically inactive and in a vicious circle of disuse and weakness. His physical inactivity results in muscle and cardiovascular deconditioning and obesity. Altogether, he is at risk for cardiovascular disease. Seven years ago, he experienced an ischemic cerebrovascular incident, which further increased his experienced fatigue.

TREATMENT OF FATIGUE

Most treatment studies in patients with muscular dystrophy do not describe the efficacy of the intervention in terms of decreasing muscle fatigue or experienced fatigue. Nevertheless, we will provide a critical overview of the possible treatment options with respect to fatigue in these patients. Treatment strategies that will be reviewed include physical exercise training, drug treatment and cognitive behaviour therapy.

Training studies in animals

Extrapolating data from animal studies to humans must be done with caution, because there are large differences in biomechanical properties and phenotypic expression of the dystrophic disorder between humans and, for example, the mdx mouse. Nevertheless, it may still be valuable to consider animal studies first, since unique information can be obtained (Abresch et al., 1998; Wineinger et al., 2002). Exercise training in animals mainly consisted of high-repetition aerobic-type activities like swimming, treadmill running or voluntary-wheel running. Two reviews described that dystrophic animals had a normal (and beneficial) adaptation to mild, voluntary submaximal aerobic exercise, which generally included an increase in muscle strength per cross-sectional area of muscle tissue and a reduction in muscle degeneration. The oxidative capacity and the proportion of oxidative fibres were increased, especially in slow-twitch muscles and in the muscles that were not severely affected by the dystrophy (Carter et al., 2002; Wineinger et al., 2002). Aerobic training apparently increases the amount of type I muscle fibres, as hypothesised earlier in this chapter (see the section on pathophysiological studies of muscle fatigue). Younger animals tend to benefit more from exercise studies than older animals. The muscles of young dystrophic mdx mice have a greater rate of recovery of force production than those of older mdx mice. Histological and contractile studies suggest that this difference is due to an increased regenerative capacity in young, dystrophin-deficient mdx mice, which is lost in older mdx mice (Dupont-Versteegden et al., 1994; Carter et al., 1995).

Carter *et al.* reviewed studies of exercise training and contraction-induced muscle injury in animal models of muscular dystrophy (Carter *et al.*, 2002). A majority of the studies in both normal and dystrophic animals showed that untrained eccentric exercise (lengthening of the muscle during contraction) may injure the contractile and cytoskeletal components of the muscle fibres. During eccentric exercise, sarcomeres are stretched and the actin and myosin filaments are pulled apart, leading to disruption of the thick and thin filament array and subsequent damage to cytoskeletal proteins. The inability to quickly repair a disruption of the membrane causes an elevation in intracellular calcium concentration, which triggers calcium-activated degradation pathways and further structural damage. This damage results in fibre degeneration, followed by inflammation and, eventually, fibre regeneration. Probably because of their increased regenerative capacity, muscles of younger mdx mice recovered more rapidly than those of older mdx mice (McArdle *et al.*, 1992; Sacco *et al.*, 1992; Dick and Vrbova, 1993; Brussee *et al.*, 1997). Based on these animal studies, one can conclude that submaximal aerobic exercise training can be beneficial. However, eccentric exercise training should be avoided.

Training studies in muscular dystrophy patients

In the past, many patients with muscular dystrophy were advised not to exercise because of the belief that too much exercise might lead to overuse weakness (Johnson and Braddom, 1971; Fowler Jr, 1984; Petrof, 1998). Yet, in their Cochrane review on muscle strength training and aerobic exercise training for patients with muscle diseases, van der Kooi *et al.* concluded that moderate-intensity strength training in MD and FSHD appeared not to be harmful, although there was insufficient evidence to establish its benefit (van der Kooi *et al.*, 2005). This conclusion was based on merely two randomised clinical trials (Lindeman *et al.*, 1995; van der Kooi *et al.*, 2004). When randomised clinical trials are scarce, evidence from non-randomised studies and other designs, such as pre–post studies or case-control studies, may be particularly relevant (Hartling *et al.*, 2005). For this reason, Cup *et al.* reviewed not only randomised clinical trials, but also controlled clinical trials and other designs of sufficient quality, using the list by van Tulder *et al.* (Van Tulder *et al.*, 1997; Cup *et al.*, 2007). All types of exercise therapy and other physical therapy modalities were included for patients with muscular dystrophy, among which were patients with FSHD, LGMD, MD and DMD. Cup *et al.* also concluded that exercise training is not harmful in muscular dystrophies (Cup *et al.*, 2007). However, based on the reviewed studies, there was insufficient evidence for the effectiveness of muscle strengthening exercises, although there were some indications that aerobic exercises may have a positive effect on body functions, as well as on activities and participation.

There are several limitations to consider when reviewing training studies in muscular dystrophies. First of all, there are only very few randomised controlled trials, each small in sample size. Second, studies are not immediately comparable because they have used training protocols which differ regarding the intensity and duration of the training, targeted muscle groups, type of strength training (i.e. isometric or isokinetic) and type of controls. The majority of exercise training studies have evaluated non-supervised home programmes of relatively short duration, using submaximal, low-intensity training levels. The short duration of most strengthening studies does not allow differentiation between neural training effects versus muscle fibre hypertrophy, which generally occurs after six weeks. Third, the compliance of patients, especially during non-supervised home protocols, is a possible confounding factor in all training studies. Fourth, because of the scarcity of patients of each muscular dystrophy, studies have often grouped together several disorders. Persons with different types of muscular dystrophy may, however, respond very differently to exercise (Fowler Jr, 2002). Fifth, some studies used the contralateral non-exercised muscle as a control in muscle strengthening interventions (Aitkens et al., 1993; Kilmer et al., 1994; Tollback et al., 1999). The problem with this study design is that there may be confounding cross-over effects in the non-exercised muscles. Moreover, one can hardly expect meaningful effects of a single-limb training programme on a patient's activities, participation and well-being (Fowler Jr, 2002). Olsen et al. (2005) investigated the effect of aerobic training in eight patients with FSHD. Twelve weeks of low-intensity aerobic exercise improved maximal oxygen uptake and workload with no signs of muscle damage. The authors concluded that aerobic training is a safe method to increase exercise performance in patients with FSHD. Most importantly, only one study described the effect of strength training for experienced fatigue (van der Kooi et al., 2007; see the following section).

To conclude, aerobic exercise training appears not to be harmful in muscular dystrophies and could have a positive effect on functioning, activities and participation, but the number of high-quality studies is low.

Medication for muscle fatigue and experienced fatigue

No curative pharmacological interventions are available. Nevertheless, many agents have been proposed as a potential pharmacological treatment for decreasing muscle fatigue in muscular dystrophies. Creatine and 2-agonists have been studied most frequently. Creatine is a well-known nutritional supplement among athletes because it increases muscle force and lean body mass (Bemben and Lamont, 2005). Supplementation of creatine monohydrate might enhance muscle performance in patients with muscular dystrophy, as they tend to have lower skeletal muscle creatine levels (see the section on pathophysiological studies of muscle fatigue). A recent systematic review on creatine for treating muscle disorders concluded that short- to intermediate-term supplementation with creatine monohydrate in patients with muscular dystrophy may result in a significant, but minimal, increase in maximal isometric force of approximately 8.5% in quantitative muscle testing. Most of the potentially clinically relevant effects in muscular dystrophies were seen in dystrophinopathies (Kley et al., 2008).

Other investigators expected a positive effect of 2-agonists in decreasing muscle fatigue as high doses of 2-agonists have muscle anabolic properties. In animals and healthy volunteers, 2-adrenergic agonists, such as clenbuterol and albuterol, increase muscle strength and muscle mass, in particular when combined with strength training (Yang and McElligott, 1989; Maltin and Delday, 1992). Based on the model by Kalkman et al. (see the section on perpetuating factors of experienced fatigue) we might expect that these drugs can decrease experienced fatigue, as many of them are aimed at increasing muscle strength. However, in the study by van der Kooi et al., albuterol and a strength-training programme did not have any effect on experienced fatigue (van der Kooi et al., 2007). Moreover, 2-agonists are often associated with numerous undesirable side effects, including increased heart rate and muscle tremor, factors that have limited their therapeutical potential. Consequences of prolonged use are presently unclear (van der Kooi et al., 2004, 2007).

Thus, although creatine and albuterol appear to be effective in increasing muscle strength in healthy subjects, they seem to have little effect on muscle strength and experienced fatigue in muscular dystrophies.

Cognitive behaviour therapy

Muscular dystrophies have a large impact on psychosocial functioning as patients must continuously adapt to their progressive illness. Illness cognitions and coping styles influence the level of physical activity and, consequently, experienced fatigue and health status. Hence, changing illness cognitions and coping style may lead to a better quality of life. A cognitive behaviour approach has been proven successful in the chronic fatigue syndrome (Prins et al., 2001; Chambers et al., 2006) and for post-cancer fatigue (Gielissen et al., 2006, 2007b) and may be effective in patients with muscular dystrophy as well. Cognitive behaviour therapy in FSHD should, for instance, be focused on the known perpetuating factors of experienced fatigue as described by Kalkman et al., i.e. sleep disturbances, pain complaints and physical inactivity (Della et al., 2007; Kalkman et al., 2007b) (see the section on perpetuating factors of experienced fatigue). Therapy should be adapted to the life of each individual, resulting in an individualised treatment approach. Altogether, cognitive behaviour therapy seems a rational, promising treatment for fatigue in muscular dystrophies.

> **Clinical case: treatment of fatigue**
>
> Mr A has never used any drug for alleviating his fatigue. It has simply not been mentioned by any clinician. He believes that physical exercise is beneficial for decreasing his fatigue. In this perspective, he regrets that it is very difficult for him to exercise because of his obesity and poor cardiovascular condition. He is convinced that exercise at a maximum level will lead to overuse. He regards cognitive behaviour therapy as a promising intervention for him.
>
> **Clinical case: experienced fatigue in clinical setting**
>
> Mr A regrets that his experienced fatigue has never been asked for, nor treated by health workers. He regards his fatigue as an essential disability, whereas his physician has focused on the genetic aspects and diagnosis of the disease. He is happy that fatigue is now on the research and on the medical agenda.

CONCLUSION

Fatigue is not only a frequent, but also a very relevant symptom in patients with muscular dystrophy. Based on the content of this chapter, several recommendations for clinical practice and research can be made. Clinicians should actively ask their patients about the presence of fatigue and its individual characteristics. The nature of the experienced fatigue gives directions to its primary cause. In particular, affective disorders need to be diagnosed or ruled out. The CIS-fatigue questionnaire can be used to measure the severity of experienced fatigue.

The history taking should cover the possible perpetuating factors of fatigue and the perceived disabilities in daily life. Based on these perpetuating factors, a therapeutic intervention can be proposed. Both in MD and FSHD, specific attention should be paid to sleep disturbances. Specifically in FSHD, the level of physical activity should improve.

Aerobic exercise training appears to be a safe intervention to increase physical activity. It may also be effective in maximising functional ability and preventing chronic physical complications of inactivity. High-resistance and eccentric strength training should be avoided, particularly in muscular dystrophies caused by defects in structural proteins in the dystrophin–glycoprotein complex, i.e. LGMD, BMD and DMD.

In the future, research on fatigue in muscular dystrophies needs to investigate each type of muscular dystrophy separately, both pathophysiologically and therapeutically. All treatment strategies will ultimately depend on pathophysiology, but the absence of hard evidence should not prevent clinicians and researchers from investigating treatment options based on credible hypotheses.

FIVE KEY PAPERS THAT SHAPED THE TOPIC AREA

Study 1. Wineinger, M.A., Walsh, S.A. and Abresch, T. (2002). Muscle fatigue in animal models of neuromuscular disease. *American Journal of Physical Medicine and Rehabilitation*, 81, S81–98.

A review of Wineinger *et al.* showed that difference in fatigability between dystrophic animal muscles and muscles of healthy animals can largely be explained by differences in muscle fibre types. The dystrophic soleus muscle, which is largely composed of slow-twitch type I oxidative muscle fibres, fatigued more slowly or at the same rate as that of healthy animals. Yet, the dystrophic EDL, which has mainly type IIB fibres, was weaker than in healthy animals and generally more fatigable. No differences in fatigability were found between healthy and dystrophic diaphragm muscles. The diaphragm is composed of fast oxidative IIA muscle fibres, which are relatively fatigue resistant (Table 10.2).

Study 2. Kalkman, J.S., Schillings, M.L., Zwarts, M.J., van Engelen, B.G. and Bleijenberg, G. (2007b). The development of a model of fatigue in neuromuscular disorders: a longitudinal study. *Journal of Psychosomatic Research*, 62, 571–579.

Kalkman *et al.* investigated the perpetuating factors of experienced fatigue in patients with FSHD (n = 60) and MD (n = 70) (Figure 10.2). Only in FSHD did the level of physical (in)activity have a central place in the model. Lower levels of physical activity contributed to higher levels of experienced fatigue and, through that, to a lower health status. The level of physical activity was directly and negatively influenced by loss of muscle strength. In addition, pain complaints influenced levels of experienced fatigue, both directly and indirectly, by decreasing physical activity.

Study 3. Vignos, P.J. and Watkins, M.P. (1966). The effect of exercise in muscular dystrophy. *The Journal of the American Medical Association*, 197, 843–848.

The first controlled study of strengthening exercise in muscular dystrophies used a regimen with gradual increase in weight resistance in patients with DMD (n=14), FSHD (n=4) and LGMD (n=6) over a one-year period. The authors reported strength improvement throughout the first four months of exercise regardless of type of dystrophy. The degree of improvement was related to the initial strength of the exercised muscle. Therefore, they concluded that exercise programmes should begin early in the course of the disease.

Study 4. Florence, J.M. and Hagberg, J.M. (1984). Effect of training on the exercise responses of neuromuscular disease patients. *Medicine and Science in Sports and Exercise*, 16, 460–465.

Because the hallmark of muscular dystrophy is motor weakness, more studies have been conducted looking at muscle strength training than at aerobic exercise training. Only in 1984, Florence *et al.* first described a positive effect of aerobic exercise training on the exercise responses of neuromuscular disease patients among which patients with LGMD (n = 3) and FSHD (n = 1). All patients completed a 12-week cycle ergometry training programme, three days per week. The increase in $\dot{V}O_{2max}$ in the patients was almost the same as that of the healthy subjects (n = 4). The authors concluded that patients can develop relatively normal adaptations to training. No definitive deleterious effects of training were demonstrated in these patients.

Study 5. Lindeman, E., Leffers, P., Spaans, F., Drukker, J., Reulen, J., Kerckhoffs, M. and Koke, A. (1995). Strength training in patients with myotonic dystrophy and hereditary motor and sensory neuropathy: a randomized clinical trial. *Archives of Physical Medicine and Rehabilitation*, 76, 612–620.

The number of recent studies on the effect of training in muscular dystrophy lacking a randomised controlled design is striking. Lindeman *et al.* first conducted a randomised clinical trial on the effects of strength training in muscular dystrophy. This trial compared the effect of 24 weeks of strength training of the thigh muscles versus no training in 36 adult patients with myotonic dystrophy. The participants trained three times per week for 24 weeks, with weights adjusted to their force. In the MD patients, none of the outcomes showed any training effect. No serious side effects of the training occurred. Training loads could be gradually increased in all patients because the repetition maximum improved. Three rather weak MD patients were unable to perform the exercises according to the training instructions. Most of the differences in muscle strength outcomes (isometric, dynamic and endurance) between groups showed small, non-significant positive effects in favour of the training group. Only changes in the endurance measure (13.1 s longer maximum duration of an isometric contraction; 95% CI 2.2 to 24.0) reached statistical significance. No signs of overuse, such as a decline in strength or a rise in parameters of muscle membrane permeability were seen. However, this study imposed merely a controlled strain for a relatively short period.

GLOSSARY OF TERMS

AFQ	abbreviated fatigue questionnaire
AP	action potential
BDI-PC	Beck depression inventory primary care
BMD	Becker muscular dystrophy
CAF	central activation failure
CIS	Checklist Individual Strength
CK	creatine phosphokinase
DMD	Duchenne muscular dystrophy
DMPK gene	dystrophia myotonica protein kinase gene
EDL	extensor digitorum longus
FFM	fat-free mass
FSHD	facioscapulohumeral dystrophy
LGMD	limb girdle muscular dystrophies
MD	myotonic dystrophy
MVC	maximal voluntary contraction
PROMM	proximal myotonic myopathy
REE	resting energy expenditure
SCL-90	symptom checklist-90
SF-36	Short Form 36
SIP	sickness impact profile
TEE	total daily energy expenditure
$\dot{V}O_{2max}$	maximal oxygen uptake

REFERENCES

Aartsma-Rus, A., van Deutekom, J.C., Fokkema, I.F., van Ommen, G.J. and den Dunnen, J.T. (2006) Entries in the Leiden Duchenne muscular dystrophy mutation database: an overview of mutation types and paradoxical cases that confirm the reading-frame rule. *Muscle and Nerve*, 34, 135–144.

Abresch, R.T., Walsh, S.A. and Wineinger, M.A. (1998). Animal models of neuromuscular diseases: pathophysiology and implications for rehabilitation. *Physical Medicine and Rehabilitation Clinics of North America*, 9, 285–299.

Aitkens, S.G., McCrory, M.A., Kilmer, D.D. and Bernauer, E.M. (1993). Moderate resistance exercise program: its effect in slowly progressive neuromuscular disease. *Archives of Physical Medicine and Rehabilitation*, 74, 711–715.

Alberts, M., Smets, E.M., Vercoulen, J.H., Garssen, B. and Bleijenberg, G. (1997). "Abbreviated fatigue questionnaire": a practical tool in the classification of fatigue. *Nederlands Tijdschrift voor Geneeskunde*, 141, 1526–1530.

Bemben, M.G. and Lamont, H.S. (2005). Creatine supplementation and exercise performance: recent findings. *Sports Medicine*, 35, 107–125.

Bogdanovich, S., Perkins, K.J., Krag, T.O. and Khurana, T.S. (2004). Therapeutics for Duchenne muscular dystrophy: current approaches and future directions. *Journal of Molecular Medicine*, 82, 102–115.

Brussee, V., Tardif, F. and Tremblay, J.P. (1997). Muscle fibers of mdx mice are more vulnerable to exercise than those of normal mice. *Neuromuscular Disorders*, 7, 487–492.

Bushby, K., Norwood, F. and Straub V. (2007). The limb-girdle muscular dystrophies: diagnostic strategies. *Biochimica et Biophysica Acta*, 1772, 238–242.

Carter, G.T., Abresch, R.T. and Fowler Jr, W.M. (2002). Adaptations to exercise training and contraction-induced muscle injury in animal models of muscular dystrophy. *American Journal of Physical Medicine and Rehabilitation*, 81, S151–161.

Carter, G.T., Wineinger, M.A., Walsh, S.A., Horasek, S.J., Abresch, R.T. and Fowler Jr, W.M. (1995). Effect of voluntary wheel-running exercise on muscles of the mdx mouse. *Neuromuscular Disorders*, 5, 323–332.

Chambers, D., Bagnall, A.M., Hempel, S. and Forbes, C. (2006). Interventions for the treatment, management and rehabilitation of patients with chronic fatigue syndrome/myalgic encephalomyelitis: an updated systematic review. *Journal of the Royal Society of Medicine*, 99, 506–520.

Cossu, G. and Sampaolesi, M. (2007). New therapies for Duchenne muscular dystrophy: challenges, prospects and clinical trials. *Trends in Molecular Medicine*, 13, 520–526.

Culebras, A. (1996). Sleep and neuromuscular disorders. *Neurologic Clinics*, 14, 791–805.

Cup, E.H., Pieterse, A.J., Ten Broek-Pastoor, J.M., Munneke, M., van Engelen, B.G., Hendricks, H.T., van der Wilt, G.J. and Oostendorp, R.A. (2007). Exercise therapy and other types of physical therapy for patients with neuromuscular diseases: a systematic review. *Archives of Physical Medicine and Rehabilitation*, 88, 1452–1464.

Daniele, N., Richard, I. and Bartoli, M. (2007). Ins and outs of therapy in limb girdle muscular dystrophies. *The International Journal of Biochemistry and Cell Biology*, 39, 1608–1624.

Day, J.W., Ricker, K., Jacobsen, J.F., Rasmussen, L.J., Dick, K.A., Kress, W., Schneider, C., Koch, M.C., Beilman, G.J., Harrison, A.R., Dalton, J.C. and Ranum, L.P. (2003). Myotonic dystrophy type 2: molecular, diagnostic and clinical spectrum. *Neurology*, 60, 657–664.

De Die-Smulders, C.E., Howeler, C.J., Thijs, C., Mirandolle, J.F., Anten, H.B., Smeets, H.J., Chandler, K.E. and Geraedts, J.P. (1998). Age and causes of death in adult-onset myotonic dystrophy. *Brain*, 121, 1557–1563.

De Graaf, Bijl, R.V., Smit, F., Ravelli, A. and Vollebergh, W.A. (2000). Psychiatric and sociodemographic predictors of attrition in a longitudinal study: the Netherlands Mental Health Survey and Incidence Study (NEMESIS). *American Journal of Epidemiology*, 152, 1039–1047.

De Groot, I. (2006). Guideline on the use of corticosteroids in Duchenne muscular dystrophy from paediatric neurologists, neurologists and rehabilitation physicians. *Nederlands Tijdschrift voor Geneeskunde*, 150, 684–685.

Della, M.G., Frusciante, R., Vollono, C., Dittoni, S., Galluzzi, G., Buccarella, C., Modoni, A., Mazza, S., Tonali, P.A. and Ricci, E. (2007). Sleep quality in facioscapulohumeral muscular dystrophy. *Journal of the Neurological Sciences*, 263, 49–53.

Dick, J. and Vrbova, G. (1993). Progressive deterioration of muscles in mdx mice induced by overload. *Clinical Science (London)*, 84, 145–150.

Dupont-Versteegden, E.E., McCarter, R.J. and Katz, M.S. (1994). Voluntary exercise decreases progression of muscular dystrophy in diaphragm of mdx mice. *Journal of Applied Physiology*, 77, 1736–1741.

Eagle, M., Bourke, J., Bullock, R., Gibson, M., Mehta, J., Giddings, D., Straub, V. and Bushby, K. (2007). Managing Duchenne muscular dystrophy: the additive effect of spinal surgery and home nocturnal ventilation in improving survival. *Neuromuscular Disorders*, 17, 470–475.

Emery, A.E. (1991). Population frequencies of inherited neuromuscular diseases: a world survey. *Neuromuscular Disorders*, 1, 19–29.

Emery, A.E. (2002). The muscular dystrophies. *Lancet*, 359, 687–695.

Engel, A.G. and Franzini-Armstrong, C. (2004). *Myology*, Vol. II. New York: McGraw-Hill.

Florence, J.M. and Hagberg, J.M. (1984). Effect of training on the exercise responses of neuromuscular disease patients. *Medicine and Science in Sports and Exercise*, 16, 460–465.

Fowler Jr, W.M. (1984). Importance of overwork weakness. *Muscle and Nerve*, 7, 496–499.

Fowler Jr, W.M. (2002). Role of physical activity and exercise training in neuromuscular diseases. *American Journal of Physical Medicine and Rehabilitation*, 81, S187–195.

Gandevia, S.C. (2001). Spinal and supraspinal factors in human muscle fatigue. *Physiological Reviews*, 81, 1725–1789.

George, A., Schneider-Gold, C., Zier, S., Reiners, K. and Sommer, C. (2004). Musculoskeletal pain in patients with myotonic dystrophy type 2. *Archives of Neurology*, 61, 1938–1942.

Gielissen, M.F., Knoop, H., Servaes, P., Kalkman, J.S., Huibers, M.J., Verhagen, S. and Bleijenberg, G. (2007a). Differences in the experience of fatigue in patients and healthy controls: patients' descriptions. *Health and Quality of Life Outcomes*, 5, 36.

Gielissen, M.F., Verhagen, C.A. and Bleijenberg, G. (2007b). Cognitive behaviour therapy for fatigued cancer survivors: long-term follow-up. *British Journal of Cancer*, 97, 612–618.

Gielissen, M.F., Verhagen, S., Witjes, F. and Bleijenberg, G. (2006). Effects of cognitive behavior therapy in severely fatigued disease-free cancer patients compared with patients waiting for cognitive behavior therapy: a randomized controlled trial. *Journal of Clinical Oncology*, 24, 4882–4887.

Griffin, J.L., Williams, H.J., Sang, E., Clarke, K., Rae, C. and Nicholson, J.K. (2001). Metabolic profiling of genetic disorders: a multi-issue (1)H nuclear magnetic resonance spectroscopic and pattern recognition study into dystrophic tissue. *Analytical Biochemistry*, 293, 16–21.

Griggs, R.C. and Bushby, K. (2005). Continued need for caution in the diagnosis of Duchenne muscular dystrophy. *Neurology*, 64, 1498–1499.

Harper, P.S., Van Engelen, B.G.M.E.B. and Wilcox, D.E. (2004). *Myotonic Dystrophy: Present Management and Future Therapy*. New York: Oxford University Press.

Hartling, L., McAlister, F.A., Rowe, B.H., Ezekowitz, J., Friesen, C. and Klassen, T.P. (2005). Challenges in systematic reviews of therapeutic devices and procedures. *Annals of Internal Medicine*, 142, 1100–1111.

Howeler, C.J., Busch, H.F., Geraedts, J.P., Niermeijer, M.F. and Staal, A. (1989). Anticipation in myotonic dystrophy: fact or fiction? *Brain*, 112, 779–797.

Ingjer, F. (1979). Effects of endurance training on muscle fibre ATP-ase activity, capillary supply and mitochondrial content in man. *Journal of Physiology*, 294, 419–432.

Johnson, E.R., Abresch, R.T., Carter, G.T., Kilmer, D.D., Fowler Jr, W.M., Sigford, B.J. and Wanlass, R.L. (1995). Profiles of neuromuscular diseases: myotonic dystrophy. *American Journal of Physical Medicine and Rehabilitation*, 74, S104–116.

Johnson, E.W. and Braddom, R. (1971). Over-work weakness in facioscapulohumeral muscular dystrophy. *American Journal of Physical Medicine and Rehabilitation*, 52, 333–336.

Kalkman, J. (2006). From prevalence to predictors of fatigue in neuromuscular disorders, the building of a model. Unpublished PhD thesis, Radboud University Nijmegen.

Kalkman, J.S., Schillings, M.L., van der Werf, S.P., Padberg, G.W., Zwarts, M.J., van Engelen, B.G. and Bleijenberg, G. (2005). Experienced fatigue in facioscapulohumeral dystrophy, myotonic dystrophy, and HMSN-I. *Journal of Neurology, Neurosurgery and Psychiatry*, 76, 1406–1409.

Kalkman, J.S., Schillings, M.L., Zwarts, M.J., van Engelen, B.G. and Bleijenberg, G. (2007a). Psychiatric disorders appear equally in patients with myotonic dystrophy, facioscapulohumeral dystrophy, and hereditary motor and sensory neuropathy type I. *Acta Neurologica Scandinavica*, 115, 265–270.

Kalkman, J.S., Schillings, M.L., Zwarts, M.J., van Engelen, B.G. and Bleijenberg, G. (2007b). The development of a model of fatigue in neuromuscular disorders: a longitudinal study. *Journal of Psychosomatic Research*, 62, 571–579.

Kilmer, D.D., Abresch, R.T., McCrory, M.A., Carter, G.T., Fowler Jr, W.M., Johnson, E.R. and McDonald, C.M. (1995). Profiles of neuromuscular diseases: facioscapulohumeral muscular dystrophy. *American Journal of Physical Medicine and Rehabilitation*, 74, S131–139.

Kilmer, D.D., McCrory, M.A., Wright, N.C., Aitkens, S.G. and Bernauer, E.M. (1994). The effect of a high resistance exercise program in slowly progressive neuromuscular disease. *American Journal of Physical Medicine and Rehabilitation*, 75, 560–563.

Kirschner, J. and Bonnemann, C.G. (2004). The congenital and limb-girdle muscular dystrophies: sharpening the focus, blurring the boundaries. *Archives of Neurology*, 61, 189–199.

Kley, R.A., Tarnopolsky, M.A. and Vorgerd, M. (2008). Creatine treatment in muscle disorders: a meta-analysis of randomised controlled trials. *Journal of Neurology, Neurosurgery and Psychiatry*, 79, 366–367.

Kohler, J., Rohrig, D., Bathke, K.D. and Koch, M.C. (1999). Evaluation of the facioscapulohumeral muscular dystrophy (FSHD1) phenotype in correlation to the concurrence of 4q35 and 10q26 fragments. *Clinical Genetics*, 55, 88–94.

Lapidos, K.A., Kakkar, R. and McNally, E.M. (2004). The dystrophin glycoprotein complex: signaling strength and integrity for the sarcolemma. *Circulation Research*, 94 1023–1031.

Lewis, S.F. and Haller, R.G. (1989). Skeletal muscle disorders and associated factors that limit exercise performance. *Exercise and Sport Sciences Reviews*, 17, 67–113.

Lindeman, E., Leffers, P., Spaans, F., Drukker, J., Reulen, J., Kerckhoffs, M. and Koke, A. (1995). Strength training in patients with myotonic dystrophy and hereditary motor and sensory neuropathy: a randomized clinical trial. *Archives of Physical Medicine and Rehabilitation*, 76, 612–620.

McArdle, A., Edwards, R.H. and Jackson, M.J. (1992). Accumulation of calcium by normal and dystrophin-deficient mouse muscle during contractile activity in vitro. *Clinical Science (London)*, 82, 455–459.

McCrory, M.A., Kim, H.R., Wright, N.C., Lovelady, C.A., Aitkens, S. and Kilmer, D.D. (1998). Energy expenditure, physical activity, and body composition of ambulatory adults with hereditary neuromuscular disease. *American Journal of Clinical Nutrition*, 67, 1162–1169.

McDonald, C.M. (2002). Physical activity, health impairments, and disability in neuromuscular disease. *American Journal of Physical Medicine and Rehabilitation*, 81, S108–120.

McDonald, C.M., Johnson, E.R., Abresch, R.T., Carter, G.T., Fowler Jr, W.M. and Kilmer, D.D. (1995). Profiles of neuromuscular diseases: limb-girdle syndromes. *American Journal of Physical Medicine and Rehabilitation*, 74, S117–130.

Machuca-Tzili, L., Brook, D. and Hilton-Jones, D. (2005). Clinical and molecular aspects of the myotonic dystrophies: a review. *Muscle and Nerve*, 32, 1–18.

Maltin, C.A. and Delday, M.I. (1992). Satellite cells in innervated and denervated muscles treated with clenbuterol. *Muscle and Nerve*, 15, 919–925.

Manzur, A.Y., Kuntzer, T., Pike, M. and Swan, A. (2008). Glucocorticoid corticosteroids for Duchenne muscular dystrophy. *Cochrane Database of Systematic Reviews*, CD003725.

Mathieu, J., Allard, P., Potvin, L., Prevost, C. and Begin, P. (1999). A 10-year study of mortality in a cohort of patients with myotonic dystrophy. *Neurology*, 52, 1658–1662.

Monaco, A.P., Bertelson, C.J., Liechti-Gallati, S., Moser, H. and Kunkel, L.M. (1988). An explanation for the phenotypic differences between patients bearing partial deletions of the DMD locus. *Genomics*, 2, 90–95.

Muntoni, F. (2001). Is a muscle biopsy in Duchenne dystrophy really necessary? *Neurology*, 57, 574–575.

Muntoni, F. (2003). Cardiomyopathy in muscular dystrophies. *Current Opinion in Neurology*, 16, 577–583.

Muntoni, F., Torelli, S. and Ferlini, A. (2003). Dystrophin and mutations: one gene, several proteins, multiple phenotypes. *Lancet Neurology*, 2, 731–740.

Norwood, F., de Visser, M., Eymard, B., Lochmuller, H. and Bushby, K. (2007). EFNS guideline on diagnosis and management of limb girdle muscular dystrophies. *European Journal of Neurology*, 14, 1305–1312.

Olsen, D.B., Orngreen, M.C. and Vissing, J. (2005). Aerobic training improves exercise performance in facioscapulohumeral muscular dystrophy. *Neurology*, 64, 1064–1066.

Padberg, G.W. (1982). Facioscapulohumeral disease. Unpublished thesis, University of Leiden.

Pagala, M.K., Venkatachari, S.A., Nandakumar, N.V., Ravindran, K., Kerstein, J., Namba, T. and Grob, D. (1991). Peripheral mechanisms of fatigue in muscles of normal and dystrophic mice. *Neuromuscular Disorders*, 1, 287–298.

Petrof, B.J. (1998). The molecular basis of activity-induced muscle injury in Duchenne muscular dystrophy. *Molecular and Cellular Biochemistry*, 179, 111–123.

Petrof, B.J., Shrager, J.B., Stedman, H.H., Kelly, A.M. and Sweeney, H.L. (1993). Dystrophin protects the sarcolemma from stresses developed during muscle contraction. *Proceedings of the National Academy of Sciences USA*, 90, 3710–3714.

Phillips, M.F., Steer, H.M., Soldan, J.R., Wiles, C.M. and Harper, P.S. (1999). Daytime somnolence in myotonic dystrophy. *Journal of Neurology*, 246, 275–282.

Prins, J.B., Bleijenberg, G., Bazelmans, E., Elving, L.D., de Boo, T.M., Severens, J.L., van der Wilt, G.J., Spinhoven, P. and van der Meer, J.W. (2001). Cognitive behaviour therapy for chronic fatigue syndrome: a multicentre randomised controlled trial. *Lancet*, 357, 841–847.

Reardon, W., Newcombe, R., Fenton, I., Sibert, J. and Harper, P.S. (1993). The natural history of congenital myotonic dystrophy: mortality and long term clinical aspects. *Archives of Disease in Childhood*, 68, 177–181.

Ricci, E., Galluzzi, G., Deidda, G., Cacurri, S., Colantoni, L., Merico, B., Piazzo, N., Servidei, S., Vigneti, E., Pasceri, V., Silvestri, G., Mirabella, M., Mangiola, F., Tonali, P. and Felicetti, L. (1999). Progress in the molecular diagnosis of facioscapulohumeral muscular dystrophy and correlation between the number of KpnI repeats at the 4q35 locus and clinical phenotype. *Annals of Neurology*, 45, 751–757.

Rose, M.R. and Tawil, R. (2004). Drug treatment for facioscapulohumeral muscular dystrophy. *Cochrane Database of Systematic Reviews*, CD002276.

Sacco, P., Jones, D.A., Dick, J.R. and Vrbova, G. (1992). Contractile properties and susceptibility to exercise-induced damage of normal and mdx mouse tibialis anterior muscle. *Clinical Science (London)*, 82, 227–236.

Schara, U. and Schoser, B.G. (2006). Myotonic dystrophies type 1 and 2: a summary on current aspects. *Seminars in Pediatric Neurology*, 13, 71–79.

Schillings, M.L., Kalkman, J.S., Janssen, H.M., van Engelen, B.G., Bleijenberg, G. and Zwarts, M.J. (2007). Experienced and physiological fatigue in neuromuscular disorders. *Clinical Neurophysiology*, 118, 292–300.

Schrag, A. (2004). Psychiatric aspects of Parkinson's disease: an update. *Journal of Neurology*, 251, 795–804.

Schulte-Mattler, W.J., Muller, T., Deschauer, M., Gellerich, F.N., Iaizzo, P.A. and Zierz, S. (2003). Increased metabolic muscle fatigue is caused by some but not all mitochondrial mutations. *Archives of Neurology*, 60, 50–58.

Shahrizaila, N. and Wills, A.J. (2005). Significance of Beevor's sign in facioscapulohumeral dystrophy and other neuromuscular diseases. *Journal of Neurology, Neurosurgery & Psychiatry*, 76, 869–870.

Sharma, U., Atri, S., Sharma, M.C., Sarkar, C. and Jagannathan, N.R. (2003a). Biochemical characterization of muscle tissue of limb girdle muscular dystrophy: an 1H and 13C NMR study. *NMR in Biomedicine*, 16, 213–223.

Sharma, U., Atri, S., Sharma, M.C., Sarkar, C. and Jagannathan, N.R. (2003b). Skeletal muscle metabolism in Duchenne muscular dystrophy (DMD): an in-vitro proton NMR spectroscopy study. *Magnetic Resonance Imaging*, 21, 145–153.

Sharpe, M. and Wilks, D. (2002). Fatigue. *British Medical Journal*, 325, 480–483.

Surtees, P.G., Wainwright, N.W., Khaw, K.T. and Day, N.E. (2003). Functional health status, chronic medical conditions and disorders of mood. *British Journal of Psychiatry*, 183, 299–303.

Tawil, R. and van der Maarel, S.M. (2006). Facioscapulohumeral muscular dystrophy. *Muscle and Nerve*, 34, 1–15.

Tollback, A., Eriksson, S., Wredenberg, A., Jenner, G., Vargas, R., Borg, K. and

Ansved, T. (1999). Effects of high resistance training in patients with myotonic dystrophy. *Scandinavian Journal of Rehabilitation Medicine*, 31, 9–16.

Van der Kooi, E.L., de Greef, J.C., Wohlgemuth, M., Frants, R.R., van Asseldonk, R.J., Blom, H.J., van Engelen, B.G., van der Maarel, S.M. and Padberg, G.W. (2006). No effect of folic acid and methionine supplementation on D4Z4 methylation in patients with facioscapulohumeral muscular dystrophy. *Neuromuscular Disorders*, 6, 766–769.

Van der Kooi, E.L., Kalkman, J.S., Lindeman, E., Hendriks, J.C., van Engelen, B.G., Bleijenberg, G. and Padberg, G.W. (2007). Effects of training and albuterol on pain and fatigue in facioscapulohumeral muscular dystrophy. *Journal of Neurology*, 254, 931–940.

Van der Kooi, E.L., Lindeman, E. and Riphagen, I. (2005). Strength training and aerobic exercise training for muscle disease. *Cochrane Database of Systematic Reviews*, CD003907.

Van der Kooi, E.L., Vogels, O.J., van Asseldonk, R.J., Lindeman, E., Hendriks, J.C., Wohlgemuth, M., van der Maarel, S.M. and Padberg, G.W. (2004). Strength training and albuterol in facioscapulohumeral muscular dystrophy. *Neurology*, 63, 702–708.

Van der Werf, S.P. (2003). Determinants and consequences of experienced fatigue in chronic fatigue syndrome and neurological conditions. Unpublished thesis, University of Nijmegen.

Van Deutekom, J.C., Janson, A.A., Ginjaar, I.B., Frankhuizen, W.S., Artsma-Rus, A., Bremmer-Bout, M., den Dunnen, J.T., Koop, K., van der Kooi, A.J., Goemans, N.M., de Kimpe, S.J., Ekhart, P.F., Venneker, E.H., Platenburg, G.J., Verschuuren, J.J. and van Ommen, G.J. (2007). Local dystrophin restoration with antisense oligonucleotide PRO051. *New England Journal of Medicine*, 357, 2677–2686.

Van Tulder, M.W., Assendelft, W.J., Koes, B.W. and Bouter, L.M. (1997). Method guidelines for systematic reviews in the Cochrane Collaboration Back Review Group for Spinal Disorders. *Spine*, 22, 2323–2330.

Vercoulen, J.H., Swanink, C.M., Fennis, J.F., Galama, J.M., van der Meer, J.W. and Bleijenberg, G. (1994). Dimensional assessment of chronic fatigue syndrome. *Journal of Psychosomatic Research*, 38, 383–392.

Vignos, P.J., and Watkins, M.P. (1966). The effect of exercise in muscular dystrophy. *The Journal of the American Medical Association*, 197, 843–848.

Vitelli, F., Villanova, M., Malandrini, A., Bruttini, M., Piccini, M., Merlini, L., Guazzi, G. and Renieri, A. (1999). Inheritance of a 38-kb fragment in apparently sporadic facioscapulohumeral muscular dystrophy. *Muscle and Nerve*, 22, 1437–1441.

Walter, M.C., Lochmuller, H., Reilich, P., Klopstock, T., Huber, R., Hartard, M., Hennig, M., Pongratz, D. and Muller-Felber, W. (2000). Creatine monohydrate in muscular dystrophies: a double-blind, placebo-controlled clinical study. *Neurology*, 54, 1848–1850.

Wineinger, M.A., Walsh, S.A. and Abresch, T. (2002). Muscle fatigue in animal models of neuromuscular disease. *American Journal of Physical Medicine and Rehabilitation*, 81, S81–98.

Yang, Y.T. and McElligott, M.A. (1989). Multiple actions of beta-adrenergic agonists on skeletal muscle and adipose tissue. *Biochemical Journal*, 261, 1–10.

Zatz, M., de Paula, F., Starling, A. and Vainzof, M. (2003). The 10 autosomal recessive limb-girdle muscular dystrophies. *Neuromuscular Disorders*, 13, 532–544.

CHAPTER 11

FATIGUE AND WEAKNESS IN PATIENTS WITH INFLAMMATORY MYOPATHIES: DERMATOMYOSITIS, POLYMYOSITIS AND INCLUSION BODY MYOSITIS

Jane H. Park, Brittany C. Lee and Nancy J. Olsen

OBJECTIVES

The objectives of this chapter are to:

- characterise fatigue and weakness in myositis patients with dermato-myositis, polymyositis or inclusion body myositis;
- consider the aetiology of muscle dysfunction and abnormalities in patients with myositis;
- suggest methods for alleviation of fatigue and weakness in order to improve muscle strength, endurance and overall quality of life.

INTRODUCTION

Idiopathic inflammatory myopathies, including dermatomyositis (DM), polymyositis (PM) and inclusion body myositis (IBM), are chronic, debilitating diseases which are primarily characterised by weakness, fatigue and myalgia (Dalakas, 1991). The underlying reasons for these disabilities are not clearly understood. Symptoms can be partially alleviated by treatment with prednisone and immunosuppressive medications like methotrexate and/or azathioprine. However, most adult patients retain some disability and do not return to their original state of strength and endurance (Harris-Love, 2003). Thus, inflammatory myopathies have long-term effects on patients' lifestyles and employment possibilities. Since muscle strength and fatigue are the clinical features most readily followed during treatment, extensive investigations have provided applicable clinical tests and basic physiological insights into these aspects of inflammatory myopathies. Weakness or strength is evaluated with the ability to lift weights, and fatigue or endurance is measured by patient self-assessments such as the health assessment questionnaire (HAQ), or by testing the effects of prolonged exercise (Dastmalchi *et al.*, 2007).

CLINICAL EVALUATION

Dermatomyositis and polymyositis show symmetrical proximal muscle weakness in the thighs, shoulders and neck, whereas inclusion body myositis demonstrates weakness and atrophy in both proximal and distal muscles. These three diseases are diagnosed by clinical examination and laboratory tests. Elevated levels of serum enzymes, namely creatine phosphokinase (CK), aldolase, lactic dehydrogenase (LDH) and aspartate and alanine aminotransferases (AST and ALT), suggest muscle degeneration and fragmentation. However, serum enzyme levels, particularly CK, can be normal or elevated by non-myopathic factors such as CK inhibitors in the serum, intense exercise, gender and race. In addition, muscle enzyme levels from biopsies have documented defects in respiratory chain enzyme activities and increased succinic dehydrogenase (Rifai *et al.*, 1995; Chariot *et al.*, 1996). Muscle damage is further verified by abnormal EMG and biopsy. Distinguishing histological features of the diseases as demonstrated by biopsy are well-recognised. DM shows immune processes of perivascular distributions of inflammatory cells, primarily B-lymphocytes and T4+ cells, against capillaries and secondarily against muscle necrosis. PM involves a cell-mediated cytotoxicity by T-lymphocytes, especially T8+ cytotoxic cells and macrophages. IBM myotoxicity is mediated by T-cells producing intranuclear tubular inclusions (Wortmann, 2005). Biopsy has an advantage over magnetic resonance imaging (MRI) in that it can distinguish the type of inflammatory infiltration and the features of the vascular system.

MAGNETIC RESONANCE IMAGING OF MUSCLE MORPHOLOGY

Since biopsy and EMG are painful and not suitable for long-term evaluation of chronic muscle diseases, MRI is now used to evaluate muscle morphology as related to weakness and fatigue (Park *et al.*, 1990; Dalakas, 1991). MRI provides an image of the entire thigh or shoulder, and thereby avoids the error of a negative biopsy within a small, restricted area. As a non-invasive technique, longitudinal MRI examinations are utilised for therapeutic evaluations of individual thigh muscles (Park *et al.*, 1994). Examples of T1- and T2-weighted images are shown in Figure 11.1 to illustrate inflammation and progressing stages of fat infiltration in affected patients.

T1-weighted images (left-hand column) are rapidly acquired images and do not show inflammation. This can be seen in the similarity of the signal intensity in the thigh of a normal control (A) and a DM patient (C). By contrast, T2-weighted images (right-hand column) are acquired four times more slowly and demonstrate high signal intensity (brightness) in the DM vastus muscles (D), presumably the mobile water of inflammation. For the PM patient (E), T1-weighted images show infiltrated fat with rapidly rotating methyl protons, as illustrated in the vastus muscles of quadriceps and also in the superficial fat of all subjects. T2-weighted images (F) show fat as somewhat darkened, but the vastus muscle itself is lightened by overlying inflammation. This can be definitively observed in the STIR images (Figure 11.2), which were acquired with a sequence that suppresses (blackens) the fat signal and thereby enhances the bright signal of the inflammation (Dalakas, 1991). The chronic PM patient (G) shows total fat replacement in the quadriceps, but preservation of the rectus femoris, biceps femoris, semimembranosus, gracillus and sartorius. This 70-year-old patient went to the gym five times per week and was able to walk and exercise with his preserved muscles at an unexpectedly high level of function. IBM patients were not included in our discussion as they represent only 15% of this myositis group and are a different clinical phenotype (Alexanderson *et al.*, 2007).

MRI quantification with T1 and T2 relaxation times

Signal intensity differences between the patients can be quantitatively evaluated by calculation of T1 and T2 relaxation times within a defined region of a given muscle. Inflammation within a muscle produces high T1 and T2 values (Table 11.1) (Park *et al.*, 1990, 1994). By contrast, fat infiltration shows a substantially lower T1 value and a high T2 value. These calculations are not only useful for characterising the severity and extent of inflammation or fat infiltration, but also for evaluating responses to

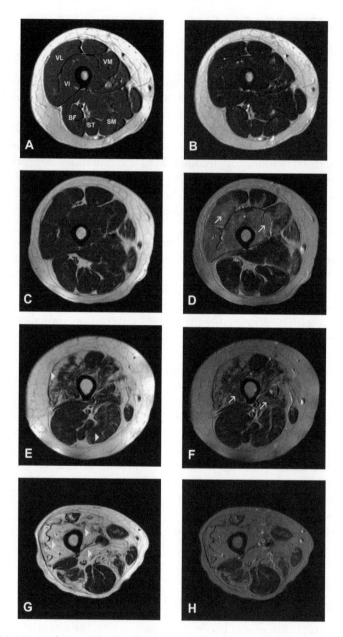

Figure 11.1 T1- and T2-weighted images of the thigh of control, DM and PM patients. T1- and T2-weighted images of a control subject (A, B) and DM (C,D), and PM (E, F) and chronic PM patients (G, H). The left column illustrates T1-weighted images (A, C, E,G), and the right column shows corresponding T2-weighted images (B, D, F, H). Arrowheads on T1-weighted images indicate fat, and arrows on T2-weighted images indicate inflammation. VL: vastus lateralis; VI: vastus intermedius; VM: vastus medialis; BF: biceps femoris; ST: semitendinosus; SM: semimembranosus; MRI: T1-weighted images; TR/TE=500/20, 80; T2-weighted images, TR/TE=2000/20, 80

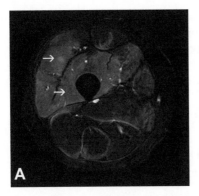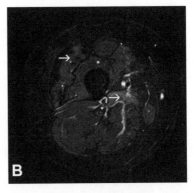

Figure 11.2 STIR images of DM and PM thighs. The STIR image of DM patient (A) shows inflammation in the quadriceps muscles. For the PM patient (B), both suppression of the fat signal and the enhancement of the inflammation in the vastus lateralis and adductor magnus (arrows)

ongoing therapies. When T1 and T2 calculations are confounded with infiltrated fat in the presence of inflammation, a STIR image can be quantitatively measured by increased signal intensity to show selected areas of inflammation (Figure 11.2) (Dalakas, 1991). Self-assessment measurements of fatigue and weakness on a 10 cm visual analogue scale (VAS) indicate severe problems with functional activities (Table 11.1). Although the perception of fatigue and weakness varies among patients, longitudinal evaluation by an individual patient may be particularly revealing.

The focal characteristics of inflammation and/or fat infiltration are strikingly demonstrated in MRI images (Figure 11.1). In general, the anterior quadriceps muscles are more affected than the posterior hamstring muscles, which may have normal relaxation values (Figure 11.1: D, F). This accounts for the proximal weakness observed in rising from a chair or getting out of a car. Although the pattern seems well-defined in the muscles of these three patients, selection of different muscles and distribution of inflammation has been noted. DM cases have been described in which

Table 11.1 T1 and T2 relaxation times and fatigue and weakness scores of control and diseased muscles

Subject (images)	Vastus lateralis	T1 (ms)	T2 (ms)	Fatigue	Weakness
Control (A, B)	Normal	1,396	29	0	0
DM (C, D)	Inflammation	1,806	49	5.8	3.9
PM (E, F)	Fat infiltration	569	44	8.5	7.6
Chronic PM (H, I)	Fat replacement	370	46	6.6	7.1

Note
High T1 and T2 relaxation times are indicative of inflammation (DM). Fat infiltration and replacement are shown by low T1 and high T2 relaxation times. Fatigue and weakness were quantified on a visual analog scale, with 0 meaning no fatigue or weakness and 10 meaning severe fatigue or weakness.

inflammation has been suppressed by immune therapy, but the patient still reported weakness and fatigue (Newman and Kurland, 1992; Park *et al.*, 1994). These findings demonstrate that functional ability does not always correspond with image abnormalities; therefore, investigations with magnetic resonance spectroscopy (^{31}P MRS) have been conducted to evaluate metabolic abnormalities.

METABOLIC ABNORMALITIES RELATED TO WEAKNESS AND FATIGUE

Non-invasive ^{31}P MRS is useful for characterisation of metabolic abnormalities in the muscles of patients with inflammatory myopathies. ^{31}P MRS monitors metabolic status by determining levels of the high-energy phosphate compounds, ATP and phosphocreatine (PCr), which are required for muscle contraction (Park *et al.*, 1990, 1994; Lodi *et al.*, 1998; Cea *et al.*, 2002; Pfleiderer *et al.*, 2004). Figure 11.3 shows ^{31}P spectra from a myositis patient and control subject at rest. Patients have lower P$_i$, PCr and ATP peaks compared to control subjects. The concentration of substrates is calculated from the determination of the area under the curve and the application of appropriate correction constants (Park *et al.*, 1990, 1994).

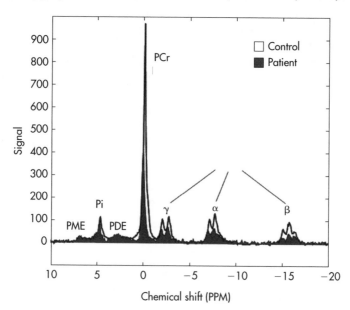

Figure 11.3 Comparison of ^{31}P MRS spectra of the quadriceps muscles of a normal control and an inflammatory myopathy patient at rest. The positions of the P$_i$ and PCr peaks, along with the α-, β- and γ-phosphate peaks of ATP, PDE and PME are shown. The muscles of the patient had lower peaks for PCr and ATP than did those of the control subject

Studies have shown that at rest, PCr and ATP are approximately 38% lower in both DM (Park *et al.*, 1995, 2008) and PM (Park *et al.*, 2008) patients compared to controls.

During the course of exercise, PCr levels in patients decrease in order to maintain constant ATP levels (Figure 11.4). During exercise at 25% MVC, P_i/PCr ratios are 56% greater in DM patients compared to controls, and 52% greater at 50% MVC. Elevated P_i/PCr ratios indicate a deficiency in DM muscles for generation and maintenance of ATP. PM patients and controls have similar ratios at 25% MVC exercise, and at 50% MVC, PM patients' ratios are 20% less than controls. The similar ratios between controls and PM patients do not mean that the two groups are equally efficient. The fact that PM patients lifted less than half the weight by controls (8 lbs versus 21.5 lbs) means that diseased muscles required less PCr for contraction. The calculations of ratios may be misleading unless the quantitative values for the metabolites are determined. The work/energy cost ratios (weight lifted/[P_i/PCr]) of DM and PM patients are lower than those of normal controls, but ratios between patient groups are not different. Though DM patients have higher P_i/PCr ratios, they lifted more weight than PM patients (15 lbs versus 8 lbs), thus making the work/energy cost ratios approximately the same. These MRS findings suggest that, with regard to muscle function, DM patients are more metabolically active than PM patients.

^{31}P MRS can also determine magnesium (Mg^{2+}) levels in muscle. Mg^{2+} is important because it is a required cofactor for all enzymatic reactions involving ATP, and for mitochondrial membrane stability. Also, Mg deficiency can lead to muscle weakness (Cronin and Knochel, 1983). ^{31}P MRS measurements of Mg levels in muscles are determined by the chemical shift of the β-phosphate peak of ATP when it binds to free Mg^{2+} and forms the enzymatically active MgATP complex (Figure 11.5). Free Mg^{2+} levels represent the biologically active form of Mg *in vivo*.

In previous studies, MgATP levels in DM and PM patients were found to be approximately 40% below control values at rest (Figure 11.6) (Niermann *et al.*, 2002; Qi *et al.*, 2007, 2008a). During exercise, DM patients lose about 15% of their MgATP, while PM patients with a very low workload were able to maintain their levels throughout the exercise protocol. Loss of MgATP during exercise represents a serious defect in metabolism. ATP is degraded to ADP and then to AMP, which can leave the muscle cell as adenosine. To subsequently rebuild ATP without the adenine starting blocks can require 2–3 hours, resulting in a prolonged period of significant fatigue. Juvenile DM (JDM) patients have MgATP levels that are 32% below juvenile controls at rest, but, unlike the adult DM patients, exercise does not decrease levels of this compound in children (Niermann *et al.*, 2002).

Levels of free Mg^{2+} in resting DM or PM muscles were not significantly different from control muscles. During exercise, free Mg^{2+} in the muscles

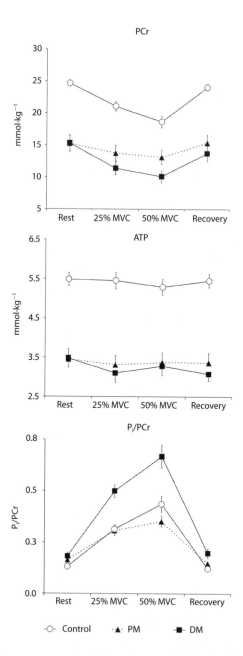

Figure 11.4 PCr and ATP concentrations and P_i/PCr ratios in control subjects and DM and PM patients during rest, exercise and recovery. Values are mean ± SEM mmol·kg^{-1} wet weight muscle. DM and PM patients had less PCr and ATP compared to control subjects throughout the exercise protocol. DM patients had higher P_i/PCr ratios compared to controls, indicating problems with oxidative phosphorylation. PM patients had similar or lower P_i/PCr ratios compared to controls because they lifted substantially less weight and therefore had less PCr utilisation

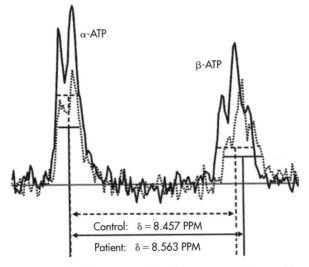

Figure 11.5 Chemical shift of the β-ATP peak in the spectrum of the quadriceps muscles of an inflammatory myopathy patient. The β-ATP peak in the spectrum for the control subject (solid line) exhibits a symmetrical triplicate shape, indicative of MgATP complex. In the patient's spectrum (dotted line), the β-ATP peak is less symmetrical and shifted to the right, indicating decreased Mg binding. The distance (δ) from the half-height midpoint of the α-peak to the quarter-height midpoint of the β-peak was utilised for the calculations of Mg

of adult and juvenile controls rose by 17% or 30%, respectively. In contrast, free Mg^{2+} in DM and JDM patients decreased significantly to approximately 40% of their respective control values and remained low, even during recovery. These low levels of free Mg^{2+} may be related to weakness and fatigue since this cation is the biologically active magnesium in muscle (Welch and Ramadan, 1995). It is required for optimisation of enzyme reactions, maintenance of membrane integrity, stabilisation of mitochondria, regulation of ATP-dependent ion channels of K^+, Na^+, and Ca^{2+} transport (Welch and Ramadan, 1995; Irish *et al.*, 1997; Blazev and Lamb, 1999). Low levels of Mg^{2+} may result in hypercontractility of smooth muscle and arteriolar vasoconstriction, which may produce hypoperfusion of oxygen and substrates (Blazev and Lamb, 1999). Thus, muscle dysfunction could be induced by magnesium deficiencies through multiple mechanisms.

Treatment of myositis patients with prednisone and immunosuppressive drugs results in increases in levels of MgATP and free Mg^{2+} in muscle tissue. Increases in magnesium are concordant with improved strength and endurance (Niermann *et al.*, 2002). Therefore, MgATP and free Mg^{2+} may be significant factors in the pathophysiology of myositic diseases.

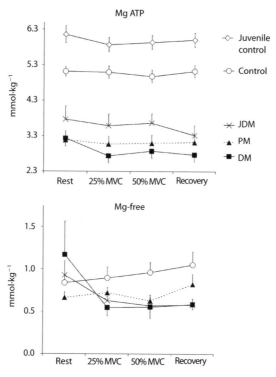

Figure 11.6 Differences in MgATP and free Mg^{2+} concentrations in control subjects and DM, JDM and PM patients during rest, exercise and recovery. Values are mean ± SEM mmol·kg wet weight muscle. Control subjects have higher concentrations of MgATP than DM, JDM and PM patients throughout the exercise protocol. Juvenile controls also had greater concentrations of MgATP compared to their adult control counterparts. Free Mg^{2+} concentrations increased over the course of exercise in control subjects, but decreased in DM, JDM and PM patients

ADDITIONAL FACTORS RELATED TO FATIGUE AND WEAKNESS

Lung involvement

In dermatomyositis and polymyositis, extramuscular organ involvement is common, such as a typical erythematous skin rash in dermatomyositis, as well as involvement with the lungs, heart, and joints. Lung involvement has been found in two separate studies to affect approximately 75% DM/PM population, primarily as interstitial lung disease (ILD) (Fathi *et al.*, 2007, 2008). The high frequency of ILD in this population suggest that chest X-ray, high resolution computed tomography and lung function tests be performed in DM/PM patients. ILD responds to treatment with prednisone and adjunct immunosuppressive drugs such as methotrexate and azathioprine. The hypoventilation associated with ILD and weakness of respiratory muscles will clearly decrease oxidative phosphorylation in the mitochondria of muscles. In extreme cases, progressive lung disease may result in atrophy and fibrosis of diseased muscles. Lung disorders may lead to decreased patient activity which, in turn, may result in muscle dysfunction. However, cardiac involvement is rare in this group of patients, and its effects on exercise are therefore not discussed.

Relationships between carbohydrate and fat metabolism

Regulation of carbohydrate and fat utilisation in muscle was studied in 1963 by Randle *et al.*, who introduced glucose–fatty acid cycle relationships in metabolism (Randle *et al.*, 1963). They proposed that an increase in fatty acid would result in increased fat metabolism and inhibition of glucose utilisation. The detailed enzymatic pathways and relationships of these data to insulin sensitivity has resulted in extensive present day studies of the glucose–fatty acid cycle in type 2 diabetes, and also in muscle exercises at various intensities. With ^{31}P MRS experiments, mitochondrial oxidative phosphorylation in the skeletal muscle of diabetic patients was shown to be substantially decreased (Phielix and Mensink, 2008). Although it was difficult to determine whether mitochondria were impaired by insulin resistance or were a primary point of pathogenesis, the data suggest a starting point for myopathies with fat infiltration. Mitochondrial damage and fat infiltration in polymyositis patients have been observed, but the disease mechanisms were not determined (Schroder and Molnar, 1997). Damaged mitochondria may, in part, explain the more severe weakness and exercise abnormalities in polymyositis patients compared to dermatomyositis patients (Figures 11.4 and 11.5).

Hypoperfusion as a contributor to muscle deficiency

Numerous histological studies have documented abnormalities in the arteries and capillaries of DM/PM muscles (Emslie-Smith and Engel, 1990; Grundtman *et al.*, 2008). The pathological findings include reduced capillary lumen, disrupted capillary membrane and reduced capillary density. These abnormalities may represent the known hypoperfusion in muscle, resulting in reduced substrate concentration and oxygen metabolism. The elegant ^{31}P MRS kinetic studies with DM and PM patients placed hypoperfusion as the primary factor in defective oxidative phosphorylation in mitochondria (Cea *et al.*, 2002). Abnormalities in perfusion and diffusion in DM/PM patients were demonstrated with MRI methods of diffusion-weighted imaging (DWI), and diffusion tensor imaging (DTI) which provide a microscopic picture of water movement required for delivery of substrates and oxygenation (Qi *et al.*, 2008b; Park *et al.*, 2009). DTI fibre tracking shows fibre fragmentation and spatial displacement that affect diffusion. Clearly, many factors could contribute to the reduced metabolism which produces fatigue and weakness. There are real difficulties in prioritising these factors as the interplay between parameters is substantial.

TREATMENTS FOR ALLEVIATION OF WEAKNESS AND FATIGUE

Exercise therapy to improve strength and fatigue

Patients with inflammatory myositis show improvement in muscle function with prednisone treatment and adjunct immunosuppressive agents, such as methotrexate and azathioprine. However, most adult patients are left with residual weakness and fatigue, which requires continued medication (Harris-Love, 2003). Despite the early objections to exercise for dermatomyositis and polymyositis patients, it has now been shown that moderate exercise programmes improve both muscle strength and endurance (Hicks *et al.*, 1993; Wiesinger *et al.*, 1998; Alexanderson *et al.*, 2007; Dastmalchi *et al.*, 2007). The reasons for improvements are revealed in the well-documented study of biopsy data which show that fibre type is altered (Dastmalchi *et al.*, 2007). After 12 weeks of exercise, nine patients (four DM and five PM) showed fibre type changes. Type I fibres, which contain mitochondria for oxidative phosphorylation and slow myosin heavy chains, increased from 32% to 42% of the total fibre count. This increase was due to a change in type IIC fibres to type I fibres. The relative proportion of fast fibres, type IIA and IIB, was unchanged after training. Moreover, the mean cross-sectional area of type II fibres increased by 25%, whereas type I fibre cross-sectional area did not change. In addition to increased strength and decreased fatigue, the health assessment scores for lifestyle characteristics (SF-36) indicate substantial improvements which correlate with fibre type change and increased muscle fibre area. This is the first study to show changes in fibre types in inflammatory muscle diseases.

Creatine supplementation for improved energy metabolism

Creatine supplements are known to have modest effects on highly trained athletes during periods of training (Kreider, 2003). Muscular dystrophies are also improved with creatine supplementation (Tarnopolsky *et al.*, 2004). However, creatine did not improve all exercise tasks equally. Benefits of creatine are thought to be due to changes in the muscle bioenergetics through increased PCr levels.

For inflammatory diseases, an excellent double-blind, randomised, placebo-controlled trial was conducted for dermatomyositis and polymyositis patients (Chung *et al.*, 2007). The purpose of this creatine trial was to determine whether improvements with regular exercise could be further promoted by creatine supplementation over a period of six months. In this investigation, 37 DM or PM patients were assigned to an exercise programme. Of these, 19 patients received creatine and 18 received a placebo. Statistical analyses showed, as stated in the previous paragraph, that both groups demonstrated increased muscle endurance. Creatine-treated patients demonstrated further improvement in endurance and the ability to undertake high-intensity exercise. ^{31}P MRS studies showed that the additional exercise improvements in the creatine group correlated with increased PCr/β-NTP ratios, as demonstrated in other studies of myositis patients (Rico-Sanz *et al.*, 1999; Schunk *et al.*, 1999; Park *et al.*, 2000; Scott and Kingsley, 2004). Thus, the improvements are, in part, due to metabolic changes associated with increased PCr. There were no changes in measures such as serum CK levels and health status. The authors conclude that exercise provides a key to rehabilitation, and oral creatine supplements could enhance exercise benefits at a reasonable cost to patients. Creatine does not, however, substitute for the immunosuppressive drugs that treat the basic myositic diseases.

Creatine and magnesium supplementations

Since DM and PM patients have decreased levels of PCr, ATP and Mg^{2+}, a preliminary study was proposed to determine whether creatine and Mg^{2+} would be tolerated and enhance muscle performance. Five patients (one DM and four PM) were enrolled in a six-month open-label trial. Creatine was given for 30 days, and, for the following five months, a daily dose of either 200 mg or 400 mg per day of Mg^{2+} as magnesium aspartate was added. All five patients adhered to the therapeutic regimes from their physicians and completed the trial. PCr, ATP and Mg^{2+} levels, strength and function improved significantly during the trial. Four out of five patients had decreased fatigue and pain, as measured on the VAS ($P=$NS) and by HAQs ($P=0.04$). Patients showed functional improvement by their physician's global assessment. Due to technical difficulties, only three of the patients completed follow-up MRI exams. In these patients, T2 values decreased from 38.2 to 34.6 ms ($P=0.04$), suggesting partial resolution of inflammation. Creatine and Mg^{2+} supplementation may be a safe and effective treatment for DM and PM patients. Therefore, a larger double-blind trial should be considered. The above results are consistent with the data from the larger creatine trial discussed above (Park *et al.*, 2005; Qi *et al.*, 2005; Chung *et al.*, 2007).

CASE HISTORIES SHOWING CLINICAL ASPECTS

Case histories of 11 dermatomyositis patients were studied longitudinally at the request of the physicians (Park *et al.*, 1994). The progression and/or regression of disease indicated a wide variety of clinical results. Therefore, two cases were selected to illustrate the excellent recovery of a juvenile dermatomyositis patient with all clinical factors improving, and an adult woman with a disease flare followed by clinical improvement.

Patient 1: A 14-year-old Caucasian boy had a sudden onset of weakness, pain and typical erythematous rash. He was examined with MRI and MRS prior to the initiation of prednisone treatment, three weeks after a dosage of $60 \text{mg} \cdot \text{d}^{-1}$ prednisone was started, and at five and seven months, when the dosage had been reduced to $35 \text{mg} \cdot \text{d}^{-1}$. During this course of treatment, he showed decreases in serum CK levels and increases in strength (MVC) (Table 11.2). Initial T2-weighted images showed abnormal regions of high signal intensity in the quadriceps muscles, indicative of inflammation, whereas the posterior hamstring muscles appeared normal with a dark homogeneous signal (Figure 11.7B). T2 images taken in the seventh month of treatment show the resolution of inflammation with both the quadriceps and hamstring muscles having dark homogeneous signal intensity. These results were verified by the calculation of elevated T1 values in the initial images of the quadriceps muscle (1,890 ms) and a return to normal T1 levels in the seven-month evaluation period (1,265 ms). Biochemical abnormalities in the initial ^{31}P MRS were detected as an approximately 40% decrease in both ATP and PCr. The P_i/PCr ratio was initially elevated, indicating inefficient generation and/or utilisation of ATP and PCr. Over the 5–7-month period, the P_i/PCr ratio decreased. This metabolic improvement was consistent with the increase in MVC from 15 lbs to 40 lbs. During exercise at 25% MVC, normalisation of P_i/PCr ratios, despite the higher workload, demonstrated improvement in bioenergetic status (work/energy-cost ratios increased from 7.1 to 32.3).

These data corresponded with the physician's global evaluation of improved status. However, mild residual fatigue persisted, as evaluated in the activity questionnaire, and a restriction on excessive exercise in physical education classes was suggested.

Patient 2: A 67-year-old woman presented with mild weakness and was examined with MRI and ^{31}P MRS prior to receiving medication. She was treated with a low dose of prednisone and showed considerable improvement. Hence, medication was terminated. A flare in disease activity occurred at 15.5 months, and she was sent for re-evaluation with MRI/MRS. Treatment was restarted with daily doses of 60 mg prednisone, and considerable improvement was noted. Initial PCr values of $20 \text{mmol} \cdot \text{kg}^{-1}$ muscle dropped to $11 \text{mmol} \cdot \text{kg}^{-1}$ at 15.5 months, but with the resumption of prednisone treatment, returned at 33 months to $18 \text{mmol} \cdot \text{kg}^{-1}$.

Table 11.2 Improvements in Patient 1 after seven months of prednisone treatment

Month	CPK (IU/ml)	MVC (lbs)	T1 (ms)	ATP (mmol·kg^{-1})	PCr (mmol·kg)	P$_i$/PCr ratio	Work/energy cost
1	4,962	15	1,890	4.2	13.6	0.49	7.1
7	63	40	1,265	6.9	23.6	0.31	32.3
Normal juvenile	30–210	32 ± 5	1,083 ± 26	6.6 ± 0.3	25.5 ± 1.2	0.27 ± 0.02	29.9 ± 5.3

Notes
Laboratory and metabolic values for Patient 1. Normal values are mean ± SEM for six healthy control subjects. T1 values are taken from VL.

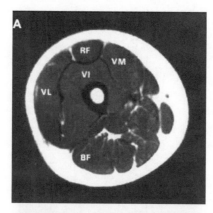

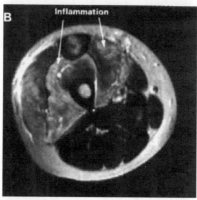

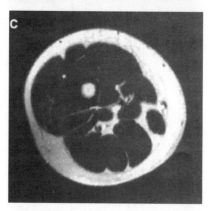

Figure 11.7 T1- and T2-weighted MRI images of the thigh of Patient 1. (A) Initial T1-weighted image, showing homogeneous signal intensity in major muscle groups. VL: vastus lateralis; VI: vastus intermedius; VM: vastus medialis; RF: rectus femoris; BF: biceps femoris. (B) Initial T2-weighted image, showing inflammation, as indicated by increased signal intensity (brightness), in RF, VL, VI, and VM muscles. (C) T2-weighted image after five months of prednisone treatment shows homogeneous signal intensity, indicating inflammation has resolved

When P_i/PCr ratios were considered, the initial ratio at rest was comparable to normal values at 0.15, but with exercise, increased to an abnormally high level of 0.6 (Figure 11.8). During the disease flare at 15.5 months, all ratios at rest, exercise and recovery were elevated, indicating a severe deterioration of the bioenergetic status in the muscles. When prednisone treatment was restarted, all metabolic parameters were improved. However, her T1 values did not return to normal levels. Her slightly elevated T1 values (1,439 ms) were essentially unchanged over 33 months.

This patient's history is interesting because the morphology of the MR images did not correlate with clinical tests, and in this case, the biochemical abnormalities of the MRS evaluations provided the most sensitive indicators of disease activity.

CONCLUSION

The major symptoms of inflammatory myositis, weakness and fatigue are closely intertwined. Pharmacological treatment depresses the immune

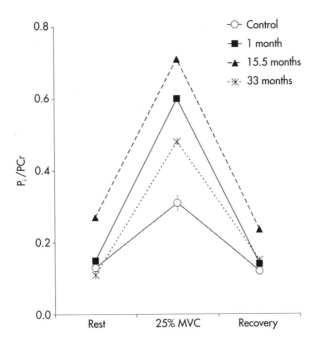

Figure 11.8 P_i/PCr ratios for Patient 2 during relapse and remission. Initially, patient ratios are similar to control values at rest and recovery and only elevated with exercise. During disease relapse at 15.5 months, ratios are elevated at all stages of the protocol. After treatment at 33 months, a trend toward normalization of P_i/PCr ratios is observed

responses and improves both strength and persistence during tasks. Over time and with exercise through physical therapy, greater muscle fibre size and increased numbers of type I oxidative fibres improve strength and, concordantly, endurance for selected exercise tasks. Creatine, or a combination of creatine and Mg supplements, demonstrate that biochemical intervention, which increases high energy compounds, PCr and ATP, and also Mg, will enhance strength, endurance and the lifestyles of patients with DM and PM. Substantial advances in clinical treatment and understanding of basic aetiology promise to evoke substantial opportunities for future investigations of idiopathic inflammatory myopathy (IIM).

FIVE KEY PAPERS THAT SHAPED THE TOPIC AREA

Study 1. Park, J.H., Vital, T.L., Ryder, N.M., Hernanz-Schulman, M., Partain, C.L., Price, R.R. and Olsen, N.J. (1994). Magnetic resonance imaging and ^{31}P magnetic resonance spectroscopy provide unique quantitative data useful in the longitudinal management of patients with dermatomyositis. *Arthritis and Rheumatism*, 37, 736–746.

This paper explored the utility of using MRI and ^{31}P MRS to evaluate disease progression/improvements in DM patients over an extended period of time. Monitoring of DM activity level is commonly performed with a strength evaluation or physical exam, and by measuring serum levels of muscle enzymes such as CPK. However, increased CPK is not directly indicative of disease status, as not all patients show an elevation in this enzyme. Invasive tests such as EMG and muscle biopsy are also not desirable for long-term monitoring. MRI and MRS are non-invasive and can be repeated multiple times. This paper presents long-term evaluations of 11 DM patients, both with and without useful laboratory findings. T1 and T2 relaxation times were calculated from MRI images, and concentrations of metabolites such as P_i, PCr and ATP were determined from resonance areas of the ^{31}P MRS spectra. Of the 11 DM patients, seven had consistently normal CPK levels, nine showed statistical changes in T1 and T2 relaxation values following the clinical course of the disease, and nine exhibited changes in biochemical parameters that were also consistent with clinical evaluation. However, these MRI and MRS changes were not temporally concordant, indicating that inflammation and metabolic abnormalities are not necessarily resolved at the same rate. Overall, this study shows the absence or resolution of inflammation does not appear sufficient for some patients to return to normal metabolic status. Conversely, metabolic changes can also be an early indication of muscles abnormalities. MRI and MRS examinations proved very useful for initial diagnosis of the extent and severity of DM and subsequent evaluation of therapy.

Study 2. Harris-Love, M.O. (2003). Physical activity and disablement in the idiopathic inflammatory myopathies. *Current Opinions of Rheumatology*, 15, 679–690.

This review focuses on the enablement–disablement model of disability applied to idiopathic inflammatory myopathies and discusses the role of physical activity in the enablement process. First proposed in 1965, the model has been revised and expanded by the Institute of Medicine to better convey the multidirectional and dynamic relationship between the domains of disablement. Their Enabling–Disabling Process focuses on the role of the enabling process, separation of disability from the individual and emphasis on person–environment interaction. Pathological factors such as estimates of disease activity and damage may be used to guide exercise intervention and lend insight into the potential for muscle to exhibit adaptive response to strength training. Impairments of IIM are muscle weakness, excessive fatigue and diminished aerobic capacity that can be targeted by exercise therapy. Disability is the consequence of functional limitations in a social context and often measured by self-report. Cardiovascular fitness and muscle strength can be improved with various modes of physical activity and are a valuable adjunct therapy to pharmacological treatment in PM/DM patients. These aspects can decrease functional limitations and disability as part of the enablement process. Conversely, decreased physical activity as a result of IIM can cause a decline in health and independent functioning. Overall, the Enablement–Disablement model takes into account the continuum of health and functioning instead of exclusively focusing on disability, and provides a better conceptual framework to understand the effectiveness of exercise interventions in the treatment of IIM. A comprehensive description of clinical activities in future trials will help discern how much exercise helps with physiological adaptations, the appropriate timing and the type of exercise done to promote clinical guidelines.

Study 3. Dastmalchi, M., Alexanderson, H., Loell, I., Stahlberg, M., Borg, K., Lundberg, I.E. and Esbjornsson, M. (2007). Effect of physical training on the proportion of slow-twitch type I muscle fibres, a novel nonimmune-mediated mechanism for muscle impairment in polymyositis or dermatomyositis. *Arthritis and Rheumatism*, 57, 1303–1310.

This paper compared fibre type composition and fibre area in healthy controls and patients with chronic PM and DM and sought to determine if 12 weeks of physical training changes these muscle characteristics. The relative frequency of fast (type II) and slow (type I) twitch fibres in the muscle determines the muscle's endurance capacity. Muscle fibre type, cross-sectional area and amount of regenerating fibres were determined from biopsies. The Functional Index (FI) in myosites was used to evaluate muscle strength and endurance. The Swedish version of the Medical Outcomes Study Short Form 36 (SF-36) was used to evaluate subject's perceived health-related quality of life. Initially, patients had fewer type I fibres and more type IIB and IIC fibres compared to controls. After 12 weeks of physical training, the patients had increased type I fibres with training and increased type II muscle fibre cross-sectional area, thus making trained muscles more similar to those of controls. Muscle endurance and quality of life scores also increased after training. Positive correlations were found between the increase in type I fibres and physical function (subset of SF-36), as well as function and type II fibre cross-sectional area. The low percentage of oxygen-dependent type I fibres found in patients before the exercise programme was hypothesised to be due to adaptation to muscle tissue hypoxia. The exact mechanism(s) still needs to be determined, but the authors proposed it could be due to improved microcirculation, lowered total peripheral resistance, and reduced skeletal muscle ischemia. The increase in cross-sectional area of type II fibres was attributed to the type of weight training performed, which is known to increase muscle protein synthesis preferably in type II muscles fibres. The clinical relevance of the changes in fibre type composition and fibre area were supported by the improvement of physical capacity measured by both the FI test and the reported improvement of physical functioning in the SF-36 questionnaire.

Study 4. Chung, Y.L., Alexanderson, H., Pipitone, N., Morrison, C., Dastmalchi M., Stahl-Hallengren, C., Richards, S., Thomas, E.L., Hamilton, G., Bell, J.D., Lundberg, I.E. and Scott, D.L. (2007). Creatine supplements in patients with idiopathic inflammatory myopathies who are clinically weak after conventional pharmacologic treatment: six-month, double-blind, randomized, placebo-controlled trial. *Arthritis and Rheumatism*, 57, 694–702.

This paper presents one of the largest reported randomised, placebo-controlled trials in IIM. Oral creatine is generally considered to improve athletic and sporting performance. This study was designed to test the hypothesis that patients with established, treated IIM and persisting weakness would have additional benefits from exercise combined with oral creatine therapy. In a six-month, two-centre, double-blind, randomised, controlled trial, a total of 29 patients were randomised to receive oral creatine supplements (eight days, $20\,mg\cdot d^{-1}$ then $3\,mg\cdot d^{-1}$) or placebo. All patients followed a home exercise programme which consisted of resistive, range of motion and stretching exercises, as well as a 15 min daily walk. The primary outcome measure was aggregate functional performance time (AFPT), which was a functional assessment measured in times to complete four simple tasks involving walking and stair climbing. ^{31}PMR spectroscopy was used to evaluate muscle bioenergetics as determined by the PCr/β-NTP ratio. Assessments were taken before and at the third and sixth months of the trial. AFPT showed a median decrease of 11.4% in the creatine group versus 3.7% decrease with placebo from initial performance. PCr/β-NTP ratios increased in the creatine group, but remained unchanged in the placebo group. Creatine increased the benefits of exercise on endurance and improved ability to undertake high-intensity exercise, an effect maintained over five months. Therefore, creatine could be a useful adjunct therapy to extend the benefits of exercise on muscle weakness and fatigue.

Study 5. Niermann, K.J., Olsen, N.J. and Park, J.H. (2002). Magnesium abnormalities of skeletal muscle in dermatomyositis and juvenile dermatomyositis. *Arthritis and Rheumatism*, 46, 475–488.

This study utilised ^{31}P MRS to characterise muscle magnesium (Mg) levels in DM and JDM patients and to evaluate how effective immunosuppressive therapy was in increasing free and ATP-bound Mg (Mg^{2+} and MgATP). Spectra were obtained during rest, two graded levels of exercise and recovery. Levels of free ATP, biologically active free Mg^{2+} and enzymatically active MgATP were calculated from these spectra at each of the four stages of the exercise protocol. MgATP levels in DM and JDM myopathic muscles were at least 37% lower than those in normal muscles during rest, exercise and recovery ($P < 0.0005$). Free Mg^{2+} levels were normal in DM and JDM myopathic muscles at rest, but were significantly lower than control values during exercise and recovery ($P < 0.029$ and $P < 0.005$ for DM and JDM, respectively). Prednisone and immunosuppressive therapy partially reversed the magnesium abnormalities, as evidenced by elevation of the levels of MgATP and free Mg^{2+}. Increased levels of MgATP and free Mg^{2+} were concordant with decreased weakness and fatigue.

GLOSSARY OF TERMS

^{31}P MRS	phosphorus magnetic resonance spectroscopy
ADP	adenosine diphosphate
ALT	alanine aminotransferase
AMP	adenosine monophosphate
AST	aspartate aminotransferase
ATP	adenosine triphosphate
BF	biceps femoris
CK	creatine phosphokinase
CTL	adult control subject
DM	dermatomyositis
EMG	electromyography
HAQ	health assessment questionnaire
IBM	inclusion body myositis
IIM	idiopathic inflammatory myopathy
JCTL	juvenile control subject
JDM	juvenile dermatomyositis
LDH	lactate dehydrogenase

Mg	magnesium
MgATP	magnesium adenosine triphosphate
MRI	magnetic resonance imaging
MVC	maximum voluntary contraction
PCr	phosphocreatine
PDE	phosphodiester
P_i	inorganic phosphate
PM	polymyositis
PME	phosphomonoester
SM	semimembranosus
ST	semitendinosus
TE	echo time
TR	repetition time
VAS	visual analogue scale
VI	vastus intermedius
VL	vastus lateralis
VM	vastus medialis

REFERENCES

Alexanderson, H., Dastmalchi, M., Esbjornsson-Liljedahl, M., Opava, C.H. and Lundberg, I.E. (2007). Benefits of intensive resistance training in patients with chronic polymyositis or dermatomyositis. *Arthritis and Rheumatism*, 57, 768–777.

Blazev, R. and Lamb, G.D. (1999). Low [ATP] and elevated [Mg^{2+}] reduce depolarized-induced Ca^{2+} release in rat skinned skeletal muscle fibres. *Journal of Physiology (London)*, 520, 203–215.

Cea, G., Bendahan, D., Manners, D., Hilton-Jones, D., Lodi, R., Styles, P. and Taylor, D.J. (2002). Reduced oxidative phosphorylation and proton efflux suggest reduced capillary blood supply in skeletal muscle of patients with dermatomyositis and polymyositis: a quantitative ^{31}P-magnetic resonance spectroscopy and MRI study. *Brain*, 125, 1635–1645.

Chariot, P., Ruet, E., Authier, F.J., Labes, D., Poron, F. and Gherardi, R. (1996). Cytochrome c oxidase deficiencies in the muscle of patients with inflammatory myopathies. *Acta Neuropathology*, 91, 530–536.

Chung Y.L., Alexanderson, H., Pipitone, N., Morrison, C., Dastmalchi, M., Stahl-Hallengren, C., Richards, S., Thomas, E.L., Hamilton, G., Bell, J.D., Lundberg, I.E. and Scott, D.L. (2007). Creatine supplements in patients with idiopathic inflammatory myopathies who are clinically weak after conventional pharmacologic treatment: six-month, double-blind, randomized, placebo-controlled trial. *Arthritis and Rheumatism*, 57, 694–702.

Cronin, R.E. and Knochel, J.P. (1983). Magnesium deficiency. *Advances in Internal Medicine*, 28, 509–533.

Dalakas, M.C. (1991). Polymyositis, dermatomyositis and inclusion-body myositis. *The New England Journal of Medicine*, 325, 1487–1498.

Dastmalchi, M., Alexanderson, H., Loell, I., Stahlberg, M., Borg, K., Lundberg, I.E. and Esbjornsson, M. (2007). Effect of physical training on the proportion of slow-twitch type I muscle fibres, a novel nonimmune-mediated mechanism for muscle impairment in polymyositis or dermatomyositis. *Arthritis and Rheumatism*, 57, 1303–1310.

Emslie-Smith, A.M. and Engel, A.G. (1990). Microvascular changes in early and advanced dermatomyositis: a quantitative study. *Annals of Neurology*, 27, 343–356.

Fathi, M., Lundberg, I.E. and Tornling, G. (2007). Pulmonary complications of polymyositis and dermatomyositis. *Seminars in Respiratory and Critical Care in Medicine*, 28, 451–458.

Fathi, M., Vikgren, J., Boijsen, M., Tylen, U., Jorfeldt, L., Tornling, G. and Lundberg, I.E. (2008). Interstitial lung disease in polymyositis and dermatomyositis: longitudinal evaluation by pulmonary function and radiology. *Arthritis and Rheumatism*, 59, 677–685.

Grundtman, C., Tham, E., Ulfgren, A. and Lundberg, I.E. (2008). Vascular endothelial growth factor is highly expressed in muscle tissue of patients with polymyositis and patients with dermatomyositis. *Arthritis and Rheumatism*, 58, 3224–3238.

Harris-Love, M.O. (2003). Physical activity and disablement in the idiopathic inflammatory myopathies. *Current Opinions of Rheumatology*, 15, 679–690.

Hicks, J.E., Miller, F., Plotz, P., Chen, T.H. and Gerber, L. (1993). Isometric exercise increases strength and does not produce sustained creatine phosphokinase increases in a patient with polymyositis. *Journal of Rheumatology*, 20, 1399–1401.

Irish, A.B., Thompson, C.H., Kemp, G.J, Taylor, D.J. and Radda, G.K. (1997). Intracellular free magnesium concentrations in skeletal muscle in chronic uraemia. *Nephron*, 76, 20–25.

Kreider, R.B. (2003). Effects of creatine supplementation on performance and training adaptations. *Molecular and Cellular Biochemistry*, 244, 89–94.

Lodi, R., Taylor, D.J., Tabrizi, S.J., Hilton-Jones, D., Squier, M.V., Seller, A., Styles, P. and Schapira, A.H. (1998). Normal in vivo skeletal muscle oxidative metabolism in sporadic inclusion body myositis assessed by ^{31}P-magnetic resonance spectroscopy. *Brain*, 121 (11), 2119–2126.

Newman, E.D. and Kurland, R.J. (1992). P-31 magnetic resonance spectroscopy in polymyositis and dermatomyositis: altered energy utilization during exercise. *Arthritis and Rheumatism*, 35, 199–203.

Niermann, K.J., Olsen, N.J. and Park, J.H. (2002). Magnesium abnormalities of skeletal muscle in dermatomyositis and juvenile dermatomyositis. *Arthritis and Rheumatism*, 46, 475–488.

Park, J.H., Lee, B.C. and Qi, J. (2009). Magnetic resonance imaging (DTI) of abnormalities in the thigh muscles of polymyositis patients. *Proceedings of the International Society of Magnetic Resonance in Medicine*, 17, 1903.

Park, J.H., Niermann, K.J., Ryder, N.M., Nelson, A.E., Das, A., Lawton, A.R., Hernanz-Schulman, M. and Olsen, N.J. (2000). Muscle abnormalities in juvenile dermatomyositis patients: P-31 magnetic resonance spectroscopy studies. *Arthritis and Rheumatism*, 43, 2359–2367.

Park, J.H., Olsen, N.J., King Jr, L., Vital, T., Buse, R., Kari, S., Hernanz-Schulman, M. and Price, R.R. (1995). Use of magnetic resonance imaging and P-31 magnetic resonance spectroscopy to detect and quantify muscle dysfunction in the amyopathic and myopathic variants of dermatomyositis. *Arthritis and Rheumatism*, 38, 68–77.

Park, J.H., Qi, J., Kroop, S.F. and Olsen, N.J. (2005). Preliminary trial of creatine and magnesium supplementation in myositis patients. *Arthritis and Rheumatism*, 52, S316.

Park, J.H., Qi, J., Lee, B.C. and Olsen, N.J. (2008). Comparisons of metabolic defects in exercising muscles of patients with polymyositis and dermatomyositis. *Arthritis and Rheumatism*, 58, S230.

Park, J.H., Vansant, J.P., Kumar, N.G., Gibbs, S.J., Curvin, M.S., Price, R.R., Partain, C.L. and James Jr, A.E. (1990). Dermatomyositis: correlative MR imaging and P-31 MR spectroscopy for quantitative characterization of inflammatory disease. *Radiology*, 177, 473–479.

Park, J.H., Vital, T.L., Ryder, N.M., Hernanz-Schulman, M., Partain, C.L., Price, R.R. and Olsen, N.J. (1994). Magnetic resonance imaging and P-31 magnetic resonance spectroscopy provide unique quantitative data useful in the longitudinal management of patients with dermatomyositis. *Arthritis and Rheumatism*, 37, 736–746.

Pfleiderer, B., Lange, J., Loske, K.D. and Sunderkotter, C. (2004). Metabolic disturbances during short exercises in dermatomyositis revealed by real-time functional ^{31}P magnetic resonance spectroscopy. *Rheumatology*, 43, 696–703.

Phielix, E. and Mensink, M. (2008). Type 2 diabetes mellitus and skeletal muscle metabolic function. *Physiology and Behaviour*, 84, 252–258.

Qi, J., Lee, B.C., Olsen, N.J. and Park, J.H. (2007). Abnormalities in magnesium (Mg) and ATP levels correlate with muscle dysfunction in polymyositis. *Proceedings of the International Society of Magnetic Resonance in Medicine*, 15, 2606.

Qi, J., Park, J.H., Kroop, S.F. and Olsen, N.J. (2005). P-31 MRS measures the metabolic changes in a creatine and magnesium supplementation trial for myositis patients. *Radiological Society of North America, Radiology*, Supplement 237, 400.

Qi, J., Niermann, K.J., Lee, B.C., Olsen, N.J. and Park, J.H. (2008a). Metabolic deficiency in skeletal muscle of dermatolyositis (DM) and polymyositis (PM) patients detected with quantitative P-31 MRS. *Radiological Society of North America, Radiology*, Supplement 249, 628.

Qi, J., Olsen, N.J., Price, R.R., Winston, J.A. and Park, P.H. (2008b). Diffusion-weighted imaging of inflammatory myopathies: polymyositis and dermatomyositis. *Journal of Magnetic Resonance Imaging*, 27, 212–217.

Randle, P.J., Hales, C.N., Garland, P.B. and Newsholme, E.A. (1963). The glucose-fatty acid cycle: its role in insulin sensitivity and the metabolic disturbances of diabetes mellitus. *Lancet*, 1, 785–789.

Rico-Sanz, J., Zehnder, M., Buchli, R., Kuhne, G. and Boutellier, U. (1999). Non-invasive measurement of muscle high-energy phosphates and glycogen concentrations in elite soccer players by ^{31}P- and ^{13}C-MRS. *Medicine and Science in Sports and Exercise*, 31, 1580–1586.

Rifai, Z., Welle, S., Kamp, C. and Thornton, C.A. (1995). Ragged red fibres in normal aging and inflammatory myopathy. *Annals of Neurology*, 37, 24–29.

Schroder, J.M. and Molnar, M. (1997). Mitochondrial abnormalities and peripheral neuropathy in inflammatory myopathy, especially inclusion body myositis. *Molecular and Cellular Biochemistry*, 174, 277–281.

Schunk, K., Pitton, M., Duber, C., Kersjes, W., Schadmand-Fischer, S. and Thelen, M. (1999). Dynamic phosphorus-31 magnetic resonance spectroscopy of the quadriceps muscle: effects of age and sex on spectroscopic results. *Investigations in Radiology*, 34, 116–125.

Scott, D.L. and Kingsley, G.H. (2004). Use of imaging to assess patients with muscle disease. *Current Opinions in Rheumatology*, 16, 678–683.

Tarnopolsky, M.A., Mahoney, D.J., Vajsar, J., Rodriguez, C., Doherty, T.J., Roy, B.D. and Biggar, D. (2004). Creatine monohydrate enhances strength and body composition in Duchenne muscular dystrophy. *Neurology*, 62, 1771–1777.

Welch, K.M.A. and Ramadan, N.M. (1995). Mitochondria, magnesium and migraine. *Journal of Neurological Sciences*, 134, 9–14.

Wiesinger, G.F., Quittan, M., Graninger, M., Seeber, A., Ebenbichler, G., Sturm, B., Kerschan, K., Smolen, J. and Graninger, W. (1998). Benefit of 6 months long-term physical training in polymyositis/dermatomyositis patients. *British Journal of Rheumatology*, 37, 1338–1342.

Wortmann, R.L. (2005). Inflammatory Diseases of muscle and other myopathies. In E.D. Harris Jr., R.C. Budd, M.C. Genovese, G.S. Firestein, J.S. Sargent and C.B. Sledge (eds), *Kelley's Textbook of Rheumatology*, 7th edn, Vol. 2. Philadelphia: Elsevier Saunders.

CHAPTER 12

MUSCLE FATIGUE IN METABOLIC MYOPATHIES

Ronald G. Haller and John Vissing

OBJECTIVES

The aim of this chapter is to provide an overview on:

- the biochemical basis and key clinical features of two classes of human metabolic myopathy;
- studies that have illuminated the mechanism of muscle fatigue when the key pathways of muscle energy production are limited;
- the insights provided by these energy defects for the role of these metabolic pathways in normal muscle fatigue.

INTRODUCTION

In this chapter we will review the implications of two classes of inborn errors of muscle metabolism for muscle fatigue during exercise: muscle mitochondrial defects and disorders of muscle glycogenolysis or glycolysis, utilising as an example of a complete block in muscle glycogenolysis, muscle glycogen phosphorylase deficiency (GSD V, McArdle disease).

MITOCHONDRIAL MYOPATHIES

Mitochondrial myopathies are typically attributable to genetic defects that impair the function of the respiratory chain in skeletal muscle (DiMauro and Schon, 2003). A major clinical consequence of significant respiratory chain defects is a reduced capacity for aerobic exercise, i.e. exercise that is fuelled by the combustion of fats and carbohydrates to carbon dioxide and water, with ATP produced by oxidative phosphorylation. A characteristic of exercise that is fully supported by oxidative phosphorylation is its ability to be sustained for prolonged periods of time, in contrast to exercise that depends upon anaerobic metabolism (phosphocreatine hydrolysis and anaerobic glycogenolysis) to fuel muscle contractions, which rapidly results in muscle fatigue (Allen *et al.*, 2008). Characteristic of anaerobic metabolism is its ability to support rates of ATP turnover that are more than two-fold that of oxidative phosphorylation, so anaerobic metabolism is required for muscle contractions at maximal effort, and when the rate of ATP demand exceeds maximal rates of oxidative phosphorylation (Sahlin, 1986). Accordingly, the limits of an individual's aerobic capacity is marked by the activation of anaerobic metabolism and is associated with a steep increase in lactate production and in the lactate/pyruvate ratio in working muscle and in blood. In healthy humans, an important variable in determining aerobic capacity is the level of physical conditioning. Trained individuals have a greater aerobic capacity by virtue of having developed the circulatory and muscle mitochondrial adaptations that increase the capacity for oxygen delivery and utilisation (Gollnick and Saltin, 1983). In habitually sedentary individuals, the opposite is true. Hence the range of "normal" aerobic capacity in healthy humans is substantial.

Aerobic capacity is commonly measured in metabolic equivalents termed METS, representing multiples of resting metabolic rate, which by convention is 3.5 ml of oxygen per minute per kilogram of body weight (Åstrand and Rodahl, 1986). In our experience, patients with mitochondrial myopathies have an average peak aerobic capacity of approximately 4.5 METS, with a large percentage having a peak capacity of less than 4 METS (Taivassalo *et al.*, 2003). As a result, affected patients have a greatly restricted range of physical activities that can be supported by oxidative phosphorylation (Figure 12.1). Accordingly, activities considered minor by healthy humans are sufficient to accelerate anaerobic metabolism and cause tachycardia and muscle fatigue in association with marked increases in blood lactate (Figure 12.2A and 12.2B).

Interestingly, the character of muscle fatigue in patients with mitochondrial myopathy is similar to that of healthy subjects (Wiles *et al.*, 1981). Patients do not develop distinctive symptoms of exercise intolerance, but simply have overall low tolerance of aerobic exercise that commonly includes symptoms of exertional shortness of breath and tachycardia. The

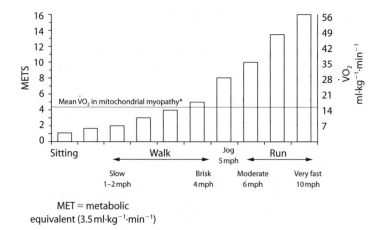

MET = metabolic
equivalent (3.5 ml·kg⁻¹·min⁻¹)

Figure 12.1 Correlation between oxygen utilisation expressed in ml per kg per minute (right axis), METS (left axis) and levels of physical activity (X axis)

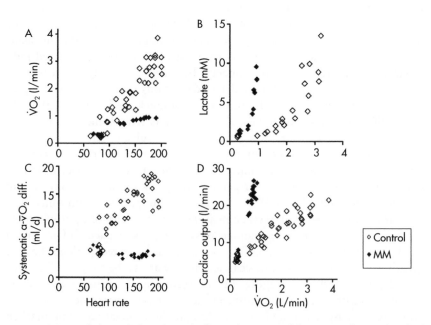

Figure 12.2 Metabolic and physiological effects of a severe defect in muscle mitochondrial metabolism associated with deficiency of multiple iron–sulphur cluster-containing proteins (including succinate dehydrogenase and aconitase and subunits of complex I and III of the respiratory chain (Haller *et al.*, 1991) – a disorder recently shown to be due to a mutation in the iron–sulphur cluster scaffold, ISCU, gene (Mochel *et al.*, 2008)). Results during cycle exercise for an affected patient (filled diamonds) and for control subjects (open diamonds) are shown for: A: $\dot{V}O_2$ versus heart rate; B: lactate relative to $\dot{V}O_2$; C: systemic arterio-venous O_2 difference relative to heart rate; and D: cardiac output relative to $\dot{V}O_2$

non-specific nature of muscle fatigue that occurs with physical activities that would be expected to be easily tolerated often leads to the mistaken belief that patient symptoms derive from being in poor physical condition.

The metabolic correlates of skeletal muscle fatigue in mitochondrial myopathy include changes in metabolites that are associated with muscle fatigue in healthy subjects. These include depletion of phosphocreatine, increases in muscle and blood lactate and lactate/pyruvate, a decline in muscle pH and increases in muscle levels of ADP and diprotonated phosphate (Matthews et al., 1991). In healthy humans, interventions that increase the oxygen-carrying capacity of blood by increasing red cell mass (blood doping, erythropoietin administration) increase exercise capacity consistent with the notion that oxygen availability rather than levels of functional mitochondria limits oxidative phosphorylation during exercise involving large muscle groups. In contrast, in mitochondrial myopathies, the restriction in oxidative metabolism resides in a defective respiratory chain that limits the ability of mitochondria to utilise available oxygen.

The distinctive pathophysiology of exercise in muscle mitochondrial defects is a mismatch between muscle oxygen utilisation and the circulatory and ventilatory responses to exercise that are normally closely geared to muscle oxygen utilisation (Taivassalo et al., 2003). This mismatch is illustrated by the effects of muscle mitochondrial dysfunction upon the components of oxygen utilisation contained in the Fick equation:

oxygen utilisation $(\dot{V}O_2)$ = systemic oxygen delivery (cardiac output) × systemic arterio-venous (a-$\bar{v}$) O_2 difference

Assuming normal O_2-carrying capacity (normal levels and function of haemoglobin), arterialised blood contains about 20 ml O_2, and mixed venous blood 15 ml O_2 per 100 ml of blood, so resting systemic a-$\bar{v}$ O_2 difference is about 5 ml. From rest to peak leg exercise, systemic a-$\bar{v}$ O_2 difference in healthy individuals increases three-fold from 5 ml to about 15 ml·dl^{-1}, and O_2 saturation of femoral venous blood falls from about 75% to 25%. This increase in O_2 uptake is attributable to the avid mitochondrial utilisation of oxygen, and to the low Km of cytochrome c oxidase for O_2 (Richardson et al., 2006). From rest up to moderate levels of exercise, the activation of mitochondrial oxidative phosphorylation lowers intracellular O_2 tensions from a pO_2 of approximately 30 mmHg to a level of 3–5 mmHg, establishing a steep O_2 gradient from capillaries (pO_2 ~35 mmHg) to respiring mitochondria (Richardson et al., 2006).

When mitochondrial metabolism limits oxidative phosphorylation, working muscle is unable to normally extract available oxygen from blood, so muscle oxygen levels remain abnormally high and systemic arterio-venous O_2 levels remain low during peak exercise (Haller et al., 1991; Bank and Chance, 1994; Taivassalo et al., 2002; Taivassalo et al., 2003). In the most severe mitochondrial defects, systemic a-$\bar{v}$ O_2 difference remains at or below resting levels (Figure 12.2C) (Taivassalo *et al.*, 2003). Correspondingly, the circulatory delivery of oxygen (cardiac output and blood flow) is greatly exaggerated relative to oxygen utilisation. Normally, the delivery of O_2 by the circulation is closely matched to oxygen utilisation by working muscle; but when impaired mitochondrial metabolism limits oxidative metabolism, the increase in cardiac output relative to oxygen utilisation may be 3–5 times normal or even greater (Figure 12.2D) (Haller and Vissing, 1999; Taivassalo *et al.*, 2003). This "hyperkinetic" circulatory response to exercise in mitochondrial myopathy indicates that intact mitochondrial metabolism is required for the normal matching of O_2 utilisation and delivery during exercise and suggests that blocked muscle oxidative phosphorylation in working muscle activates circulatory reflexes to drive the circulation during exercise (Haller and Vissing, 1999). Defects in muscle oxidative metabolism also result in exaggerated ventilatory responses to exercise in which ventilation is abnormally high in relation to oxygen utilisation and carbon dioxide production (Flaherty *et al.*, 2001; Taivassalo *et al.*, 2003). Correspondingly, affected patients often experience dyspnea during exercise. In fact, patients with severe mitochondrial myopathies may experience dyspnea as the major symptom limiting physical exertion, despite the fact that the primary metabolic defect is impaired muscle oxidative phosphorylation. The mechanism underlying the prominence of exertional dyspnea in patients with severe oxidative defects that are restricted to skeletal muscle is unknown.

MUSCLE GLYCOLYTIC DEFECTS: McARDLE DISEASE AND RELATED DISORDERS

Myophosphorylase deficiency (glycogen storage disease type V, McArdle disease) is an autosomal-recessive disorder attributable to mutations in the gene for the muscle form of glycogen phosphorylase on chromosome 11q13 (Aquaron *et al.*, 2007; Deschauer *et al.*, 2007). The effect of all described pathogenic mutations to date is a complete absence of biochemical activity of myophosphorylase, and thus a complete block in muscle glycogen breakdown. Accordingly, McArdle disease provides unique insight into the role of glycogen in muscle energy metabolism.

The classical clinical features of McArdle disease are premature muscle fatigue, electrically silent muscle contractures and rhabdomyolysis (lysis of muscle fibres) produced by exercise that normally would be powered by anaerobic glycogenolysis. Muscle contractures should not be confused with ordinary muscle cramps. They are electrically silent, that is, they are not maintained by muscle or nerve excitation, but rather mimic muscle rigor. With a contracture, muscle is effectively locked in contraction and attempted stretching of the muscle causes severe pain. Recovery commonly is delayed for half an hour or more. Available evidence suggests that muscle contractures are mediated by energy limitations that result in sustained abnormal levels of intracellular calcium (Ruff, 1996). Exertional muscle contractures are routinely associated with muscle fibre necrosis and a breach in sarcolemmal integrity that results in a liberation of muscle cytoplasmic contents into the bloodstream. One of the most dramatic consequences is the release of the muscle haem protein myoglobin from necrotic muscle fibres, resulting in myoglobinuria (causing dark, commonly cola-coloured urine attributable to myoglobin excreted in the urine). Myoglobinuria may result in renal injury, acute tubular necrosis and acute renal failure. These dramatic consequences of exercise in patients with McArdle disease are mimicked by other inborn errors of muscle carbohydrate metabolism, including muscle phosphofructokinase deficiency, muscle phosphoglycerate mutase deficiency, phosphoglycerate kinase deficiency and muscle lactate dehydrogenase deficiency. In each of these disorders, muscle contractions at maximal effort result in catastrophic muscle failure, in which muscle not only fatigues, but develops contractures and muscle necrosis, which, in severe cases, leads to myoglobinuria. These clinical observations indicate that muscle glycogenolysis/glycolysis is not only crucial for supplying energy for muscle contractions during peak exercise, but is also necessary for normal muscle fatigue. That is to say, anaerobic glycogenolysis not only is able to briefly support the most powerful of muscle contractions, but it also is necessary for producing the metabolic milieu in working muscle that safely sets limits to maximal muscle power output so that, after a period of rest and recovery, fatigued skeletal muscle is able to resume normal muscle contractions. Thus, muscle glycogenolysis may be viewed as both the normal engine and the normal breaking system of *machina carnis*.

In the classical clinical description of McArdle disease, ischemic forearm exercise was employed to demonstrate the characteristic block in muscle lactate production. Ischemic exercise also produces a characteristic pattern of abnormal muscle fatigability (Figure 12.3). The ischemic forearm test is performed by inflating a blood pressure cuff above systolic blood pressure on the arm, and then performing maximal handgrip contractions, commonly with contractions every other second for 60 s. In healthy subjects, a gradual decline in maximal voluntary muscle contractions (MVC) occurs so that maximal contractile force declines to about 70% of MVC at the end of the test (Figure 12.3). A similar fatigue pattern is elicited by electrical stimulation of muscle, indicating that contractile force is not limited by voluntary effort. Perhaps not surprisingly, the pattern of fatigue during ischemic exercise in patients with mitochondrial myopathy is similar to healthy subjects (Figure 12.3). In contrast, in patients with muscle phosphorylase deficiency, the pattern of fatigue is highly abnormal. During such testing, contractile force is similar to healthy subjects for the first 5–7 muscle contractions, suggesting that phosphocreatine hydrolysis is able to support muscle energy requirements during this period. Thereafter, a dramatic drop off of contractile force occurs, so that by 15 maximal contractions, force has declined to about 30% of initial MVC, and patients rarely can continue beyond 40 s of testing (20 muscle contractions) because of the development of a muscle contracture. Electrically stimulated muscle contractions in affected patients produces a similar pattern of fatigue (Wiles *et al.*, 1981).

The metabolic mechanisms of muscle fatigue, contractures and muscle necrosis when glycogenolysis or glycolysis are impaired have not been established with certainty, but a variety of metabolic abnormalities accompany blocked muscle glycogenolysis (Figure 12.4). These include the characteristic failure of lactate production (Figure 12.4A). Whereas healthy subjects increase blood lactate 4–5-fold during ischemic exercise, venous lactate levels actually fall during forearm exercise in patients with McArdle disease. As a result of the block in lactate production, the normal fall in muscle pH (increase in [H^+]) does not occur (Figure 12.4B) (Argov *et al.*, 1987). In healthy subjects, maximal exercise is associated with a steep fall in muscle pH from resting levels of approximately 7.0 to about 6.4. In contrast, such exercise in McArdle disease results in an *increase* in muscle pH, denoting a *decrease* in [H^+] (Figure 12.4B). The abnormal pH of working phosphorylase-deficient muscle may have a number of deleterious effects. Among the most important of these is its effect on the creatine kinase reaction. Creatine kinase catalyses the reversible reaction by which

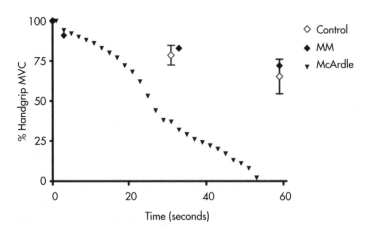

Figure 12.3 Development of fatigue during ischemic handgrip exercise recorded as percentage of original MVC in which subjects are requested to perform a handgrip exercise consisting of 1 s MVC followed by 1 s of relaxation for 60 s. The mean ±SD for 12 healthy control subjects (open diamonds) are compared with a patient with a mitochondrial myopathy (closed diamond) and a patient with McArdle disease (closed triangle). The fatigue pattern of patients with other defects of muscle glycolysis is comparable to McArdle disease (not shown)

phosphocreatine (PCr) hydrolysis is coupled to the phosphorylation of ADP. This is a proton-consuming reaction that is normally stimulated by [H⁺] production via glycolysis. Low levels of [H⁺] caused by blocked glycogenolysis causes relative inhibition of PCr hydrolysis and results in dramatically elevated levels of ADP during exercise (Radda, 1986). As a consequence, the level of AMP production via myokinase is increased, and the level of AMP deamination via myoadenylate deaminase is accelerated, resulting in high levels of production of IMP and ammonia (Haller and Vissing, 2004a). Accordingly, impaired muscle lactate production in inborn errors of muscle glycogenolysis or glycolysis is associated with an exaggerated increase in ammonia production (Figure 12.4C) that both mirrors the abnormal accumulation of muscle ADP and AMP in affected patients, and is helpful clinically to confirm that blunted lactate production during diagnostic exercise testing is due to a metabolic block rather than poor effort (when the increase in both lactate and ammonia are blunted) (Haller and Vissing, 2004a).

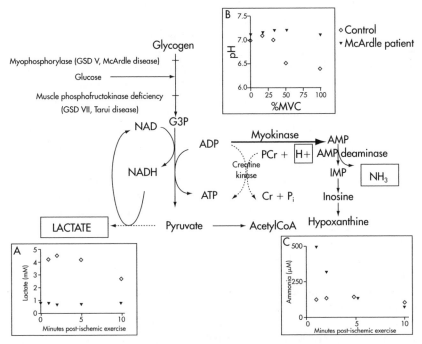

Figure 12.4 Outline of muscle glycogen and glucose metabolism, and the site of metabolic block in McArdle disease and in muscle phosphofructokinase deficiency. Inserts A–C reveal the results of ischemic exercise for control subjects (open diamonds) and a patient with McArdle disease (closed triangle) with respect to lactate production (A) and ammonia production (C). Insert B compares the change in muscle pH during aerobic exercise (measured using ³¹P magnetic resonance spectroscopy) at different percentages of MVC. Results for such testing in patients with muscle phosphofructokinase (not shown) duplicate those shown for the McArdle patient

An additional important consequence of blocked glycogenolysis is deficient substrate-level phosphorylation of ADP, i.e. blocked ATP production at the level of phosphoglycerate kinase and pyruvate kinase. Substrate level ATP production has been postulated to preferentially supply ATP for the membrane sodium–potassium ATPase, so impaired glycolytic flux might be expected to impair membrane pump function and contribute to the abnormal fatigue and sarcolemmal failure that occurs with intense exercise in McArdle disease and related disorders. The regenerative action potential that underlies sarcolemmal excitability results in an increase in extracellular (and decrease in intracellular) potassium, as well as an increase in intracellular sodium. Maintaining membrane excitability during repeated muscle contraction depends upon the ability of sodium–potassium ATPase, the sodium–potassium pump, to maintain ion gradients and prevent the excessive accumulation of extracellular potassium (Clausen, 1986). Physiological studies have established that abnormal fatigability in McArdle disease is associated with impaired excitation–contraction coupling/sarcolemmal inexcitability, and that the rise in blood and presumably extracellular potassium is exaggerated in relation to the level of muscle contractions in these patients (Cooper et al., 1989). These results implicate sarcolemmal inexcitability and impaired sodium–potassium pump function in the pathophysiology of abnormal muscle fatigue in McArdle disease. Possible mechanisms include: the block in substrate-level phosphorylation that is the preferred source of energy for the sodium–potassium pump (Han et al., 1992; James et al., 1996); absence of acidosis, which may be important for maintaining muscle membrane excitability by decreasing chloride permeability (Pedersen et al., 2004); loss of the osmotic effect from lactate accumulation, which may account for absence of the normal increase in water content of exercised muscle (as monitored by magnetic resonance imaging), and may promote higher than normal concentrations of extracellular potassium in exercising muscle (Fleckenstein et al., 1991); and the exaggerated accumulation of ADP during exercise that may inhibit ATPases, including sodium–potassium ATPase (Ruff and Weissman, 1995). In addition, we have found that sodium–potassium pump levels in muscle samples from McArdle patients, measured by ^{3}H-ouabain binding were only about 70% of control levels (Haller et al., 1998). Sodium–potassium pump levels increase with aerobic training and decrease with deconditioning, so the relative decrease in pump levels could be attributable to relative deconditioning in patients. However, we have found that the increase in pump numbers with aerobic conditioning may be blunted in McArdle patients undergoing exercise training (Haller, Vissing and Clausen, unpublished observation). This suggests that additional factors may be involved, such as disruption of the normal close functional relationship between glycolytic, substrate-level phosphorylation and the activity of the sodium–potassium pump (James et al., 1996).

The fact that fatiguing exercise in McArdle disease and related glycolytic disorders results in muscle contractures and muscle fibre necrosis in addition to impaired excitation–contraction coupling suggests that additional pathogenic factors are involved. A valuable model for studying these issues has been iodoacetate-poisoned muscle that was shown by Lundsgaard to produce muscle contraction and block lactate production in frog skeletal muscle. Iodoacetate blocks glycolysis by inhibiting the enzyme, glyceraldehyde-3-phosphate dehydrogenase. Using iodoacetate-poisoned rat skeletal muscle, Ruff et al. found that muscle contractures in this model are associated with a marked increase in intracellular calcium and with an increase in myofibrillar calcium sensitivity in association with a marked increase in cellular ADP (Ruff and Weissman, 1991; Ruff and Weissman, 1995; Ruff, 1996). Muscle ATP was not substantially reduced, suggesting that ATP depletion per se was not responsible for these results. Interestingly, in this model, the increase in inorganic phosphate that normally parallels the decrease in muscle PCr is blunted compared to normal muscle, owing to the accumulation of phosphorus-containing glycolytic intermediates behind the metabolic block. A similar blunted increase in inorganic phosphate due to a build-up of phosphomonoesters in glycolytic intermediates behind the metabolic block occurs in muscle phosphofructokinase deficiency and in distal glycolytic defects. The accumulation of diprotonated inorganic phosphate (owing to the increase in [H^+] when glycolysis is intact) has been implicated in mediating muscle fatigue in healthy humans (Miller et al., 1988), but cannot have a role in the abnormal exertional fatigue, contractures and necrosis in McArdle disease and glycolytic defects, since diprotonated phosphate does not accumulate in the absence of acidosis. Also, inorganic phosphate seems effectively eliminated as a cause of the exertional muscle fatigue in McArdle disease and glycolytic defects, since the pathophysiology of exertional muscle failure in both conditions seems identical, despite marked differences in inorganic phosphate levels (Duboc et al., 1987). A remaining common denominator in the mediation of exertional muscle failure in these conditions is an exaggerated increase in muscle ADP levels, which could operate by inhibiting multiple ATPases.

The foregoing has focused upon the importance of anaerobic glycogenolysis and glycolysis in supplying energy for maximal effort muscle contractures, and the catastrophic consequences for muscle function and integrity when muscle contractions at maximal effort are undertaken when muscle glycogenolysis is blocked. This is in keeping with the common view that blocked glycogenolysis in McArdle disease impairs only anaerobic metabolism and that those symptoms of exercise intolerance in affected patients relate solely to the consequences of this energy limitation. However, the metabolism of glycogen is also necessary to support normal muscle oxidative metabolism, and patients with complete blocks in muscle glycogenolysis (i.e. McArdle disease) or glycolysis (muscle phosphofructokinase (PFK) deficiency, GSD VII, Tarui disease) also have distinctive symptoms attributable to impaired muscle oxidative metabolism (Haller et *al.*, 1985; Haller and Lewis, 1991). Blocked muscle glycogenolysis results in oxidative phosphorylation being limited by substrate availability, underscoring the critical need for glycogen to support normal peak rates of oxidative metabolism. Available evidence is consistent with the view that blocked glycogenolysis restricts the rate of production of acetyl-CoA and/or the level of anaplerosis necessary to increase 4-carbon tricarboxylic acid (TCA) cycle intermediates (e.g. fumarate, malate and, ultimately, oxaloacetate) to support normal peak flux of the TCA cycle, thus restraining the rate of generation of reducing equivalents, NADH and FADH, necessary for normal function of the respiratory chain and normal rates of oxidative phosphorylation (Sahlin *et al.*, 1990; Sahlin *et al.*, 1995). As a result, peak levels of oxidative metabolism in patients with McArdle disease and muscle PFK deficiency are about half or less that of normal (Haller and Lewis, 1991; Haller and Vissing, 2004b). The physiological hallmark of this limitation in muscle oxidative phosphorylation is a restricted capacity of working muscle to extract available oxygen from blood as indicated by a low peak systemic a-$\bar{v}$ O_2 difference, which is associated with an exaggerated increase in the transport of oxygen by the circulation such that the increase in cardiac output relative to O_2 utilisation is 2–3 times normal (Haller *et al.*, 1985; Haller and Lewis, 1991).

A central feature of substrate-limited oxidative metabolism is the variation in oxidative capacity that accompanies changing availability of extramuscular fuels in response to exercise and diet. This is exemplified by the "second wind" phenomenon that is characteristic of McArdle disease (Braakhekke *et al.*, 1986; Haller and Vissing, 2002; Vissing and Haller, 2003a). In the first 5–8 min of exercise, McArdle patients have a markedly restricted exercise capacity, so relatively minor exertion such as walking at a moderate pace or pedalling a bicycle at a low workload causes fatigue and tachycardia (Haller and Vissing, 2002). However, when exercise is sustained or resumed after a brief rest, patients experience a dramatic improvement in exercise capacity and perceived exertion in association with a marked reduction in exercise heart rate, so that exercise that previously caused fatigue is easily tolerated. Our studies indicate that the markedly low oxidative capacity early in exercise is due to the fact that glycogen is the fuel that normally enables oxidative phosphorylation to increase to peak levels within the first 3–5 min of the transition from rest to exercise. Compounding the oxidative limitation is a relatively low availability of extramuscular fuels – primarily glucose and free fatty acids – during the first minutes of exercise (Haller and Vissing, 2002). Thus, peak oxygen utilisation within the first 5–7 min of exercise in patients with McArdle disease is similar to that of patients with mitochondrial myopathies (Figure 12.5). Like patients with mitochondrial myopathies, McArdle patients have a restricted capacity to extract available oxygen from blood with a peak systemic a-$\bar{v}$ O_2 difference that is about half that of healthy subjects tested under the same conditions (Figure 12.5) (Haller *et al.*, 1985; Haller and Vissing, 2002). The dramatically improved exercise capacity accompanied by a steep drop in exercise heart rate that occurs at 7–10 min of exercise is attributable to a sudden, large increase in oxidative capacity, marked by an average increase in peak $\dot{V}O_2$ of approximately 25%, which is associated with a similar percentage increase in peak a-$\bar{v}$ O_2 difference (Haller and Vissing, 2002). The drop in heart rate mirrors the fall in exertion perceived by the patient, as heart rate and exercise intensity, expressed as a percentage of an individuals peak $\dot{V}O_2$, are proportional. For a 35-year-old patient, a heart rate of 175 bpm at 5 min of exercise indicates the patient was exercising at about 95% of their maximal oxidative capacity. The fall in heart rate to 125 bpm at the same workload after the onset of the second wind indicates exercise at that point is only about 70% of maximal oxidative capacity. Accordingly, after the second wind, the patient is able to exercise at a substantially higher workload because of a higher peak $\dot{V}O_2$ and a-$\bar{v}$ O_2 difference (Haller and Vissing, 2002). Our studies indicate that increased hepatic glucose production and muscle uptake of blood glucose are crucial for the development of a second wind (Vissing *et al.*, 1992;

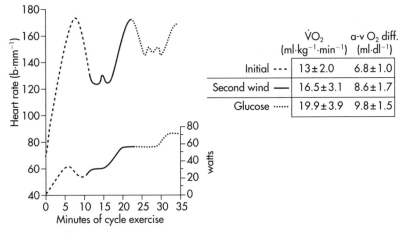

	$\dot{V}O_2$ (ml·kg^{-1}·min^{-1})	a-v O_2 diff. (ml·dl^{-1})
Initial - - -	13±2.0	6.8±1.0
Second wind —	16.5±3.1	8.6±1.7
Glucose ······	19.9±3.9	9.8±1.5

Figure 12.5 Heart rate (upper trace) and corresponding workrate (lower trace) for a patient with McArdle disease during sustained cycle exercise after an overnight fast. In the initial 5–7 min of exercise, a 30-watt work level is near maximal and elicits a heart rate approaching 180 bpm. Between the seventh and tenth minutes, the patient develops a spontaneous second wind so that exercise that was near maximal before is now easily tolerated at a heart rate that is almost 60 bpm lower than before the second wind. Between minutes 15 and 20 the workload is increased, demonstrating that the patient is able to exercise at 55 watts at the same heart rate that accompanied 30 watts before the second wind. Between minutes 23 and 25, the patient is given an infusion of glucose that doubles blood glucose levels. The glucose infusion is associated with a glucose-induced second wind with improved exercise tolerance and a more than 20 bpm drop in heart rate. After glucose, the patient's maximal workload increases to 70 watts. The table at the right shows the corresponding $\dot{V}O_2$ and systemic a-v O_2 difference for the patient's initial, spontaneous second wind, and glucose infusion condition showing a progressive increase in peak $\dot{V}O_2$ that is parallelled by progressive increases in systemic a-$\bar{v}$ O_2 difference consistent with increasing substrate availability for oxidative phosphorylation under second wind and glucose conditions

Haller and Vissing, 2002). Evidence supporting this conclusion includes the fact that patients with muscle PFK deficiency, who are unable to oxidise muscle glycogen or blood glucose, do not develop a second wind under exercise conditions that consistently produce one in McArdle disease (Haller and Vissing, 2004b).

Although McArdle patients have substantially improved exercise capacity after the onset of the spontaneous second wind, oxidative metabolism remains substrate-limited. This is evident in the fact that increasing blood glucose concentrations by intravenous glucose administration produces another second wind, with a drop in exercise heart rate and an increase in peak $\dot{V}O_2$ and peak a-$\bar{v}$ O_2 difference (Figure 12.5) (Haller and Vissing, 2002). Although muscle glucose utilisation appears to be necessary for both the spontaneous second wind and glucose-induced, second, second wind, the metabolic mechanism has not been fully elucidated. In addition to direct combustion via the production of acetyl-CoA, glucose may enhance oxidative capacity by augmenting levels of oxaloacetate, thereby promoting the oxidation of fatty acids.

Diet is also an important variable in determining oxidative capacity in McArdle disease. Oral ingestion of sucrose markedly improves exercise capacity, lowers exercise heart rate and abolishes the spontaneous second wind in McArdle disease attributable to increased blood levels of glucose and lactate (Vissing and Haller, 2003b; Andersen et al., 2008; Andersen and Vissing, 2008). In contrast, in muscle PFK deficiency, glucose infusion or a high-carbohydrate meal has the opposite effect, and reduces muscle oxidative capacity (Haller and Lewis, 1991). This relates to the fact that glucose cannot be metabolised by PFK-deficient muscle, and thus has no beneficial effect upon energy production, and to the fact that glucose causes an insulin-mediated reduction in blood levels of free fatty acids, the key oxidative fuel when the metabolism of muscle glycogen and glucose are blocked.

CASE STUDY

Clinical history

A 39-year-old man was referred for evaluation of the cause of symptoms of exertional muscle fatigue, muscle tightness and pigmenturia. The patient's first recollection of exercise limitations occurred during standardised exercise testing in junior high and high school. Particularly, upper body exercise such as pull-ups, push-ups and rope climbing caused abnormally rapid fatigue, often accompanied by nausea. For example, he could never perform more than 1–2 pull-ups. If such exertion were continued he would develop cramps, in which the affected limb would be locked up, unable to be straightened for minutes to hours, and the attempt to do so would cause severe pain. His legs have less frequently been affected, but maximal effort running, particularly if he had not warmed up could cause similar cramps in his legs. On many occasions, exertional cramping and muscle pain have been followed by dark urine, typically hours after the bout of exertional muscle cramps. He has continued to experience these symptoms over the years with a general downhill course marked by developing symptoms after lesser degrees of exercise.

Despite these problems, the patient has been relatively athletic. He tried out for track, running the mile in high school, though he was never good enough to make the team. His best time for the mile was about 6.5 min. In college, he participated in a physical education class, for which he ran three miles per day at a 6.5–7 min per mile pace. At age 18, he rode across country on a bicycle tour. He had not cycled significantly before and was one of the slower riders in his group for the first part of the trip, but by the end he was near the leaders. After a hiatus from participating in running or cycling, he notes that he is prone to develop exertional cramps and dark urine until he has trained for a week or two. After high school he spent several years as a "ski bum". There he noted he fatigued more rapidly than his peers in difficult runs. Particularly steep slopes with moguls that required frequent turns would cause cramping in his thighs.

Diet has no major effect on exercise capacity, though he thinks he may cramp more easily on an empty stomach. No circumstances other than exercise have triggered muscle symptoms. He characterises his muscle as weaker than normal, but illustrates this as tiring out quickly with heavy exertion.

Evaluation: rationale

The history is strongly suggestive of a metabolic muscle disorder, with multiple episodes of pigmenturia suggesting myoglobinuria following maximal effort exercise. Such heavy exercise is described as producing muscle tightness, "cramps" and "locking up" in which the muscle remains contracted from "minutes to hours", and attempting to stretch the muscle produces pain. These episodes are highly suggestive of muscle contractures, and the maximal effort exercise that triggers them normally is supported by anaerobic metabolism, particularly anaerobic glycogenolysis. However, the history suggests that his aerobic exercise capacity is much higher than in muscle phosphorylase deficiency (McArdle disease), the most common muscle glycolytic disorder. Specifically, the history of being able to run 1–3 miles at a 6.5–7 min per mile pace predicts an aerobic capacity far higher than is typical of McArdle disease (Figure 12.1).

In order to confirm the history of a relatively normal aerobic capacity and to determine if the patient has evidence of a spontaneous second wind, a cycle exercise test would be appropriate – ideally with the capacity to measure cardiac output as well as $\dot{V}O_{2max}$ in order to be able to calculate systemic a-$\bar{v}$ O_2 difference to determine if oxygen extraction by working muscle is normal.

In addition, it is important to evaluate the patient's capacity for anaerobic exercise using an ischemic forearm test or a maximal effort aerobic forearm test.

Exercise testing: results

The patient underwent cycle exercise testing which revealed a work capacity of 170 watts, a $\dot{V}O_2$ of 2.71 ($30\,ml\cdot kg^{-1}\cdot min^{-1}$), peak heart rate of 165 bpm, peak cardiac output of 17 l, and peak a-$\bar{v}$ O_2 difference of $16\,ml\cdot dl^{-1}$ O_2 blood – all values within the normal range (compare with normal values in Figure 12.2) in contrast to typical values found with such testing for patients with McArdle disease (Figure 12.5). In addition, prolonged cycle exercise testing performed on a separate day revealed no evidence of a second wind.

Ischemic forearm testing in which the patient performed a maximal handgrip contraction every 2 s revealed the results shown in Figure 12.6. The patient completed the first eight maximal effort contractions normally, but thereafter began to rapidly fatigue and stopped after 17 handgrips (33 s) with a contracture affecting the finger flexors (Figure 12.6A). Venous lactate rose from a resting level of approximately 1 mM to 3.2 mM at 1 min post-exercise (Figure 12.6B), or about 70% of the increase seen in healthy control subjects (Figure 12.6B). However, the ammonia response was highly abnormal, increasing to a level almost three times that of control subjects (Figure 12.6D). This abnormal ammonia response was dramatically apparent when ammonia relative to lactate values were plotted (Figure 12.6C). The results show that the lactate–ammonia response is linear in healthy subjects, but that the increase in ammonia was dramatically higher relative to the increase in lactate, indicating an abnormally high level of AMP deamination consistent with abnormally high levels of ADP during maximal effort exercise in the patient. These results suggest that the patient has an unusual muscle glycolytic disorder, in which some capacity for glycolysis is preserved. Further evaluation of this patient, employing ^{31}P MRS revealed normal resting spectra. During handgrip exercise, the fall in PCr is normally matched by a proportional increase in inorganic phosphate. In this patient, the increase in inorganic phosphate during exercise was markedly blunted, and there is an anomalous large phosphomonoester (PME) peak indicating the accumulation of phosphorus containing glycolytic intermediates and confirming the fact that the patient has a block in muscle glycolysis (Figure 12.7). Noteworthy as well is the fact that despite the ability to increase lactate during forearm exercise, muscle in the patient does not acidify (Figure 12.7).

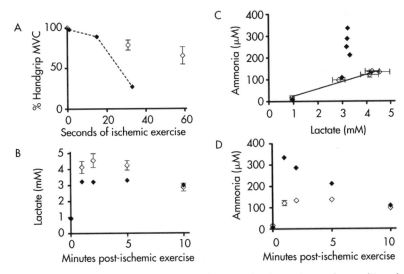

Figure 12.6 Ischemic forearm testing results for control subjects (open diamonds) and the patient described in the case study (closed diamonds) showing: A: rapid muscle fatigue in the patient; B: blunted lactate production in the patient; C: lactate and corresponding ammonia level at rest and during exercise, demonstrating a linear relationship between lactate and ammonia in control subjects and dramatically elevated levels of ammonia relative to lactate in the patient; and D: exaggerated ammonia production in the patient

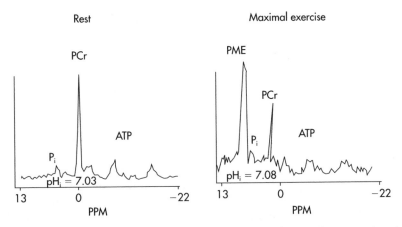

Figure 12.7 ^{31}P MRS at rest and during maximal aerobic handgrip in the patient described in the case report. Rest spectra are normal. During heavy exercise, the increase in inorganic phosphate (P_i) is markedly blunted and there appears a large phosphomonoester peak, representing the accumulation of phosphorus containing intermediates of glycolysis behind a metabolic block. Note that muscle pH is abnormally high during heavy exercise (compare with the results for healthy subjects in Figure 12.4)

Interpretation/implications

The diagnosis, in this instance, was achieved by muscle biopsy with muscle enzyme analysis, which indicated that the patient had phosphoglycerate kinase deficiency (5% of normal enzyme activity). The clinical and exercise results in this patient are typical for so-called distal glycolytic defects in which some level of residual enzyme activity is preserved. Remarkably, although a substantial level of increase in lactate occurs with exercise, the level of residual glycolysis is insufficient to cause a substantial fall in muscle pH, and the pattern of fatigue with maximal effort anaerobic exercise in patients with this class of glycolytic defect is virtually the same as in patients with a complete block in glycogenolysis or glycolysis (compare Figure 12.3 and Figure 12.6A). This suggests that the deficit in energy availability relative to energy demand and the metabolic milieu of working muscle during maximal effort anaerobic exercise are similar to those in patients with a complete metabolic block with a characteristic exaggerated increase in ammonia relative to lactate, indicating high levels of ADP, AMP and AMP deamination.

However, there is a significant difference in muscle oxidative capacity in patients with partial, compared to complete, blocks in muscle glycogen or glycogen and glucose utilisation with minor levels of residual glycolysis supporting essentially normal oxidative capacity in patients with partial glycolytic defects.

CONCLUSION

Human genetic disorders of muscle energy metabolism provide unique perspectives from which to view the function of metabolic pathways in muscle physiology and metabolism. Muscle respiratory chain disorders that limit muscle oxidative phosphorylation illuminate the necessity of oxidative metabolism for powering sustained muscle contractions and reveal a key role for mitochondrial metabolism in regulating the delivery of oxygen to working muscles. Disorders of muscle glycogenolysis and glycolysis reveal the crucial role of these metabolic processes for supplying both anaerobic and aerobic energy for muscle contraction. The pathological fatigue that occurs when glycogenolysis and/or glycolysis is blocked implies an important role for these metabolic pathways in normal muscle fatigue.

FIVE KEY PAPERS THAT SHAPED THE TOPIC AREA

Study 1. Wiles, C.M., Jones, D.A. and Edwards, R.H. (1981). Fatigue in human metabolic myopathy. *Ciba Foundation Symposium*, 82, 264–282.

This is a classical study of muscle fatigue in mitochondrial myopathy and McArdle disease from the laboratory of Professor Richard Edwards, a pioneer in the physiological investigation of human muscle diseases. These studies employ neural stimulation of muscle to eliminate voluntary effort in the investigation of muscle fatigue in these disorders in order to define specific defects of muscle excitation or excitation–contraction coupling.

Study 2. Ruff, R.L. (1996). Elevated intracellular Ca^{2+} and myofibrillar Ca^{2+} sensitivity cause iodoacetate-induced muscle contractures. *Journal of Applied Physiology*, 81, 1230–1239.

This is one of a series of excellent studies utilising the iodoacetate-inhibition of glyceraldehyde-3-phosphate dehydrogenase to illuminate the metabolic mechanism of muscle fatigue and muscle contractures when muscle glycolysis or glycogenolysis is blocked.

Study 3. Haller, R.G. and Vissing, J. (2002). Spontaneous second wind and glucose-induced second, "second wind" in McArdle disease: oxidative mechanisms. *Archives in Neurology*, 59, 1395–1402

This study reveals that the unavailability of muscle glycogen causes muscle oxidative phosphorylation to be limited by substrate availability and that this oxidative limitation is most severe in the first 5–7 min of exercise; that the spontaneous second wind that occurs after 7–10 min of sustained exercise, associated with a fall in exercise heart rate and perceived exertion, is due to a ~25% increase in oxidative capacity; and that muscle oxidative phosphorylation remains substrate-limited after the spontaneous second wind as indicated by the fact that a glucose infusion produces another second wind due to a ~20% further increase in muscle oxidative capacity.

Study 4. Taivassalo, T., Jensen, T.D., Kennaway, N., DiMauro, S., Vissing, J. and Haller R.G. (2003). The spectrum of exercise tolerance in mitochondrial myopathies: a study of 40 patients. *Brain*, 126, 413–423.

This study evaluated patients with a broad range of impaired muscle oxidative phosphorylation, showing that the key physiological limitation for oxidative metabolism in muscle respiratory chain defects is a limited capacity to extract oxygen from blood (low peak a-$\bar{v}$ O_2 difference) and that peak a-$\bar{v}$ O_2 difference represents a surrogate marker of the severity of the respiratory chain defect. Furthermore, the study reveals a close, inverse relationship between the severity of impaired muscle oxidative phosphorylation (as indicated by peak a-$\bar{v}$ O_2 difference in cycle exercise) and the level of exaggerated circulatory and ventilatory responses to exercise in these patients.

Study 5. Allen, D.G., Lamb, G.D. and Westerblad, H. (2008). Skeletal muscle fatigue: cellular mechanisms. *Physiological Review* 88, 287–332.

This is a comprehensive and up-to-date review of studies of cellular mechanisms of muscle fatigue. While fatigue in metabolic myopathies is not a focus of this review, it provides an excellent assessment of the role of muscle metabolism in normal muscle fatigue that is highly relevant to the consideration of the pathophysiology of metabolic muscle disorders.

GLOSSARY OF TERMS

a-v̄ O_2 difference	arterio-venous oxygen difference
ADP	adenosine diphosphate
AMP	adenosine monophosphate
ATP	adenosine triphosphate
GSD	glycogen storage disease
GSD V	muscle phosphorylase deficiency
GSD VII	muscle phosphofructokinase deficiency
IMP	inosine monophosphate
MET	metabolic equivalent
MVC	maximal voluntary contraction
PCr	phosphocreatine
PFK	phosphofructokinase
PME	phosphomonoester
P_i	inorganic phosphate
pO_2	partial pressure of oxygen, commonly expressed in millimeters of mercury
TCA	tricarboxylic acid cycle

REFERENCES

Allen, D.G., Lamb, G.D. and Westerblad, H. (2008). Skeletal muscle fatigue: cellular mechanisms. *Physiological Reviews*, 88, 287–332.

Andersen, S.T. and Vissing, J. (2008). Carbohydrate- and protein-rich diets in McArdle disease: effects on exercise capacity. *Journal of Neurology and Neurosurgery and Psychiatry*, 79, 1359–1363.

Andersen, S.T., Haller, R.G. and Vissing J. (2008). Effect of oral sucrose shortly before exercise on work capacity in McArdle disease. *Archives in Neurology*, 65, 786–789.

Aquaron, R., Berge-Lefranc, J.L., Pellissier, J.F., Montfort, M.F., Mayan, M., Fig-arella-Branger, D., Coquet, M., Serratrice, G. and Pouget, J. (2007). Molecular characterization of myophosphorylase deficiency (McArdle disease) in 34 patients from Southern France: identification of 10 new mutations: absence of genotype-phenotype correlation. *Neuromuscular Disorders*, 17, 235–241.

Argov, Z., Bank, W.J., Maris, J. and Chance, B. (1987). Muscle energy metabolism in McArdle's syndrome by in vivo phophorus magnetic resonance spectroscopy. *Neurology*, 37, 1720–1724.

Åstrand, P.O. and Rodahl, K. (1986). *Textbook of Work Physiology: Physiological Basis of Exercise*. New York: McGraw-Hill.

Bank, W. and Chance, B. (1994). An oxidative defect in metabolic myopathies: diagnosis by non-invasive tissue oxymetry. *Annals in Neurology*, 36, 830–837.

Braakhekke, J.P., deBruin, M.I., Stegeman, D.F., Wevers, R.A., Binkhorst, R.A. and Joosten, E.M.G. (1986). The second wind phenomenon in McArdle's disease. *Brain*, 109, 1087–1101.

Clausen, T. (1986). Regulation of active Na^+–K^+ transport in skeletal muscle. *Physiological Reviews*, 66, 542–580.

Cooper, R.G., Stokes, M.J. and Edwards, R.H. (1989). Myofibrillar activation failure in McArdle's disease. *Journal of Neurological Sciences*, 93, 1–10.

Deschauer, M., Morgenroth, A., Joshi, P.R., Glaser, D., Chinnery, P.F., Aasly, J., Schreiber, H., Knape, M., Zierz, S. and Vorgerd, M. (2007). Analysis of spectrum and frequencies of mutations in McArdle disease: identification of 13 novel mutations. *Journal of Neurology*, 254, 797–802.

DiMauro, S. and Schon, E.A. (2003). Mitochondrial respiratory-chain diseases. *The New England Journal of Medicine*, 348, 2656–2668.

Duboc, D., Jehenson, P., Dinh, S., Marsac, C., Syrota, A. and Fardeau, M. (1987). Phosphorus NMR spectroscopy study of muscular enzyme deficiencies involving glycogenolysis and glycolysis. *Neurology*, 37, 663–674.

Flaherty, K.R., Wald, J., Weisman, I.M., Zeballos, R.J., Schork, M.A., Blaivas, M., Rubenfire, M. and Martinez, F.J. (2001). Unexplained exertional limitation: characterization of patients with a mitochondrial myopathy. *American Journal of Respiratory and Critical Care Medicine*, 164, 425–432.

Fleckenstein, J.L., Haller, R.G., Lewis, S.F., Bertocci, L., Payne, J., Barker, B., Payne, J., Parkey, R.W. and Peshock, R.M. (1991). Myophosphorylase deficiency impairs exercise-enhancement on MRI of skeletal muscle. *Journal of Applied Physiology*, 71, 961–969.

Gollnick, P.D. and Saltin, B. (1983). Skeletal muscle adaptability: significance for metabolism and performance. In L.D. Peachey (ed.) *Handbook of Physiology*. Bethesda: American Physiological Society, pp. 555–631.

Haller, R.G. and Lewis, S.F. (1991). Glucose-induced exertional fatigue in muscle phosphofructokinase deficiency. *The New England Journal of Medicine*, 324, 364–369.

Haller, R.G. and Vissing, J. (1999). Circulatory regulation in muscle disease. In B. Saltin, R. Boushel, N. Secker and J.H. Mitchell (eds). *Exercise and Circulation in Health and Disease*. Champaign: Human Kinetics, pp. 263–273.

Haller, R.G. and Vissing, J. (2002). Spontaneous second wind and glucose-induced second, "second wind" in McArdle disease: oxidative mechanisms. *Archives in Neurology*, 59, 1395–1402.

Haller, R.G. and Vissing, J. (2004a). Functional evaluation of metabolic myopathy.

In A.G. Engel and C. Franzini-Armstrong (eds). *Myology*, edition III, Vol. 1. New York: McGraw-Hill, pp. 665–679.

Haller, R.G. and Vissing, J. (2004b). Lack of a spontaneous second wind in muscle phosphofructokinase deficiency. *Neurology*, 62, 82–87.

Haller, R.G., Clausen, T. and Vissing, J. (1998). Reduced levels of skeletal muscle Na⁺K⁺-ATPase in McArdle disease. *Neurology*, 50, 37–40.

Haller, R.G., Henriksson, K.G., Jorfeldt, L., Hultman, E., Wibom, R., Sahlin, K., *et al.* (1991). Deficiency of skeletal muscle succinate dehydrogenase and aconitase: pathophysiology of exercise in a novel human muscle oxidative defect. *Journal of Clinical Investigations*, 88, 1197–1206.

Haller, R.G., Lewis, S.F., Cook, J.D. and Blomqvist, C.G. (1985). Myophosphorylase deficiency impairs muscle oxidative metabolism. *Annals of Neurology*, 17, 196–199.

Han, J.W., Thieleczek, R., Varsanyi, M. and Heilmeyer Jr, L.M. (1992). Compartmentalized ATP synthesis in skeletal muscle triads. *Biochemistry*, 31, 377–384.

James, J.H., Fang, C.H., Schrantz, J., Hasselgren, P.-O., Paul, R.J. and Fischer, J.E. (1996). Linkage of aerobic glycolysis to sodium–potassium transport in rat skeletal muscle. *Journal of Clinical Investigations*, 98, 2388–2397.

Matthews, P.M, Allaire, C., Shoubridge, E.A., Karpati, G., Carpenter, S. and Arnold, D.L. (1991). In vivo muscle magnetic resonance spectroscopy in the clinical investigation of mitochondrial disease. *Neurology*, 41, 114–120.

Miller, R.G., Boska, M.D., Moussavi, R.S., Carson, P.J. and Weiner, M.W. (1988). ³¹P nuclear magnetic resonance studies of high-energy phosphates and pH in human muscle fatigue. *Journal of Clinical Investigations*, 31, 1190–1196.

Mochel, F., Knight, M.A., Tong, W.H., Hernandez, D., Ayyad, K., Taivassalo, T., Andersen, P.M., Singleton, A., Rouault, T.A., Fischbeck, K.H., Haller, R.G. (2008). Splice mutation in the iron–sulfur cluster scaffold protein ISCU causes myopathy with exercise intolerance. *American Journal of Human Genetics*, 82, 652–660.

Pedersen, T.H., Nielsen, O.B., Lamb, G.D. and Stephenson, D.G. (2004). Intracellular acidosis enhances the excitability of working muscle. *Science*, 305, 1144–1147.

Radda, G.K. (1986). The use of NMR spectroscopy for the understanding of disease. *Science*, 233, 640–645.

Richardson, R.S., Duteil, S., Wary, C., Wray, D.W., Hoff, J. and Carlier, P.G. (2006). Human skeletal muscle intracellular oxygenation: the impact of ambient oxygen availability. *Journal of Physiology*, 571, 415–424.

Ruff, R.L. (1996). Elevated intracellular Ca²⁺ and myofibrillar Ca²⁺ sensitivity cause iodoacetate-induced muscle contractures. *Journal of Applied Physiology*, 81, 1230–1239.

Ruff, R.L and Weissman, J. (1991). Iodoacetate-induced contracture in rat skeletal muscle: possible role of ADP. *American Journal of Physiology*, 261, C828–836.

Ruff, R.L. and Weissman, J. (1995). Iodoacetate-induced skeletal muscle contracture: changes in ADP, calcium, phosphate, and pH. *American Journal of Physiology*, 268, C317–322.

Sahlin, K. (1986). Metabolic changes limiting muscle performance. In B. Saltin (ed.), *Biochemistry of Exercise VI*. Champaign: Human Kinetics, pp. 323–343.

Sahlin, K., Areskog, N.H., Haller, R.G., Henriksson, K.G., Jorfeldt, L. and Lewis, S.F. (1990). Impaired oxidative metabolism increases adenine nucleotide breakdown in McArdle's disease. *Journal of Applied Physiology*, 69, 1231–1235.

Sahlin, K., Jorfeldt, L., Henriksson, K.G., Lewis, S.F. and Haller, R.G. (1995). Tricarboxylic acid cycle intermediates during incremental exercise in healthy subjects and in patients with McArdle's disease. *Clinical Science (London)*, 88, 687–693.

Taivassalo, T., Abbott, A., Wyrick, P. and Haller, R.G. (2002). Venous oxygen levels during aerobic forearm exercise: an index of impaired oxidative metabolism in mitochondrial myopathy. *Annals of Neurology*, 51, 38–44.

Taivassalo, T., Jensen, T.D., Kennaway, N., DiMauro, S., Vissing, J. and Haller, R.G. (2003). The spectrum of exercise tolerance in mitochondrial myopathies: a study of 40 patients. *Brain*, 126, 413–423.

Vissing, J. and Haller, R.G. (2003a). A diagnostic cycle test in McArdle disease. *Annals of Neurology*, 54, 539–542.

Vissing, J. and Haller, R.G. (2003b). The effect of oral sucrose on exercise tolerance in McArdle's disease. *The New England Journal of Medicine*, 349, 2503–2509.

Vissing, J., Lewis, S.F., Galbo, H. and Haller, R.G. (1992). Effect of deficient muscular glycogenolysis on extramuscular fuel production in exercise. *Journal of Applied Physiology*, 72, 1773–1779.

Wiles, C.M, Jones, D.A. and Edwards, R.H. (1981). Fatigue in human metabolic myopathy. *Ciba Foundation Symposium*, 82, 264–282.

INDEX